AXX $12.50

Elementary Linear Algebra

EVAR D. NERING

Professor of Mathematics
Arizona State University

W. B. SAUNDERS COMPANY
Philadelphia · London · Toronto

W. B. Saunders Company: West Washington Square
Philadelphia, Pa. 19105

12 Dyott Street
London, WC1A 1DB

833 Oxford Street
Toronto, M8Z 5T9, Canada

Elementary Linear Algebra　　　　　　　　　　　　　　ISBN 0-7216-6755-4

© 1974 by W. B. Saunders Company. Copyright under the International Copyright Union. All rights reserved. This book is protected by copyright. No part of it may be reproduced, stored in a retrieval system, or transmitted in any form or by any means, electronic, mechanical, photocopying, recording, or otherwise, without written permission from the publisher. Made in the United States of America. Press of W. B. Saunders Company. Library of Congress catalog card number 73-90724.

Print No.:　　9　8　7　6　5　4　3　2

Preface

This book is intended primarily for use as a textbook in a linear algebra course for students in their sophomore year. The technical prerequisite is even more modest since calculus appears in only a few examples and exercises. These can be omitted without penalty since they serve only an enrichment purpose.

Something vaguely referred to as "mathematical maturity" is the ability to accept an abstract definition and go through a logical argument based on such a definition. Courses that most students are likely to have encountered before taking linear algebra are not well adapted to developing mathematical maturity. Linear algebra is ideal for this purpose since the ideas, theorems, and proofs are rather uncomplicated. The book starts at a concrete, non-abstract level and becomes more abstract as things progress. The increase in sophistication between the beginning and the end is substantial, but a great effort has gone into preparing for the increase in sophistication by giving examples and computational procedures to illustrate every abstraction.

We start with a discussion of systems of linear algebraic equations. This is a concrete problem that most students are already familiar with. Systems of linear equations also serve as a natural context in which to describe the pivot operation. The pivot operation is the principal computational step in the simplex method for linear programming, and in this context it has been around for some time. A pivot operation is merely a well-defined set of elementary row operations. There is a significant pedagogical benefit from using the pivot operation. For one thing, the pivot operation collects in one package several routine steps and leaves those steps in which decisions must be made more clearly delineated. Also, the pivot operation has an important interpretation as a change of coordinates in which one vector in a basis is replaced by another. Because of the close relationship between the pivot operation and this special type of change of basis, the pivot operation occurs in almost every situation where a computation is required because of a change of coordinates.

The pivot operation is also closely connected with the most efficient method known for solving systems of linear equations, the method of Gaussian elimination. The numerical steps are carefully explained and the amount of work required is compared with the amounts required using other methods. While not covering numerical methods as such, the treatment contained in this text gives the reader a good foundation for learning numerical methods.

Based on years of teaching linear algebra courses, I have strong opinions about the best way to study this material. There is a natural inclination for a student to spend a lot of time examining and re-examining a particular section that he finds elusive. This sort of approach may work well for some students and for some subjects, but I think it is not the best way for most students to study linear algebra.

I think the material in linear algebra is too "thin" for this kind of intensive examination. After one has looked at it for a reasonable time there isn't any more to be done with it. The important thing about linear algebra is the structure of the subject, the way it is put together and the way the ideas build on each other. I recommend the "bulldozer" approach. Dig in and try to keep going at a reasonable pace. Even if some things seem unclear, keep going. Eventually, it is not possible to go any farther. Return to a place where you feel secure, set your "blade" a little deeper and take another run at it. On subsequent passes the ideas will fit together better and make more sense.

One reason this method seems to work is that the subject contains a lot of repetition, either the repetition of a specific idea, the repetition of a technique, or the repetition of a pattern of steps. This repetition not only helps with later material but also reinforces earlier material. Grinding away at one point makes this kind of help unavailable.

I also recommend working most of the exercises. While some of the exercises can get rather involved, I have tried to minimize numerical work. You should expect a numerical problem to have a rather simple solution. The appearance of cumbersome fractions or decimals while working an exercise is usually a clue that an error has been made (perhaps by me), or that another path might be easier. However, in the material on orthogonal and unitary matrices the appearance of radicals is unavoidable.

I wish to express my deep appreciation to several people who read preliminary versions of this book. Particular acknowledgment is due to Robert E. Lynch, Purdue University, Peter Renz, Wellesley College, and Arlo W. Schurle, University of North Carolina at Charlotte. Robert E. Lynch corrected some misimpressions I had about numerical computations and suggested some significant modifications which are contained in the text. Peter Renz suggested the use of mapping diagrams. I had known about and used mapping diagrams, but it hadn't occurred to me that they would have any use in an elementary book. Although the book can be read while omitting all references to the mapping diagrams, I think most students will find them helpful. It is hard to make significant changes in a manuscript after it is almost complete. Thus, I couldn't go as far in incorporating the readers' suggestions as I would have liked. I hope they will accept the compromises that were made without feeling the need to disavow any connection with the project.

Above all I want to express singular gratitude to Albert W. Tucker, Princeton University. Several times he had urged me to consider using the pivot operation as the basis for an elementary book on linear algebra. I was already familiar with the pivot operation as it is used in linear programming, but I didn't see its advantage from a pedagogical viewpoint. In fact, I didn't respond to his suggestions until I had made two fruitless starts on this book. When I did try it, wonder of wonders, it worked. Using it, the book seemed to write itself. With it, rather complicated computational procedures are easy to describe. From the conceptual point of view it makes every tool available when it is needed. Thus, the computational and conceptual aspects of linear algebra dovetail and complement each other very nicely. There is no need to emphasize one or the other aspect, and to do so makes things more difficult rather than easier. If this book has any contribution to make to the advance of the teaching of linear algebra, the use of the pivot operation is the most significant.

EVAR D. NERING

Contents

1
LINEAR EQUATIONS 1
- 1-1 Linear Equations and Matrices 2
- 1-2 Solutions of Systems of Linear Equations 10
- 1-3 Basic Form 14
- 1-4 Equivalent Systems of Equations 17
- 1-5 Elementary Operations 25
- 1-6 Gaussian Elimination 35

2
VECTORS AND VECTOR SPACES 42
- 2-1 Coordinate Spaces 42
- 2-2 Vectors 50
- 2-3 Solution Sets as Sets of Vectors 57
- 2-4 Subspaces and Homogeneous Equations 62
- 2-5 Linear Combinations 67
- 2-6 Linear Independence 76
- 2-7 Bases and Dimension 82
- 2-8 Bases and Equations 92

3
BASES AND COORDINATE SYSTEMS 99
- 3-1 Introducing Coordinates 99
- 3-2 Change of Basis 107
- 3-3 Elementary Operations and Change of Coordinates 116
- 3-4 Elementary Matrices 126
- 3-5 A Useful Factorization of a Matrix 132

4

LINEAR TRANSFORMATIONS 146

- 4-1 Linear Transformations 147
- 4-2 Matrix Representation of Linear Transformations 153
- 4-3 General Properties of Linear Transformations 165
- 4-4 Composed Transformations and Inverses 173
- 4-5 Geometric Examples 183
- 4-6 Change of Basis 191
- 4-7 Linear Transformations and Linear Problems. 199

5

DETERMINANTS 208

- 5-1 The Determinant as a Volume 208
- 5-2 Permutations 217
- 5-3 The Formula for the Determinant 223
- 5-4 Elementary Operations and Determinants 230
- 5-5 Cofactors 234

6

EIGENVECTORS AND EIGENVALUES 240

- 6-1 Algebra of Subspaces 241
- 6-2 Eigenvectors and Invariant Subspaces 246
- 6-3 The Characteristic Matrix and Characteristic Equation 256
- 6-4 Equivalence and Similarity 264
- 6-5 Spectral Resolution 268
- 6-6 The Hamilton-Cayley Theorem 273

7

ORTHOGONAL AND UNITARY SPACES 279

- 7-1 Inner Products 280
- 7-2 Orthogonality and Orthonormal Bases 288
- 7-3 Unitary and Orthogonal Similarity 295
- 7-4 Orthogonal and Unitary Transformations 303
- 7-5 Upper Triangular Form 306
- 7-6 Normal Matrices 309
- 7-7 The Situations Over the Real Numbers 316

APPENDIX 325
    A–1   Sets and Logic... 325
    A–2   Indexed Sets 329
    A–3   Fields, Rings, and Groups 338
    A–4   Alphabet.. 342

ANSWERS TO EXERCISES... 343

INDEX 373

1
LINEAR EQUATIONS

Linear algebra is a rather "clean" mathematical subject. The definitions and theorems are satisfyingly simple and precise. Most of the proofs are illuminating rather than irrelevant, and they are usually short and direct. As a mathematical subject, linear algebra should be among the easiest since the technical prerequisites are very modest. But still many students have difficulty with it. The most important reason for this is that many of the most interesting models and applications of linear algebra occur much later in a student's mathematical experience. Linear algebra is fundamental to many different mathematical subjects and is, therefore, somewhat abstract. If a student studies linear algebra early in his career he has little in the way of models to hold up as an illustration of the thing being described.

There is one part of linear algebra that most students are likely to have already met. This concerns the problem of solving a system of simultaneous linear algebraic equations. For this reason we shall take up this problem as a means of introducing the ideas and concepts of linear algebra. These concepts will generalize to have meaning and application in a far greater context. Therefore, we shall deal with this problem, and methods for solving it, in some detail and with greater care than would be the case if our interest did not extend beyond it.

1-1 LINEAR EQUATIONS AND MATRICES

We assume that the reader is familiar with linear equations and with systems of linear equations as they are treated in high school and college algebra courses. A typical problem of this form is to find any and all solutions to the system

$$\begin{aligned} 2x + 3y &= 4 \\ 5x - 2y &= -9. \end{aligned} \qquad (1.1)$$

We shall have to allow for the possibility of systems with many more equations and many more unknowns. We wish to adopt notation more convenient for such larger systems. Instead of using $x, y, \ldots$, etc., for the unknowns, we shall usually use one letter and a set of subscripts to distinguish the various unknowns. Thus we prefer to write the system (1.1) in the form

$$\begin{aligned} 2x_1 + 3x_2 &= 4 \\ 5x_1 - 2x_2 &= -9. \end{aligned} \qquad (1.2)$$

Conceptually, there is no difference between (1.1) and (1.2).

We are interested in systems of linear equations in general: Do such systems always have solutions, how do we describe or specify the solution, what techniques will produce solutions effectively and efficiently?

We shall generally write a system of m equations in n unknowns in the form

$$\begin{aligned} a_{11}x_1 + a_{12}x_2 + \cdots + a_{1n}x_n &= b_1 \\ a_{21}x_2 + a_{22}x_2 + \cdots + a_{2n}x_n &= b_2 \\ &\vdots \\ a_{m1}x_1 + a_{m2}x_2 + \cdots + a_{mn}x_n &= b_m. \end{aligned} \qquad (1.3)$$

We refer to a system like (1.3) as an $m \times n$ (spoken "m by n") system of linear equations. The example given in (1.2) is a two-by-two system in which $m = n = 2$.

The space required to write these equations can be reduced considerably by using standard summation notation,

$$\sum_{j=1}^{n} a_{1j}x_j = a_{11}x_1 + a_{12}x_2 + \cdots + a_{1n}x_n. \qquad (1.4)$$

With this notation, the entire system (1.3) can be written in the form

$$\sum_{j=1}^{n} a_{ij} x_j = b_i : \quad i = 1, 2, \ldots, m. \quad (1.5)$$

By leaving out all plus signs, equal signs, and unknowns in (1.3) we can obtain the array

$$\begin{bmatrix} a_{11} & a_{12} & \cdots & a_{1n} & b_1 \\ a_{21} & a_{22} & \cdots & a_{2n} & b_2 \\ \cdot & \cdot & & \cdot & \cdot \\ \cdot & \cdot & & \cdot & \cdot \\ \cdot & \cdot & & \cdot & \cdot \\ a_{m1} & a_{m2} & \cdots & a_{mn} & b_m \end{bmatrix} \quad (1.6)$$

If we agree to write a system of equations in a standardized form, like (1.3), then the correspondence between (1.3) and (1.6) is clearly understood and (1.3) can always be reconstructed from (1.6). Thus (1.6) contains as much information as (1.3). We intend to solve the system (1.3) by obtaining a sequence of slightly different systems of equations, each with the same set of solutions, until we arrive at one that is particularly easy to solve. The saving obtained by removing superfluous notation is considerable, and (1.6) is particularly convenient for our purposes. The square brackets used in (1.6) are not essential for representing systems of equations, but they are used in anticipation of the following definition.

Definition *A rectangular array of numbers is called a* **matrix**. *The numbers that appear in the array are called the* **elements** *of the matrix. In this book a matrix will be indicated by enclosing the array in square brackets. Two matrices are considered to be equal if and only if they have the same number of rows and columns and have the same numbers in corresponding positions of the array. A matrix with m rows and n columns is called an m × n (spoken, "m by n") matrix. A square matrix with n rows and n columns is said to be of* **order** *n.*

A specific matrix must simply be written out and displayed in full. But most of our discussion will concern matrices with unspecified elements, as in (1.6). To discuss matrices in general we need more convenient notation. We shall usually use a notation like a_{ij} to denote the element in row i and column j of the array.

The matrix with a_{ij} as a typical element will be denoted by $[a_{ij}]$. We shall consistently use the first index for the row index and the second index for the column index. Thus, a_{rs} is the element in row r column s, but $[a_{ij}]$ and $[a_{rs}]$ denote the same matrix since both symbols stand for the entire array. Often we shall use a single capital italic letter to denote a matrix.

Let $A = [a_{ij}]$ and $B = [b_{ij}]$ be two matrices with the same number of rows and the same number of columns. The *sum* of A and B, denoted by $A + B$, is the matrix obtained by adding corresponding elements in A and B. Specifically, $a_{ij} + b_{ij}$ is the element in row i and column j of $A + B$.

Let $A = [a_{ij}]$ and let c be any number. We define $cA = [ca_{ij}]$. It is part of the special terminology of linear algebra to refer to numbers as *scalars* and to call cA a *scalar multiple* of A.

The product of two matrices is somewhat more involved. Let $A = [a_{ij}]$ be an $m \times n$ matrix and let $B = [b_{ij}]$ be an $n \times r$ matrix. The *product* of A and B, denoted by AB, is the matrix with $\sum_{k=1}^{n} a_{ik} b_{kj}$ in row i and column j. Notice that in this sum corresponding elements from row i of A and column j of B are multiplied and added. It is helpful to keep the following picture in mind.

$$\text{row } i \text{ of } A \begin{bmatrix} & & & \\ a_{i1} & a_{i2} & \cdots & a_{in} \\ & & & \end{bmatrix} \begin{bmatrix} b_{1j} \\ b_{2j} \\ \cdot \\ \cdot \\ \cdot \\ b_{nj} \end{bmatrix} \begin{array}{c} \text{column } j \text{ of } B \\ \\ \end{array}$$

At first it is a good idea to use a finger from the left hand to run along the elements of row i of A, and to use a finger from the right hand to run through the corresponding elements of column j of B. Compute the products and accumulate the sum. With very little practice this can be done with "no hands."

Even when AB is defined, the product BA might not be defined. If A is an $m \times n$ matrix, AB will be defined if and only if B has n rows. BA will be defined if and only if B has m columns. Thus, AB and BA will both be defined if and only if A is $m \times n$ and B is $n \times m$. In this case AB will be a square matrix of order m, and BA will be a square matrix of order n. There is no chance for AB to be the same as BA unless $m = n$. Even in that case it is not generally true that AB is the same as BA.

These three operations, addition of matrices, scalar multiplication, and multiplication of matrices, define an algebraic structure in the set of matrices. Certain laws, like the associative and distributive laws, can be shown to hold for this algebraic structure. To prove these laws at this point would require straightforward computations based on the definitions given. They are not difficult, but they are also not very instructive. Later (Section 4–4), when we see how matrices can be used to represent linear transformations, these proofs will become easy exercises. At this point we are content to state these properties without proof. In the following relations, remember that upper case italic letters denote matrices and lower case italic letters denote scalars.

$$(A + B) + C = A + (B + C) \tag{1.7}$$

$$A + B = B + A \tag{1.8}$$

$$(ab)C = a(bC) \tag{1.9}$$

$$a(B + C) = aB + aC \tag{1.10}$$

$$(a + b)C = aC + bC \tag{1.11}$$

$$(aB)C = B(aC) = a(BC) \tag{1.12}$$

$$(AB)C = A(BC) \tag{1.13}$$

$$A(B + C) = AB + AC \tag{1.14}$$

$$(A + B)C = AC + BC \tag{1.15}$$

The *main diagonal* of a matrix is the set of elements in the diagonal starting in the upper left-hand corner. All elements in the main diagonal have the same row and column indices. Thus, a_{ii} is the element in the i-th position of the main diagonal of $[a_{ij}]$. A square matrix in which all elements not in the main diagonal are zero is called a *diagonal matrix*.

We define δ_{ij} to be

$$\delta_{ij} = \begin{cases} 1, & \text{if } i = j, \\ 0, & \text{if } i \neq j. \end{cases} \quad (1.16)$$

This useful symbol is known as the *Kronecker delta*. We shall use it here to construct the n-th order square matrix $I_n = [\delta_{ij}]$. The matrix I_n is called the n-th order *identity matrix*. It is a diagonal matrix with 1's in the main diagonal. Usually all matrices under consideration are of the same order, or the order of I_n is known from context. Thus, usually we shall simply write I for the identity matrix. If A is an $m \times n$ matrix, then $I_m A = A$ and $A I_n = A$. It is easily seen from the definitions that $\sum_{k=1}^{m} \delta_{ik} a_{kj} = a_{ij}$, and $\sum_{k=1}^{n} a_{ik} \delta_{kj} = a_{ij}$. In particular, if A is a square matrix and I is an identity of the same order, then $IA = AI = A$.

Let A, B, and C be square matrices of the same order such that $AB = I$ and $CA = I$. Then $C = CI = C(AB) = (CA)B = IB = B$. B is called a right inverse of A, and C is called a left inverse of A. We have shown that if A has both a right inverse and a left inverse, then the left inverse and right inverse are identical and that the right inverse is also a left inverse. If A and B are square matrices of the same order such that $AB = BA = I$, then B is called an *inverse* of A. The argument above shows that a matrix cannot have two inverses. We could consider one to be a left inverse and the other to be a right inverse and show they are identical. Thus, we shall speak of *the* inverse of a matrix A and denote it by A^{-1}. Since $AA^{-1} = A^{-1}A = I$, A is also the inverse of A^{-1}. Later (Section 4-4) we will be able to show that if A and B are square matrices of the same order such that $AB = I$, then $BA = I$.

Matrix notation allows us to express the system of equations (1.3) in extremely compact form. Let A be the $m \times n$ matrix whose elements are the coefficients appearing on the left side of (1.3). That is, let

$$A = \begin{bmatrix} a_{11} & \cdots & a_{1n} \\ \vdots & & \vdots \\ a_{m1} & \cdots & a_{mn} \end{bmatrix} \quad (1.17)$$

Then let

$$X = \begin{bmatrix} x_1 \\ x_2 \\ \cdot \\ \cdot \\ \cdot \\ x_n \end{bmatrix}. \qquad (1.18)$$

Here X is an $n \times 1$ matrix. It is a matrix with only one column so that the column index can be suppressed. Let

$$B = \begin{bmatrix} b_1 \\ b_2 \\ \cdot \\ \cdot \\ \cdot \\ b_m \end{bmatrix}. \qquad (1.19)$$

AX and B are matrices, each with one column and m rows. To be equal, AX and B must have the same element in each position of the array. Thus, the m equations in (1.3) are collectively equivalent to the matrix equation

$$AX = B. \qquad (1.20)$$

The equality of two matrices of the same size imposes one condition for each position in the array. A matrix equation, like (1.20), is equivalent to a collection of simultaneous equations or conditions. Since many applications of mathematics involve a large number of simultaneous conditions, matrices are frequently used to express these conditions in compact form.

The matrix A in (1.17) is called the *coefficient matrix* of the system (1.3), and the matrix (1.6) is called the *augmented matrix* of the system (1.3).

EXERCISES 1-1

1. For $A = \begin{bmatrix} 1 & 2 & 3 \\ 4 & 5 & 6 \end{bmatrix}$, $B = \begin{bmatrix} 1 & -1 & 0 \\ 2 & 0 & -2 \end{bmatrix}$, $C = \begin{bmatrix} -1 & 1 \\ 2 & 0 \\ -1 & 3 \end{bmatrix}$,

   perform the following calculations as indicated.

   a) $2A + B$
   b) $(A + B)C$
   c) $AC + BC$
   d) $(AC)B$.

2. Show that

a) $\begin{bmatrix} 1 & 0 \\ 2 & 1 \end{bmatrix} \begin{bmatrix} 1 & 2 \\ 0 & 1 \end{bmatrix} \neq \begin{bmatrix} 1 & 2 \\ 0 & 1 \end{bmatrix} \begin{bmatrix} 1 & 0 \\ 2 & 1 \end{bmatrix}.$

(This illustrates the fact that matrix multiplication is not always commutative.)

b) $\begin{bmatrix} 1 & 2 \\ 2 & 4 \end{bmatrix} \begin{bmatrix} 2 & -4 \\ -1 & 2 \end{bmatrix} = \begin{bmatrix} 0 & 0 \\ 0 & 0 \end{bmatrix}.$

(This illustrates that a product of non-zero matrices might be zero.)

c) $\begin{bmatrix} 0 & 1 \\ 0 & 0 \end{bmatrix} \begin{bmatrix} 0 & 1 \\ 0 & 0 \end{bmatrix} = \begin{bmatrix} 0 & 0 \\ 0 & 0 \end{bmatrix}.$

(This illustrates that a power of a non-zero matrix might be zero.)

d) $\begin{bmatrix} 0 & 1 & 0 \\ 0 & 0 & 1 \\ 0 & 0 & 0 \end{bmatrix}^2 = \begin{bmatrix} 0 & 0 & 1 \\ 0 & 0 & 0 \\ 0 & 0 & 0 \end{bmatrix},$

$\begin{bmatrix} 0 & 1 & 0 \\ 0 & 0 & 1 \\ 0 & 0 & 0 \end{bmatrix}^3 = \begin{bmatrix} 0 & 0 & 0 \\ 0 & 0 & 0 \\ 0 & 0 & 0 \end{bmatrix}.$

e) $\begin{bmatrix} 1 & 2 & 3 \\ -1 & -2 & -2 \\ 1 & 2 & 1 \end{bmatrix}^3 = \begin{bmatrix} 0 & 0 & 0 \\ 0 & 0 & 0 \\ 0 & 0 & 0 \end{bmatrix}.$

f) $\begin{bmatrix} 1 & 2 \\ 3 & 7 \end{bmatrix} \begin{bmatrix} 7 & -2 \\ -3 & 1 \end{bmatrix} = \begin{bmatrix} 7 & -2 \\ -3 & 1 \end{bmatrix} \begin{bmatrix} 1 & 2 \\ 3 & 7 \end{bmatrix} = I_2.$

3. For $A = \begin{bmatrix} 1 & -2 \\ -2 & 5 \end{bmatrix}$, find a B such that $BA = I$. Verify that $AB = I$. (No systematic method for finding such a B has been suggested yet, but "trial-and-error" is not too difficult.)

4. For $A = \begin{bmatrix} 1 & 2 & 3 \\ 4 & 5 & 7 \end{bmatrix}$, $B = \begin{bmatrix} 1 & -1 \\ 2 & 0 \\ -1 & 1 \end{bmatrix}$, compute AB and BA.

5. a) For $D = \begin{bmatrix} 2 & 0 & 0 \\ 0 & -3 & 0 \\ 0 & 0 & 1 \end{bmatrix}$ and $A = \begin{bmatrix} 1 & 2 & 3 \\ 4 & 5 & 6 \\ 7 & 8 & 9 \end{bmatrix}$,

Compute AD and DA. How do the columns of AD compare with A, and how do the rows of DA compare with A?

b) Use $D' = \begin{bmatrix} -2 & 0 & 0 \\ 0 & 4 & 0 \\ 0 & 0 & 3 \end{bmatrix}$ and compute $D'A$ and AD' by using your observation in part a). Verify the accuracy of your observation by computing $D'A$ and AD' directly.

6. Let $A = \begin{bmatrix} -2 & 0 & 1 \\ -1 & -1 & 1 \end{bmatrix}$ and $B = \begin{bmatrix} 1 & 2 \\ 2 & 1 \\ 3 & 4 \end{bmatrix}$.

   Show that AB is an identity matrix, but that BA is not. Why does this not contradict the assertion made in page 6, that if $AB = I$ it follows that $BA = I$?

7. Let $D = \begin{bmatrix} 1 & 0 & 0 \\ 0 & 1 & 0 \\ 0 & 0 & 0 \end{bmatrix}$ and $A = \begin{bmatrix} 1 & 2 & 3 \\ 4 & 5 & 6 \\ 7 & 8 & 9 \end{bmatrix}$.

   Compute AD and DA. Also compute D^2. What is D^{10}?

8. Let $A = \begin{bmatrix} 1 & 2 \\ 3 & 4 \end{bmatrix}$ and $B = \begin{bmatrix} -3 & 10 \\ 15 & 12 \end{bmatrix}$.

   Show that $AB = BA$. This is an example to show that a matrix product can, under some circumstances, commute. (The "circumstance" here is that $B = A^2 - 10I$, so that $AB = A(A^2 - 10I) = A^3 - 10A = (A^2 - 10I)A = BA$.)

9. What conditions must be satisfied by A and B so that $(A + B)(A - B) = A^2 - B^2$?

10. If A is an $m \times n$ matrix and B is an $n \times r$ matrix, how many rows and columns are there in AB?

11. Write down the coefficient matrix and the augmented matrix for the system

    $$\begin{aligned} 2x_1 + 3x_2 + 3x_3 &= 16 \\ x_1 - x_2 &= 6 \\ -x_1 + x_2 + 3x_3 &= 18. \end{aligned}$$

12. Suppose a grinding machine in a machine shop must be used for every item produced by the shop. Of the three types of items produced by the shop, one unit of item A re-

quires 0.1 hours of grinding, item B requires 0.3 hours of grinding, and item C requires 0.4 hours of grinding. Suppose the manager contemplates producing x_1 units of A, x_2 units of B and x_3 units of C in each 8-hour working day. Write down a linear equation that x_1, x_2, and x_3 must satisfy if the grinding machine is to be used fully (never idle).

In using such an equation, several other conditions must be considered. For one thing, in the situation described x_1, x_2, and x_3 cannot be negative. Do we insist that x_1, x_2, and x_3 need be integers, or can we allow fractional values (by holding unfinished work until the next day)?

13. In the situation described in Exercise 12, suppose the items produced by the shop are currently ordered by one company that uses them as parts in a more complex product. Suppose, specifically, that each unit of the final product uses one of item A, 5 of item B, and 6 of item C. Assume the items are ordered and produced in that proportion. Write down two equations that the numbers x_1, x_2, and x_3 must satisfy to meet this condition.

14. Combine the equations obtained in Exercises 12 and 13, to obtain a system of three equations in x_1, x_2, and x_3. Show that if $x_1 = 2$, $x_2 = 10$, $x_3 = 12$, then each equation is satisfied.

1-2 SOLUTIONS OF SYSTEMS OF LINEAR EQUATIONS

Most students of elementary mathematics are accustomed to dealing with problems that have given just exactly the information needed to solve the problem. That is, no conditions stated are contradictory and none is superfluous. This speaks well of the person who formulated the problem, but in general one cannot expect every problem to have a unique solution, or even any solution at all.

For example, consider the system

$$x + y = 1$$
$$x + y = 2. \tag{2.1}$$

1-2 Solutions of Systems of Linear Equations • 11

It should be clear that any pair of numbers that satisfies one of the equations cannot satisfy the other. These equations are said to be *inconsistent*.

The number of equations and the number of unknowns have little to do with a system of equations being inconsistent. The simplest inconsistent system is

$$0 \cdot x = 1. \tag{2.2}$$

This single equation in one unknown may look too trivial to be seriously considered, but an equation of the same type can be obtained from the system (2.1) by subtracting the first equation from the second. For systems with many equations and many unknowns it is relatively more difficult to determine whether a system is inconsistent. But for any inconsistent system of linear equations, the equation $0 = 1$ (like equation (2.2)) can always be obtained by some combination of adding and subtracting equations. This will be shown in detail later.

Consider the system

$$\begin{aligned} x + y &= 5 \\ x - y &= 1. \end{aligned} \tag{2.3}$$

If x and y satisfy the system, then $2x = 6$ (add the two equations). Similarly, $2y = 4$ (subtract the second from the first). Thus $x = 3$, $y = 2$ is the only pair of numbers that can satisfy the system (2.3). To show that $x = 3$, $y = 2$ is a solution of the system (2.3), it is sufficient to substitute 3 for x every place that x occurs and 2 for y every place that y occurs and verify that each equation is satisfied.

Note that neither $x = 3$, nor $y = 2$, separately is a solution. The solution is the pair of numbers as a single object. Thus the system (2.3) has only one solution, a "unique" solution.

The situation where a system of linear equations has more than one solution is somewhat more involved. But the simplest system of this type,

$$x + y = 1, \tag{2.4}$$

illustrates the general idea. In equation (2.4) we can assign any value we please to one of the unknowns. An appropriate value of the other unknown can then be obtained which will satisfy the equation. In fact, this "appropriate" value can be specified by a formula. For example, from (2.4) we can obtain

$$y = 1 - x. \tag{2.5}$$

(2.5) shows how to calculate y, for whatever value is assigned to x, so that the equation (2.4) will be satisfied.

It should be clear that if x and y is any pair of numbers that satisfy (2.4), they must be related to each other by (2.5). That is, every solution of (2.4) can be obtained by choosing x arbitrarily and computing y by means of (2.5).

Given any system of equations (linear or otherwise) in n unknowns $x_1, x_2, \ldots, x_n$, a *solution* is a collection of n numbers, one for each unknown, such that when each number is substituted for its corresponding unknown in all places where the unknown occurs, all equations in the system are satisfied. The set of all solutions is the *solution set* for the system.

The solution set for the system in (2.1) is the empty set. That system has no solution. The solution set for the system in (2.4) is infinite. The solution set for the system (2.3) is finite since it contains a single element.

We wish to obtain effective methods for finding out whether a system of linear equations has a solution, and, if it has a solution, we would like to obtain the solution set. If the solution set is finite, that is a simple enough desire. We would like a list of all the elements in the solution set (and in this case it will turn out that the solution set has only one element anyway). If the solution set is infinite we cannot hope to list the elements in this set. In this case we will obtain a functional relation, similar to that in formula (2.5), by which any and all elements in the solution set can be obtained.

EXERCISES 1–2

To verify that a collection of n numbers is a solution, we substitute the numbers for the unknowns and determine that all the equations are satisfied.

1. Verify that $(-1, 1, 0)$ and $(3, -6, -1)$ are solutions of the system of equations

$$2x_1 - x_2 + 15x_3 = -3$$
$$x_1 + 2x_2 - 10x_3 = 1.$$

(We use the notation "$(-1, 1, 0)$" to denote the collection of numbers $x_1 = -1, x_2 = 1, x_3 = 0$.)

2. Verify that $(1, 1, 1)$, $(0, 3, 0)$, $(-1, 5, -1)$, and $(2, -1, 2)$ are solutions of the system of equations
$$x_1 + x_2 + x_3 = 3$$
$$x_1 + 2x_2 + 3x_3 = 6.$$

In Exercises 1 and 2 we verify that several triples of numbers are elements of the solution set. There might be more solutions than the few we are asked to verify. In Exercise 2 we might reasonably guess, from the solutions given, that $x_1 = x_3$ for every solution. Try it. The equations then become
$$x_1 + x_2 + x_3 = 2x_1 + x_2 = 3$$
$$x_1 + 2x_2 + 3x_3 = 4x_1 + 2x_2 = 6.$$

That is, *both* equations are satisfied if $x_3 = x_1$ and $x_2 = 3 - 2x_1$. This shows that
$$(x_1, 3 - 2x_1, x_1)$$
satisfies the system of equations given in Exercise 2.

3. Show that for any value of x_3, the triple $(1 + 2x_3, 3 - 2x_3, x_3)$ satisfies the system
$$x_1 + 2x_2 + 2x_3 = 7$$
$$-x_1 + x_2 + 4x_3 = 2.$$

4. Show that for any value of x_2, the triple $(5 - 7x_2, x_2, 14 + 11x_2)$ satisfies the system
$$2x_1 + 3x_2 + x_3 = 24$$
$$x_1 - 15x_2 + 2x_3 = 33.$$

5. Show that for any values of x_2 and x_3, $(4 - 2x_2 - 3x_3, x_2, x_3, 5 + x_2 - 2x_3, 6 - 4x_2 - x_3)$ satisfies the system
$$4x_1 + 2x_2 + 15x_3 + 2x_4 - x_5 = 20$$
$$-2x_1 + 3x_2 - 2x_3 + x_4 + 2x_5 = 9$$
$$x_1 - 7x_2 + 3x_3 + x_4 - 2x_5 = -3.$$

In Exercises 1 and 2 we have shown that a few triples of numbers are in the solution sets. In Exercises 3, 4, and 5 we have shown that infinitely many triples are in the solution sets. This kind of demonstration does not show how large the solution set is. It just identifies some elements in the solution set. The purpose of the following exercises is to establish limits on the solution sets.

14 • Linear Equations

6. Consider the system of equations

$$x_1 + x_2 + x_3 = 3$$
$$x_1 + 2x_2 + 3x_3 = 6.$$

Show that for any values of the unknowns that satisfy the system, we must have $x_1 = x_3$. (Hint: Assume x_1, x_2, x_3 satisfies the system and subtract twice the first equation from the second.)

7. For the system of equations given in Exercise 6, show that for any values of the unknowns that satisfy the system we must have $x_2 = 3 - 2x_1$. (Hint: Use the result of Exercise 6.)

8. Consider the system of equations

$$x_1 + 2x_2 + 2x_3 = 7$$
$$-x_1 + x_2 + 4x_3 = 2.$$

Show that for any values of the unknowns that satisfy this system, we must have $x_1 = 1 + 2x_3$ and $x_2 = 3 - 2x_3$.

9. Consider the system of equations

$$2x_1 + 3x_2 + x_3 = 24$$
$$x_1 - 15x_2 + 2x_3 = 33.$$

Show that any values of the unknowns that satisfy this system must also satisfy the conditions $x_1 = 5 - 7x_2$ and $x_3 = 14 + 11x_2$.

1-3 BASIC FORM

To see what the set of solutions of a system of linear equations looks like we shall examine a system of equations in a very special form. For a system in this special form the set of solutions is easy to obtain. In fact, an acceptable representation of the set of all solutions can be written down with no intervening steps. Later we shall show that for every system of linear equations that has a solution, it is possible by direct and simple methods to obtain a system in special form with the same solution set. This procedure is the method we will use to solve systems of linear equations: find a special system with the same solution set and read off the solution.

Suppose we have a system of linear equations in the special form:

$$\begin{aligned} x_1 & + a_{1,m+1}x_{m+1} + \cdots + a_{1n}x_n = b_1 \\ x_2 & + a_{2,m+1}x_{m+1} + \cdots + a_{2n}x_n = b_2 \\ & \\ x_m & + a_{m,m+1}x_{m+1} + \cdots + a_{mn}x_n = b_m. \end{aligned} \quad (3.1)$$

The important characteristic of this system is that for a certain set of the unknowns (equal in number to the number of equations) each appears with a non-zero coefficient in one and only one equation. Also, the coefficient of the unknown in the equation in which it appears is 1. In (3.1) these unknowns are indexed from 1 through m, but that is inessential. It is merely a matter of notational convenience. Also, the order of the equations is not important.

The form (3.1) is called a *basic form*. The unknowns that are isolated, one in each equation, are called the *basic variables*. Each equation contains only one basic variable and each basic variable appears in only one equation. This allows us to identify an equation by the basic variable it contains.

It is suggestive of the interpretations we wish to make to write the system (3.1) in a slightly different form.

$$\begin{aligned} x_1 &= b_1 - a_{1,m+1}x_{m+1} - \cdots - a_{1n}x_n \\ x_2 &= b_2 - a_{2,m+1}x_{m+1} - \cdots - a_{2n}x_n \\ & \\ x_m &= b_m - a_{m,m+1}x_{m+1} - \cdots - a_{mn}x_n. \end{aligned} \quad (3.2)$$

It should now be evident that the unknowns $x_{m+1}, x_{m+2}, \ldots, x_n$ can be assigned any numerical values whatever, and the corresponding values of $x_1, x_2, \ldots, x_m$ can easily be computed from the formulas in (3.2). The variables $x_{m+1}, \ldots, x_n$ are called the *parameters* of the solution. In (3.2) the basic variables are expressed in terms of, or functions of, the parameters.

It is easily seen that *all* solutions of the system (3.2) can be obtained by assigning values to the parameters arbitrarily and computing the corresponding values for the basic variables. Since systems (3.1) and (3.2) have the same solution set this yields all solutions to the system (3.1). Accordingly, we will consider that we have solved a system of linear equations when we obtain a parametric representation of the solution set in the general form of (3.2).

We refer to a parametric representation, such as (3.2), which gives all solutions as a *general solution*. When particular numbers are specified for the parameters, and the corresponding values of the basic variables are determined, we refer to this collection of numbers as a *particular solution*. An especially evident particular solution is obtained by giving all parameters the value zero. Thus, $x_1 = b_1, \ldots, x_m = b_m$, $x_{m+1} = x_{m+2} = \cdots = x_n = 0$ is a particular solution.

Generally, there are many different choices possible for the set of basic variables. And for each choice of the set of basic variables there is a corresponding parametric representation of the solution set. But even these choices do not limit the possible parametrizations of the solution set. An unlimited variety of changes of variables is available. For example, in (3.2) let

$$\begin{aligned} x_{n-1} &= y_{n-1} + y_n \\ x_n &= -y_{n-1} + y_n. \end{aligned} \qquad (3.3)$$

Then we obtain another parametrization,

$$\begin{aligned} x_1 &= b_1 - a_{1,m+1}x_{m+1} - \cdots - (a_{1,n-1} - a_{1n})y_{n-1} - (a_{1,n-1} + a_{1n})y_n \\ &\vdots \\ x_m &= b_m - a_{m,m+1}x_{m+1} - \cdots - (a_{m,n-1} - a_{mn})y_{n-1} - (a_{m,n-1} + a_{mn})y_n. \end{aligned} \qquad (3.4)$$

The parameters in (3.4) are $\{x_{m+1}, x_{m+2}, \ldots, x_{n-2}, y_{n-1}, y_n\}$. Implicit in (3.2) was the understanding that $n - m$ of the unknowns were taken as the parameters and they could be given arbitrary values. Actually, we should have expressed all unknowns as functions of the parameters. That is, we should have also given the relations, $x_{m+1} = x_{m+1}, \ldots, x_n = x_n$. With the change of parameters suggested we should now give the relations, $x_{m+1} = x_{m+1}, \ldots, x_{n-2} = x_{n-2}, x_{n-1} = y_{n-1} + y_n$, $x_n = -y_{n-1} + y_n$ in addition to those in (3.4). Clearly, many other parametrizations of this type are possible. In this text we shall be content to obtain a general solution resulting from a basic form. They are directly obtained, simple to write down, and for each system of equations only a finite number of basic forms can be obtained.

EXERCISES 1-3

Which of the following systems of equations are in basic form?

1. $3x_1 + x_2 - x_4 = 3$
   $2x_1 + 3x_4 + x_5 = 4$
   $-x_1 + x_3 = 5.$

2. $x_1 + x_3 = 1$
   $ x_2 + x_4 = 2$
   $x_1 + x_4 = 3$

3. $ x_3 + x_4 = 2$
   $ x_2 + x_4 = 4$
   $x_1 + x_4 = 6$

4. $x_1 + x_2 + x_3 + x_4 = 4$

5. $x_1 + x_3 = 1$
   $ x_2 + x_4 = 2$
   $ x_3 + x_5 = 3$

6. $x_1 + x_2 + x_4 + x_6 = 1$
   $x_1 + x_3 = 2$
   $x_1 + x_2 - x_4 + x_5 = 3$

7. For each of the systems in Exercises 1, 2, 3, and 6 that is in basic form, rewrite the system in parametric form (similar to (3.2)).

8. In Exercise 4, the system can be interpreted to be in four different basic forms without making any changes. Write out all four parametrizations that result.

9. In Exercise 5, the system can be interpreted to be in two different basic forms without making any changes. Write out both parametrizations that result.

1-4 EQUIVALENT SYSTEMS OF EQUATIONS

In the previous section we have shown that a system of linear equations in basic form is easy to solve. We raise the question: Given any system of linear equations can we, by simple, direct, and effective methods, find another system of equations in basic form with the same solution set? If this can be done the solution is then easily obtained from the basic form. If the original system has a solution, the answer is "yes." If the system does not have a solution a system like (3.1) cannot be obtained, since a system in that form has a solution, but the methods used to obtain the basic form will also yield the conclusion that the system fails to have a solution.

Consider the single linear equation

$$a_{11}x_1 + a_{12}x_2 + \cdots + a_{1n}x_n = b_1, \tag{4.1}$$

and the equation

$$ca_{11}x_1 + ca_{12}x_2 + \cdots + ca_{1n}x_n = cb_1 \tag{4.2}$$

obtained from (4.1) by multiplying every coefficient by the constant c. Every set of numbers that satisfies equation (4.1) also satisfies equation (4.2). If c is non-zero then (4.1) can be obtained from (4.2) by multiplying the coefficients of (4.2) by the number c^{-1}. Thus, for a non-zero number c, systems (4.1) and (4.2) have the same solutions. For a system of linear equations, *if one of the equations is multiplied in this way by a non-zero number a system is obtained with the same solution set.*

Now consider two linear equations

$$\begin{aligned}a_{11}x_1 + a_{12}x_2 + \cdots + a_{1n}x_n &= b_1 \\ a_{21}x_1 + a_{22}x_2 + \cdots + a_{2n}x_n &= b_2.\end{aligned} \tag{4.3}$$

If we multiply the first equation by the number c, and add the result to the second equation, we obtain

$$a_{11}x_1 + a_{12}x_2 + \cdots + a_{1n}x_n = b_1$$
$$(a_{21}+ca_{11})x_1 + (a_{22}+ca_{12})x_2 + \cdots + (a_{2n}+ca_{1n})x_n = b_2 + cb_1. \tag{4.4}$$

Again, every solution of system (4.3) is also a solution of (4.4). Also, if we multiply the first equation in (4.4) by $-c$ and add the result to the second equation, we obtain (4.3). Thus, for any number c, systems (4.3) and (4.4) have the same solutions. For any system of equations, *adding a multiple of one equation to another results in a system with the same solution set.*

It turns out that these two types of manipulations to obtain new systems of equations are sufficient to obtain our stated objectives. However, for convenience it is desirable to have one more type of manipulation available. We can easily see that changing the order in which the equations are listed has no effect on the solution of the system. *Interchanging two equations in a system of linear equations leads to a system with the same solution set.*

These three operations, (1) multiplying an equation by a non-zero constant, (2) adding a multiple of one equation to another, and (3) interchanging two equations, are known as *elementary operations.* These operations, or any sequence of these operations, applied to a system of linear equations, lead to a system with the same solution set. We will call two systems of equations *equivalent* if one can be obtained from the other by a sequence of elementary operations.

Now consider what can be done with a properly chosen sequence of elementary operations. Consider the general system of m linear equations in n unknowns.

$$a_{11}x_1 + \cdots + a_{1k}x_k + \cdots + a_{1n}x_n = b_1$$
$$\vdots$$
$$a_{r1}x_1 + \cdots + \boxed{a_{rk}}x_k + \cdots + a_{rn}x_n = b_r \qquad (4.5)$$
$$\vdots$$
$$a_{m1}x_1 + \cdots + a_{mk}x_k + \cdots + a_{mn}x_n = b_m.$$

Consider any variable that appears with at least one non-zero coefficient. Suppose, for example, that a_{rk} is non-zero. We have encircled this term in the array above for emphasis. We can multiply the r-th equation by a_{rk}^{-1} to obtain an equivalent system in which x_k has coefficient 1 in the r-th equation.

$$a_{11}x_1 + \cdots + a_{1k}x_k + \cdots + a_{1n}x_n = b_1$$
$$\vdots$$
$$\frac{a_{r1}}{a_{rk}}x_1 + \cdots + x_k + \cdots + \frac{a_{rn}}{a_{rk}}x_n = \frac{b_r}{a_{rk}} \qquad (4.6)$$
$$\vdots$$
$$a_{m1}x_1 + \cdots + a_{mk}x_k + \cdots + a_{mn}x_n = b_m.$$

Now add $-a_{ik}$ times the r-th equation to the i-th equation, for $i \neq r$, to obtain a new i-th equation in which x_k has a zero coefficient.

$$\left(a_{11} - \frac{a_{1k}a_{r1}}{a_{rk}}\right)x_1 + \cdots + 0 \cdot x_k + \cdots + \left(a_{1n} - \frac{a_{1k}a_{rn}}{a_{rk}}\right)x_n = b_1 - \frac{a_{1k}b_r}{a_{rk}}$$

$$\vdots$$

$$\frac{a_{r1}}{a_{rk}}x_1 + \cdots + x_k + \cdots + \frac{a_{rn}}{a_{rk}}x_n = \frac{b_r}{a_{rk}} \quad (4.7)$$

$$\vdots$$

$$\left(a_{m1} - \frac{a_{mk}a_{r1}}{a_{rk}}\right)x_1 + \cdots + 0 \cdot x_k + \cdots + \left(a_{mn} - \frac{a_{mk}a_{rn}}{a_{rk}}\right)x_n = b_m - \frac{a_{mk}b_r}{a_{rk}}$$

The complexity of the notation in the system (4.7) can be reduced by renaming the coefficients. Let

$$a_{ij} - \frac{a_{ik}a_{rj}}{a_{rk}} = a_{ij}', \quad b_i - \frac{a_{ik}b_r}{a_{rk}} = b_i' \quad (4.8)$$

for $i \neq r$ and $j \neq k$, and let

$$\frac{a_{rj}}{a_{rk}} = a_{rj}', \quad \frac{b_r}{a_{rk}} = b_r' \quad (4.9)$$

for $j \neq k$. In this notation system (4.7) becomes

$$a_{11}'x_1 + \cdots + 0 \cdot x_k + \cdots + a_{1n}'x_n = b_1'$$

$$\vdots$$

$$a_{r1}'x_1 + \cdots + x_k + \cdots + a_{rn}'x_n = b_r' \quad (4.10)$$

$$\vdots$$

$$a_{m1}'x_1 + \cdots + 0 \cdot x_k + \cdots + a_{mn}'x_n = b_m'.$$

By this sequence of operations we have obtained the system (4.10), which is equivalent to the original system (4.5) and in which x_k is a basic variable in the r-th equation.

After this is accomplished we can pass our attention to

other variables. We look for a variable that has a non-zero coefficient in an equation other than the r-th. If there is such a variable we proceed, as in the steps from system (4.5) to (4.10), to eliminate that variable from all equations but one. Since the new non-zero coefficient will be selected from an equation other than the r-th and the coefficient of x_k in that equation (in system (4.10)) is zero, the pattern of coefficients of x_k will not be disturbed. This means that x_k will remain a basic variable while the next group of steps will introduce an additional basic variable.

Let us see how these ideas work out in a particular example. Consider the system

$$\begin{aligned} 2x_1 + 3x_2 + 4x_3 &= 1 \\ 3x_1 + ②x_2 - 2x_3 &= 2 \\ 4x_1 + 2x_2 + 2x_3 &= -4. \end{aligned} \qquad (4.11)$$

Using the encircled 2 in the second equation as described we obtain x_2 as a basic variable in

$$\begin{aligned} -\frac{5}{2}x_1 \quad\quad\quad + 7x_3 &= -2 \\ \frac{3}{2}x_1 + x_2 - x_3 &= 1 \\ x_1 \quad\quad\quad + 4x_3 &= -6. \end{aligned} \qquad (4.12)$$

We could select the next basic variable in several ways, but to simplify the arithmetic we select x_1 from the third equation.

$$\begin{aligned} 17x_3 &= -17 \\ x_2 - 7x_3 &= 10 \\ x_1 \quad\quad + 4x_3 &= -6. \end{aligned} \qquad (4.13)$$

In (4.13) the only non-zero coefficient that is not in one of the equations already selected is the 17 in the first equation. Carrying out one more elimination we obtain

$$\begin{aligned} x_3 &= -1 \\ x_2 &= 3 \\ x_1 &= -2. \end{aligned} \qquad (4.14)$$

Equation (4.14) is in basic form. There are no parameters and the unique solution is easily obtained.

The process we have introduced and illustrated may be summarized in the following way. At any stage where we have completed an elimination and obtained a new basic variable, we look for a variable with a non-zero coefficient in an equation not already containing a basic variable. If we find one, we then proceed, as above, to make it an additional basic variable. The process will terminate either when we run out of equations or when we run out of variables that meet the selection conditions.

Consider first the situation in which we have more non-zero equations than basic variables but a further selection of a basic variable is not possible. Then there must be an equation of the form

$$0 = b \qquad (4.15)$$

where b is non-zero. The left side is zero because the coefficients of all variables must be zero, or else a further selection of a basic variable could be made. The right side must then be non-zero since the equation is a non-zero equation. An equation like (4.15) can have no solution, and hence the original system has no solution.

The other possibility is that the selection process continues until there are as many basic variables as there are non-zero equations. The resulting system will look like system (3.1), except that the basic variables will not necessarily be the first m variables. This is the basic form discussed in Section 1–3, and the general solution is immediately obtainable.

We see, therefore, that the process of determining a set of basic variables terminates in one of two ways: either (1) we find a set of basic variables equal in number to the number of non-zero equations remaining (perhaps less than we started with), or (2) we find at least one equation like (4.15). In case (1) a parametric general solution can be read off and this is a general solution to the original system. In case (2) the original system has no solution.

It is clear from the description of the reduction process that the choices that lead to a set of basic variables are somewhat arbitrary. To what extent is the end result dependent on these choices? Suppose that by two different sequences of choices we have selected the same set of basic variables. Notice that from a basic form we can obtain directly a parametric representation, and conversely. This is the correspondence between (3.1) and (3.2) considered in Section 1–3.

1-4 Equivalent Systems of Equations

Thus, two basic forms, with the same set of basic variables will be the same (except possibly for the order of the equations) if and only if the corresponding parametrizations are the same.

If two basic forms can be obtained from the same original system of equations, they have the same solution set as the original system of equations. We now show that if two parametrizations with the same set of basic variables and the same set of parameters are not identical they cannot describe the same solution set. For every choice of a set of values for the parameters $x_{m+1}, \ldots, x_n$ in system (3.2) we get one and only one solution in the solution set. If we have two different parametrizations with the same set of basic variables in the form of system (3.2), the parametrizations must differ in one of the b_i or in one of the a_{ij}. If they differ in one of the b_i, set all parameters equal to zero. Then for that set of parameters, x_i has one value in one solution set and a different value in the other solution set. If all the constant terms in the two parametrizations are the same and one of the a_{ij} is different in the two parametrizations, set $x_j = 1$ and set all other parameters equal to zero. Again, for that set of parameters, x_i has different values in the two solution sets. Since only one solution corresponds to each choice of values of the parameters, the two solution sets are different.

Summarizing the discussion of the last two paragraphs: if, from a given system of equations, we could obtain two different basic forms with the same set of basic variables, we would obtain two different parametrizations and two different solutions sets. Since elementary operations do not change the solutions set this is impossible. Thus, for a given set of basic variables, the basic form obtained is independent of the particular sequence of choices made in obtaining the basic form.

EXERCISES 1-4

Solve the following systems of equations. Obtain a general solution or show the system has no solutions.

1. $x_1 + x_2 = 1$
   $x_1 - 2x_2 = -5$

2. $2x_1 - 3x_2 - 8x_3 = 7$
   $-x_1 + 2x_2 + 5x_3 = -4$
   $3x_1 + 4x_2 + 5x_3 = 2$

3. $4x_1 - 8x_2 + 16x_3 = 4$
   $x_1 - 3x_2 + 6x_3 = -3$
   $2x_1 + x_2 + x_3 = 1$

4. $2x_1 + 2x_2 + 5x_3 + 6x_4 = 1$
   $x_1 + x_2 - 2x_3 + 3x_4 = 5$

5. $2x_1 + x_2 = 5$
   $5x_1 - 2x_2 = 8$
   $6x_1 + 2x_2 = 14$

6. $x_1 - 3x_2 + x_3 + x_4 = 6$
   $-x_1 + 3x_2 + 2x_3 - x_4 = 0$
   $2x_1 - 6x_2 - x_3 + 2x_4 = 6$
   $-2x_1 + 6x_2 + x_3 - x_4 = -3$

7. $2x_1 + 8x_2 + 5x_3 + 9x_4 = -4$
   $2x_1 + x_3 + 5x_4 = 4$
   $3x_1 - 7x_2 - 2x_3 + 4x_4 = 13$
   $x_1 + 5x_2 + 3x_3 + 5x_4 = -3$

8. $x_1 + x_2 + x_3 + 3x_4 = 1$
   $2x_1 + 4x_2 + 3x_3 + 7x_4 = 0$
   $3x_1 + x_2 + 2x_3 + 8x_4 = 5$
   $-2x_1 + 4x_2 + x_3 - 3x_4 = -8$

9. In Exercise 2 of Section 1-2, we considered the system
$$x_1 + x_2 + x_3 = 3$$
$$x_1 + 2x_2 + 3x_3 = 6.$$
Show that the system
$$2x_1 + x_2 = 3$$
$$-x_1 + x_3 = 0$$
is an equivalent system in basic form.

10. In Exercise 3 of Section 1-2, we considered the system
$$x_1 + 2x_2 + 2x_3 = 7$$
$$-x_1 + x_2 + 4x_3 = 2.$$
Show that the system
$$x_1 - 2x_3 = 1$$
$$x_2 + 2x_3 = 3$$
is an equivalent system in basic form.

11. In Exercise 4 of Section 1-2, we considered the system
$$2x_1 + 3x_2 + x_3 = 24$$
$$x_1 - 15x_2 + 2x_3 = 33.$$

Show that the system
$$x_1 + 7x_2 = 5$$
$$-11x_2 + x_3 = 14$$
is an equivalent system in basic form.

12. In Exercise 5 of Section 1-2, we considered the system
$$4x_1 + 2x_2 + 15x_3 + 2x_4 - x_5 = 20$$
$$-2x_1 + 3x_2 - 2x_3 + x_4 + 2x_5 = 9$$
$$x_1 - 7x_2 + 3x_3 + x_4 - 2x_5 = -3.$$
Show that the system
$$x_1 + 2x_2 + 3x_3 = 4$$
$$-x_2 + 2x_3 + x_4 = 5$$
$$4x_2 + x_3 + x_5 = 6$$
is an equivalent system in basic form.

13. The systems of equations
$$x_1 = 1$$
$$x_1 = 2,$$
and the system
$$x_1 + x_2 = 1$$
$$x_1 + x_2 = 2,$$
both have the empty set as solution sets. Why are they not equivalent systems of equations?

1-5 ELEMENTARY OPERATIONS

If one spends even a little time manipulating systems of linear equations along the lines suggested in Section 1-4, it soon becomes apparent that much of the notation (the variables and the plus and equal signs) that one writes down is repetitious and conveys little or no information. As pointed out in Section 1-1, a matrix can be used to represent a system of linear equations and the representation contains all information needed. Accordingly, it is advantageous to perform all the manipulations described in Section 1-4 on the representing matrices.

We define three types of *elementary row operations* on matrices.

Type I: Multiply a row by a non-zero scalar.
Type II: Add a multiple of one row to another row.
Type III: Interchange two rows.
(From a strictly logical point of view, a type III operation is not needed since the same effect can be achieved by a sequence of operations of types I and II. If R_1 and R_2 are two rows of a matrix, we can obtain successively R_1 and $R_1 + R_2$, $-R_2$ and $R_1 + R_2$, $-R_2$ and R_1, and then R_2 and R_1. However, our aim is to devise convenient and effective computational methods and there is no point in replacing an operation as simple as interchanging two rows by a sequence of four more complicated operations.)

The elementary operations applied to matrices correspond to the operations utilized in Section 1–4 to reduce a system of equations to a basic system. To solve a system of linear equations we first write down the augmented matrix representing the system. We then apply to this matrix a sequence of elementary operations corresponding to the sequence of operations used to reduce the system of equations to basic form.

Expressed in terms of the representing matrices, introducing a basic variable amounts to the following group of decisions and elementary operations:

1. Select a non-zero element a_{rk} in the matrix $[a_{ij}]$. This element is called the *pivot element*.
2. Divide row r by a_{rk}, so that the element in the pivot position becomes a 1.
3. Add multiples of row r to the other rows so that the other elements in column k become zeros.

This group of operations is called a *pivot operation*.

Expressed in terms of the equations, when the pivot operation is completed, x_k becomes a basic variable since it appears with a non-zero coefficient only in the r-th equation. x_k is the basic variable corresponding to the r-th equation.

When this pivot operation is completed we may proceed to select another pivot element. If the new pivot element is selected from any row other than row r, the pattern of coefficients in column k will remain undisturbed. This means x_k will remain a basic variable and the new pivot operation will introduce an additional basic variable.

When using pivot operations to solve a system of linear equations, there is no motivation to select a pivot element from a row corresponding to a basic variable. In that context we wish to increase the set of basic variables as quickly as possible until no more can be selected. However, it is desirable, for conceptual reasons, to permit selecting a pivot element from a row corresponding to a basic variable. Thus, suppose a_{rk} was a previously selected pivot element and a pivot operation was performed with respect to that element. Now we wish to select a new pivot element from row r. Suppose that a_{rj}' (in the notation of system (3.9)) is selected as a pivot element. Unless $a_{rj}' = 1$ and all other elements in column j are zero, pivoting on a_{rj}' will alter the elements in column k. If column k is changed by this pivot operation, x_k will no longer be a basic variable. But even if column k is not changed we do not want to have two basic variables corresponding to the same equation. (This kind of situation came up in Exercises 1–4 and 1–5 of Section 1–3.) Thus, regardless of the results of a pivot operation, we regard the selection of a new pivot element as introducing a new basic variable. If the new pivot element is from an equation already corresponding to a basic variable, that variable is deleted from the list of basic variables and replaced by the new one.

When a system of equations is represented by an augmented matrix the last column represents the constant terms. Thus, the pivot element should not be selected from that column, if possible, since that column does not represent the coefficients of a variable. To solve a system of linear equations we start with the augmented matrix representing the system, and sequentially introduce basic variables by selecting pivot elements from the columns containing the coefficients. The selection process terminates in one of two possible ways. The first possibility is that every non-zero row corresponds to a basic variable. In this case, the matrix obtained represents a system of equations in basic form, like (3.1). The second possibility is that there is a non-zero row that corresponds to no basic variable, but a pivot element cannot be selected from a column representing a variable. This means that the only non-zero element in that row is in the column of constant terms. But then this row represents an equation of the form $0 = b$ where $b \neq 0$. To close the matter in tidy form (and so some statements to be made later can be expressed more concisely) we shall in this case select a pivot element from the column of constants and perform one more pivot operation. In either of these two cases the resulting final matrix obtained is said to be in *basic form*. The process of obtaining the basic form by a sequence of pivot operations is called a *reduction to basic form*. The basic form obtained is also called a *reduced form*.

When the reduced, or basic, form is obtained, in one case the number of non-zero rows is equal to the number of basic variables, and the corresponding system of equations has a solution. In the other case the number of non-zero rows is one more than the number of basic variables, and the system has no solution.

The leading term of a row is the first non-zero element in that row. A matrix is in *row-echelon form* if (1) the leading term in each row is 1, (2) the leading term in each row is to the right of the leading terms of all rows above, (3) for each column containing a leading term all other elements in that column are zeros, and (4) all non-zero rows are written above all zero rows. Since each leading term corresponds to a basic variable, the row-echelon form is a basic form. Condition (2) implies that the equations are ordered in the same order as the basic variables to which they correspond. Condition (1) implies that no basic variable can be replaced by a parameter with a smaller index in a new basic form. Thus, in a row-echelon form the basic variables are selected to have the smallest indices possible.

A row-echelon form is useful because it is singled out from the several possible basic forms. We have already shown that for a given set of basic variables, the corresponding basic form is unique. For a given system of equations, the set of basic variables that gives the row-echelon form is uniquely determined. Thus, for a given system of equations, or a given matrix, the resulting row-echelon form is unique. The importance of this fact will become apparent as we make use of it.

The following matrix (5.1) is an illustration of a matrix in row-echelon form. All the 1's shown are leading elements. The elements indicated with asterisks are the only elements whose values are not determined by the locations of the leading elements.

$$\begin{bmatrix} 1 & * & 0 & * & * & 0 & 0 & * & 0 & * \\ 0 & 0 & 1 & * & * & 0 & 0 & * & 0 & * \\ 0 & 0 & 0 & 0 & 0 & 1 & 0 & * & 0 & * \\ 0 & 0 & 0 & 0 & 0 & 0 & 1 & * & 0 & * \\ 0 & 0 & 0 & 0 & 0 & 0 & 0 & 0 & 1 & * \\ 0 & 0 & 0 & 0 & 0 & 0 & 0 & 0 & 0 & 0 \end{bmatrix} \quad (5.1)$$

In matrix (5.1) the 1's which are the coefficients of basic variables are indicated in red. This is done to make the role of these numbers unambiguous. As needed, the student can underscore these coefficients to keep things straight. With experience such a device becomes unnecessary. Even where a matrix does not represent a system of equations, or an element of a matrix does not represent the coefficient of an unknown, we shall need to perform pivot operations. It is convenient to use the same terminology and refer to the 1's that result from a pivot operation as "pivot elements," "basic coefficients," or "basic elements."

Numerical Examples

(1) Consider the system of equations,

$$3x_1 + x_2 + 9x_3 = -2$$
$$3x_1 + 2x_2 + 12x_3 = -1$$
$$2x_1 + x_2 + 7x_3 = -1.$$

The augmented matrix representing this system is

$$\begin{bmatrix} 3 & \textcircled{1} & 9 & -2 \\ 3 & 2 & 12 & -1 \\ 2 & 1 & 7 & -1 \end{bmatrix}.$$

We select the encircled element as the pivot element. According to instructions, any non-zero element in the first three columns could have been selected. But choosing a 1 avoids the introduction of fractions. After the first pivot operation we have

$$\begin{bmatrix} 3 & 1 & 9 & -2 \\ -3 & 0 & -6 & 3 \\ \textcircled{-1} & 0 & -2 & 1 \end{bmatrix}.$$

Using the encircled element as a pivot element we have, after the second step in the reduction,

$$\begin{bmatrix} 0 & 1 & 3 & 1 \\ 0 & 0 & 0 & 0 \\ 1 & 0 & 2 & -1 \end{bmatrix}.$$

We have now obtained a reduced form with $\{x_1, x_2\}$ as the set of basic variables. x_3 is a parameter, and the general solution in parametric form is

$$x_1 = -1 - 2x_3$$
$$x_2 = 1 - 3x_3.$$

(2) Consider the system

$$3x_1 + x_2 + 9x_3 = -2$$
$$3x_1 + 2x_2 + 12x_3 = 1$$
$$2x_1 + x_2 + 7x_3 = -1.$$

This system has the same coefficient matrix as example (1). But notice what happens. In the following sequence of matrices the pivot elements have been encircled and no further explanation should be necessary.

$$\begin{bmatrix} 3 & ① & 9 & -2 \\ 3 & 2 & 12 & 1 \\ 2 & 1 & 7 & -1 \end{bmatrix}$$

$$\begin{bmatrix} 3 & 1 & 9 & -2 \\ -3 & 0 & -6 & 5 \\ ⊖ & 0 & -2 & 1 \end{bmatrix}$$

$$\begin{bmatrix} 0 & 1 & 3 & 1 \\ 0 & 0 & 0 & ② \\ 1 & 0 & 2 & -1 \end{bmatrix}$$

$$\begin{bmatrix} 0 & 1 & 3 & 0 \\ 0 & 0 & 0 & 1 \\ 1 & 0 & 2 & 0 \end{bmatrix}$$

In this example the system of equations has no solution.

(3) Consider the system

$$3x_1 + x_2 + 9x_3 = 10$$
$$3x_1 + 2x_2 + 10x_3 = 13$$
$$2x_1 + x_2 + 7x_3 = 9.$$

1-5 Elementary Operations

In three pivot operations we have

$$\begin{bmatrix} 3 & 1 & 9 & 10 \\ 3 & 2 & 10 & 13 \\ 2 & ① & 7 & 9 \end{bmatrix}$$

$$\begin{bmatrix} ① & 0 & 2 & 1 \\ -1 & 0 & -4 & -5 \\ 2 & 1 & 7 & 9 \end{bmatrix}$$

$$\begin{bmatrix} 1 & 0 & 2 & 1 \\ 0 & 0 & ⊖2 & -4 \\ 0 & 1 & 3 & 7 \end{bmatrix}$$

$$\begin{bmatrix} 1 & 0 & 0 & -3 \\ 0 & 0 & 1 & 2 \\ 0 & 1 & 0 & 1 \end{bmatrix}.$$

It is now easily seen that the unique solution is $x_1 = -3$, $x_2 = 1$, $x_3 = 2$.

Note: It seems that students of elementary mathematics often have the impression that a system of linear equations in which the number of equations is equal to the number of unknowns always has a unique solution. The examples given above should show that the situation is not quite so simple. In many areas of applications of linear algebra there are often reasons, based on theory, to expect the solution to be unique, as it is in example (3), or to expect the solution set to be infinite, as it is in example (1).

(4) Let us try a more complicated example:

$$\begin{aligned} 2x_1 \phantom{{}+2x_2} + 6x_3 + x_4 + 2x_5 + 4x_6 &= 5 \\ 2x_1 + 2x_2 + 10x_3 \phantom{{}+x_4} + x_5 + 3x_6 &= 8 \\ x_1 + x_2 + 5x_3 + x_4 + 3x_5 + x_6 &= 8 \\ 3x_1 \phantom{{}+2x_2} + 9x_3 + x_4 + x_5 + 7x_6 &= 4 \\ 5x_1 + 3x_2 + 21x_3 + 5x_4 + 2x_5 + 18x_6 &= 10. \end{aligned}$$

$$\begin{bmatrix} 2 & 0 & 6 & 1 & 2 & 4 & 5 \\ 2 & 2 & 10 & 0 & 1 & 3 & 8 \\ 1 & ① & 5 & 1 & 3 & 1 & 8 \\ 3 & 0 & 9 & 1 & 1 & 7 & 4 \\ 5 & 3 & 21 & 5 & 2 & 18 & 10 \end{bmatrix}$$

32 • Linear Equations

$$\begin{bmatrix} 2 & 0 & 6 & \boxed{1} & 2 & 4 & 5 \\ 0 & 0 & 0 & -2 & -5 & 1 & -8 \\ 1 & 1 & 5 & 1 & 3 & 1 & 8 \\ 3 & 0 & 9 & 1 & 1 & 7 & 4 \\ 2 & 0 & 6 & 2 & -7 & 15 & -14 \end{bmatrix}$$

$$\begin{bmatrix} 2 & 0 & 6 & 1 & 2 & 4 & 5 \\ 4 & 0 & 12 & 0 & -1 & 9 & 2 \\ -1 & 1 & -1 & 0 & 1 & -3 & 3 \\ \boxed{1} & 0 & 3 & 0 & -1 & 3 & -1 \\ -2 & 0 & -6 & 0 & -11 & 7 & -24 \end{bmatrix}$$

$$\begin{bmatrix} 0 & 0 & 0 & 1 & 4 & -2 & 7 \\ 0 & 0 & 0 & 0 & \boxed{3} & -3 & 6 \\ 0 & 1 & 2 & 0 & 0 & 0 & 2 \\ 1 & 0 & 3 & 0 & -1 & 3 & -1 \\ 0 & 0 & 0 & 0 & -13 & 13 & -26 \end{bmatrix}$$

$$\begin{bmatrix} 0 & 0 & 0 & 1 & 0 & 2 & -1 \\ 0 & 0 & 0 & 0 & 1 & -1 & 2 \\ 0 & 1 & 2 & 0 & 0 & 0 & 2 \\ 1 & 0 & 3 & 0 & 0 & 2 & 1 \\ 0 & 0 & 0 & 0 & 0 & 0 & 0 \end{bmatrix}.$$

The parameters of the general solution are x_3 and x_6, and the general solution is

$$\begin{aligned} x_1 &= 1 - 3x_3 - 2x_6 \\ x_2 &= 2 - 2x_3 \\ x_4 &= -1 - 2x_6 \\ x_5 &= 2 + x_6. \end{aligned}$$

The examples given here were contrived so that it was not necessary to handle large numbers or fractions. If such numbers were used it would add to the tedium and increase the probability of a blunder, but it would not change the pattern of the steps. In example (4) there are eleven different choices for the two variables selected as parameters, and for each choice of parameters there are many possible orders for the sequence of calculations. Some are more convenient than others, but all possible parametric solutions are considered to be equally satisfactory for our purposes.

For solving systems of linear equations in which the number of equations and unknowns is small and in which the coefficients are small integers, the technique developed in this section is quite efficient. It is called *Gauss-Jordan elimination*, or *Jordan elimination*. In terms of the number of arithmetic steps that must be performed, Gaussian elimination, which is described in the next section, is shorter. For very large systems of equations, Gaussian elimination requires about two-thirds as many steps as Jordan elimination. For systems small enough to be approachable by "paper and pencil," the difference in efficiency between the two methods is negligible.

The student is expected to obtain exact solutions for the systems given in the exercises. For systems large enough to require the use of a computing machine, or for systems in which the coefficients are approximations, it is unrealistic to require exact solutions. There is little point in trying to develop the technique for solving large systems by hand; we assume such systems will always be put on a computer. Accordingly, the exercises in the text are contrived to involve a minimal amount of arithmetic complication. Though speed is not important, accuracy is and it is within reach of every student. The answers can, and should, be checked by substituting in the given problems.

EXERCISES 1–5

1. Assuming we started with the third order identity matrix, which elementary row operations were performed to obtain the following matrices?

$$A = \begin{bmatrix} 1 & 0 & 0 \\ 0 & 3 & 0 \\ 0 & 0 & 1 \end{bmatrix},$$

$$B = \begin{bmatrix} 0 & 0 & 1 \\ 0 & 1 & 0 \\ 1 & 0 & 0 \end{bmatrix},$$

$$C = \begin{bmatrix} 1 & 0 & 0 \\ 0 & 1 & -2 \\ 0 & 0 & 1 \end{bmatrix}.$$

2. Let $D = \begin{bmatrix} 1 & 2 & 3 \\ 2 & 3 & -1 \\ -1 & 2 & 1 \end{bmatrix}.$

Compute AD, BD, and CD, where A, B, and C are the matrices in Exercise 1.

3. Reduce the following matrices to basic form by a sequence of elementary row operations grouped into pivot operations.

a) $\begin{bmatrix} 2 & 8 & 6 \\ 4 & -8 & 0 \\ 3 & 0 & 3 \end{bmatrix}$

b) $\begin{bmatrix} 2 & 4 & 2 \\ 4 & 9 & 3 \\ 2 & 3 & 3 \end{bmatrix}$

c) $\begin{bmatrix} 1 & 0 & 2 & 1 & 1 \\ 1 & 1 & 1 & 2 & 3 \\ 1 & 2 & 0 & 3 & 5 \end{bmatrix}$

d) $\begin{bmatrix} 2 & 2 & 5 & 6 \\ 1 & 1 & -2 & 2 \end{bmatrix}$

e) $\begin{bmatrix} 3 & 1 & 3 \\ 12 & 4 & 24 \\ 3 & 4 & 9 \end{bmatrix}$

4. Solve the systems of equations at the end of Section 1-4 by using matrix notation and the pivot operation to reduce to basic form.

5. The following matrices are in basic form. Identify the basic elements in each matrix.

a) $\begin{bmatrix} -1 & 0 & 1 & 2 & 0 \\ 1 & -1 & 0 & 1 & 1 \end{bmatrix}$

b) $\begin{bmatrix} 1 & 0 & 1 & 0 & 0 \\ 0 & 1 & 0 & 1 & 0 \\ 1 & 0 & 0 & 1 & 1 \end{bmatrix}$

c) $\begin{bmatrix} 0 & 0 & 0 & 0 & 0 & 0 & 1 \\ 0 & 0 & 0 & 0 & 1 & 1 & 0 \\ 0 & 0 & 1 & 2 & 0 & 2 & 0 \\ 1 & -1 & 0 & 0 & 0 & 3 & 0 \end{bmatrix}$

d) $\begin{bmatrix} 1 & 0 & 0 & 0 & 1 & 1 & 0 & 1 \\ 1 & 0 & 0 & 0 & 0 & 1 & 1 & 0 \\ 0 & 0 & -1 & 1 & 1 & 0 & 0 & 0 \\ 1 & 1 & 0 & 0 & 0 & 0 & 0 & 0 \end{bmatrix}$

1-6 GAUSSIAN ELIMINATION

With the arrival of modern high-speed computing machines has come realization that most of the pet formulas and methods of traditional mathematics are pathetically inadequate for producing numerical answers. An exception is the method of Gaussian elimination to solve a system of linear algebraic equations. This method remains today the most efficient available. The method has been highly refined in recent years and a variety of schemes based on Gaussian elimination have been developed. These variations are aimed primarily at improving the accuracy of the results or reducing the information storage requirements. In this section we will discuss the method of Gaussian elimination without getting involved with the advantages or disadvantages of any particular scheme.

Consider the system of equations

$$\begin{aligned} 3x_1 + x_2 + 9x_3 &= 10 \\ 3x_1 + 2x_2 + 10x_3 &= 13 \\ 2x_1 + x_2 + 7x_3 &= 9. \end{aligned} \qquad (6.1)$$

In the first equation, x_1 appears with a non-zero coefficient. Hence, we can solve for x_1 in the form

$$x_1 = \frac{1}{3}(10 - x_2 - 9x_3) = \frac{10}{3} - \frac{1}{3}x_2 - 3x_3. \qquad (6.2)$$

Substituting (6.2) in the second and third equations of the system (6.1) we get

$$\begin{aligned} x_2 + x_3 &= 3 \\ \frac{1}{3}x_2 + x_3 &= \frac{7}{3}. \end{aligned} \qquad (6.3)$$

This is a system with one less equation and one less unknown than we started with. We have eliminated x_1. Since x_2 appears in the first equation of (6.3) with a non-zero coefficient, we can solve for x_2 and eliminate x_2 from the remaining equation. We get

$$x_2 = 3 - x_3, \qquad (6.4)$$

$$\frac{2}{3}x_3 = \frac{4}{3}. \qquad (6.5)$$

The end result is a single equation (6.5) in one unknown which is easily solved. We obtain $x_3 = 2$. This value is substituted in (6.4) to obtain $x_2 = 1$. Finally, both $x_2 = 1$ and $x_3 = 2$ are substituted in (6.2) to obtain $x_1 = -3$.

The process leading from the system (6.1) to (6.5) is called the *forward course*, or *forward elimination*. The process of starting with a solution of (6.5) and substituting in the other equations to obtain the other unknowns is called the *backward course*, or *back substitution*.

The method described here, properly generalized to systems involving any number of equations and unknowns, is called *Gaussian elimination*. We will describe the method in terms of elementary row operations and compare Gaussian elimination with Jordan elimination in those terms.

Let

$$\begin{bmatrix} 3 & 1 & 9 & 10 \\ 3 & 2 & 10 & 13 \\ 2 & 1 & 7 & 9 \end{bmatrix} \quad (6.6)$$

be the augmented matrix representing the system (6.1). Pivoting on the first element of the first row, we obtain

$$\begin{bmatrix} 1 & \frac{1}{3} & 3 & \frac{10}{3} \\ 0 & 1 & 1 & 3 \\ 0 & \frac{1}{3} & 1 & \frac{7}{3} \end{bmatrix} \quad (6.7)$$

The first row of (6.7) represents equation (6.2), and the remaining rows represent the system (6.3).

For the next step, we pivot on the first non-zero element in the second column below the first row, except that we do *not* operate on the first row. We obtain

$$\begin{bmatrix} 1 & \frac{1}{3} & 3 & \frac{10}{3} \\ 0 & 1 & 1 & 3 \\ 0 & 0 & \frac{2}{3} & \frac{4}{3} \end{bmatrix} \quad (6.8)$$

The second row of (6.8) represents equation (6.4), and the third row represents (6.5). Solving equation (6.5) amounts to pivoting on the 2/3 in the third row without operating on the rows above. We obtain

$$\begin{bmatrix} 1 & \frac{1}{3} & 3 & \frac{10}{3} \\ 0 & 1 & 1 & 3 \\ 0 & 0 & 1 & 2 \end{bmatrix}. \tag{6.9}$$

In (6.9), each non-zero row starts with a 1, each row starts to the right of the beginning of the row above it, and the elementary operations were applied only to rows other than those containing previous pivot elements. This completes the forward course.

The backward course now amounts to completing the pivot operations by operating on the rows above the pivot elements. However, if we use the pivot elements in the reverse order from their occurrence in the forward course, there will be a saving in the number of arithmetic operations required. The next step is then

$$\begin{bmatrix} 1 & \frac{1}{3} & 0 & -\frac{8}{3} \\ 0 & 1 & 0 & 1 \\ 0 & 0 & 1 & 2 \end{bmatrix}. \tag{6.10}$$

This amounts to substituting $x_3 = 2$ in equations (6.2) and (6.4). Finally, we complete the pivot on the pivot element in the second row to obtain

$$\begin{bmatrix} 1 & 0 & 0 & -3 \\ 0 & 1 & 0 & 1 \\ 0 & 0 & 1 & 2 \end{bmatrix}. \tag{6.11}$$

In obtaining (6.11) from (6.10) it was not necessary to operate on the third element of the first row since zeros had been obtained in the third column in the previous step. It is just this fact that makes Gaussian elimination a shorter method than Jordan elimination.

For a system of any number of equations and any number of unknowns, we select pivot elements, just as with Jordan elimination. Each pivot operation is performed in two parts. In the first part, all operations are performed, when the pivot element is selected, on rows not already containing a pivot element. When all pivot elements have been selected the forward course is completed. We can then determine, just as with Jordan elimination, whether the system has a solution. The pivot operations are then completed, using the pivot elements in the reverse order from the order in which they were selected. At the completion of this, the backward course, the basic form is obtained. The end result is the same as that obtained by Jordan elimination. We shall refer to the part of the pivot operation performed in the forward course as a *forward pivot* and the part performed in the backward course as a *backward pivot*.

Let us compute the number of steps required in both Gaussian and Jordan elimination. Assume we have a system of n equations in n unknowns and that n pivots are required to solve the system (or that the system has n basic variables). We shall count each addition or subtraction as an additive step and each multiplication or division as a multiplicative step. Let $GM(n)$ denote the number of multiplicative steps required to solve an $n \times n$ system of equations by Gaussian elimination, and let $JM(n)$ denote the number of multiplicative steps required by Jordan elimination. Similarly, let $GA(n)$ and $JA(n)$ denote the number of additive steps required using Gaussian and Jordan elimination, respectively. We do not count calculations that lead to predetermined results, for example, dividing a number by itself to obtain 1.

For both Gaussian and Jordan elimination, the first step is a pivot operation. This consists of one type I elementary operation and $n-1$ type II elementary operations. The type I operation requires n divisions, and each type II operation requires n multiplications and n additions. Thus, the first pivot operation requires n^2 multiplicative steps and $n(n-1)$ additive steps.

For Jordan elimination, each successive pivot operation requires the same number of type I and type II operations, but each operation requires one less step. Thus,

$$JM(n) = n^2 + n(n-1) + n(n-2) + \cdots + n = \frac{n^2(n+1)}{2} \quad (6.12)$$

$$JA(n) = (n-1)n + (n-1)(n-1) + (n-1)(n-2)$$
$$\cdots + (n-1)$$
$$= \frac{(n-1)n(n+1)}{2}. \quad (6.13)$$

For Gaussian elimination, the elementary operations in the forward course involve the same number of steps as do the corresponding operations in Jordan elimination. In the backward course each elementary operation involves one multiplication and one addition. Thus,

$$GM(n) = n^2 + (n-1)^2 + \cdots + 1^2 + (n-1) + (n-2) + \cdots + 1$$
$$= \frac{n(n+1)(2n+1)}{6} + \frac{n(n-1)}{2} = \frac{n(n^2+3n-1)}{3}, \quad (6.14)$$

$$GA(n) = n(n-1) + (n-1)(n-2) + \cdots + 2 + (n-1)$$
$$+ (n-2) + \cdots + 1 \quad (6.15)$$
$$= \frac{n(n-1)(2n+5)}{6}.$$

In comparing Gaussian elimination with Jordan elimination, notice that every step saved by Gaussian elimination is a type II operation. Thus, the number of additive steps saved is the same as the number of multiplicative steps saved. A comparison between $GM(n)$ and $JM(n)$ is given in Table 1–6.1.

For $n \leq 5$, the difference in efficiency between Gaussian elimination and Jordan elimination is small enough to be negligible. For large n Gaussian elimination takes about two-thirds as long as Jordan elimination. At 25 microseconds per multiplication or division, $(1000)^3/3$ steps would take about two and one half hours. For large systems the difference in time is important. Not only that, but unnecessary steps contribute to the accumulated round-off errors. These considerations are so important and so involved that much energy and ingenuity have gone into finding efficient computing schemes.

To show the connections between the computations and the concepts we will develop it is more convenient to deal

TABLE 1–6.1

n	$GM(n)$	$JM(n)$
2	6	6
3	17	18
4	36	40
5	65	75
6	106	126
10	430	550
100	343,300	505,000
1000	334,333,000	500,500,000
n	(approx) $\frac{1}{3} n^3$	(approx) $\frac{1}{2} n^3$

with the pivot operation and, therefore, Jordan elimination. In the end, either method will provide the same conceptual result. For accuracy and efficiency with a computing machine, Gaussian elimination is to be preferred.

EXERCISES 1-6

1. Solve the system

$$x_1 + \frac{1}{2} x_2 = 5$$

$$\frac{1}{2} x_1 + \frac{1}{3} x_2 = 4$$

exactly. Also, represent the coefficients in decimal form using as many significant digits as you wish, and use this representation to solve the system. Compare the solutions.

2. Solve the following system of equations by performing no more than 9 multiplicative steps.

$$\begin{aligned} x_1 + 2x_2 + 7x_3 - 3x_4 + 2x_5 &= 6 \\ x_2 - x_3 + 2x_4 - x_5 &= 2 \\ x_3 + 4x_4 - 3x_5 &= 1 \\ x_5 &= 3. \end{aligned}$$

3. Solve the system

$$\begin{aligned} 2x_1 + x_2 - 2x_3 + 3x_4 &= -3 \\ 3x_1 + 2x_2 + 2x_3 + 3x_4 &= 9 \\ -2x_1 + 2x_2 + 3x_3 - 2x_4 &= 6 \\ 4x_1 + 3x_2 - x_3 + 2x_4 &= 9, \end{aligned}$$

using first Gaussian elimination and then Jordan elimination.

4. Solve the system

$$\begin{aligned} .000100 x_1 + 1.00 x_2 &= 1.00 \\ 1.00 x_1 + 1.00 x_2 &= 2.00. \end{aligned}$$

The notation for the coefficients means that each coefficient is known to three significant digits. That is, .000100 denotes a number between .0000995 and .0001005. Assume each numerical step is carried out exactly and then rounded off to three significant digits before the next numerical step is taken. Thus, $2 - 10{,}000 = -9{,}998$ would be entered as $-10{,}000$.

Do the work by performing the first pivot on .000100. Then re-do the problem by performing the first pivot on any other element. Also, solve the problem exactly by using rational numbers without roundoff. Compare the answers. (The point is that the strategy of choosing small numbers for the pivot elements is acceptable when the computation is exact, but it is not acceptable for machine computation.)

5. If you have access to a computing machine, solve the following system using Gaussian elimination.

$$11.0x_1 + 21.1x_2 - 1.10x_3 + .0703x_4 = 1.000$$
$$21.1x_1 + 25.6x_2 + 6.03x_3 - .104x_4 = 0.000$$
$$16.3x_1 - 31.7x_2 - 5.12x_3 + .0912x_4 = 0.000$$
$$19.1x_1 + 30.1x_2 + 4.14x_3 - .152x_4 = 0.000.$$

Summary

A system of linear equations is solved by obtaining a sequence of successively simpler systems with the same solution set. We finally obtain a system in which it is evident that there is no solution, or from which a parametric representation of the solution set can be read off.

The basic step in obtaining a simpler system is the pivot operation, which is more conveniently performed on the augmented matrix representing the system. Each pivot operation corresponds to introducing a basic variable. The process is terminated when a basic form is obtained.

A matrix is in *basic form* if each non-zero row contains a 1 which is the only non-zero element in its column. These 1's are called the *basic elements*. In the system represented by the matrix in basic form, each basic element is the coefficient of a basic variable. The basic variables can be expressed as functions of the other variables, which are called *parameters*.

The most efficient method known for solving a system of linear equations is Gaussian elimination. In Jordan elimination, whenever a pivot element is selected, the pivot operation is carried out in its entirety before the next pivot element is selected. In Gaussian elimination part of the pivot operation is carried out before the next pivot element is selected and part of the pivot operation is delayed. Dividing the pivot operation between the forward course and the backward course reduces the number of steps required by approximately one-third. The end result is the same by either method.

2
VECTORS AND VECTOR SPACES

In this chapter we develop a geometric-algebraic setting in which the linear equations of the previous chapter can be reinterpreted. This is not a mere pedagogical device to give the student a visual image of what we are talking about. These concepts are essential for an understanding of linear algebra as a modern mathematics subject, the keys to using linear algebra in applications.

2-1 COORDINATE SPACES

We assume the student is already accustomed to introducing coordinates in a plane or a three-dimensional space. In a plane a point is represented by a pair of coordinates, and in three dimensions a point is represented by three coordinates. Without trying to visualize what n-dimensional spaces might look like, we will extend the use of these terms and speak of a point in an n-dimensional space being represented by n coordinates. Any pictures we draw will be in two-dimensional or three-dimensional spaces. But they will be adequate to suggest the situation in spaces of higher dimension.

An *n-tuple* is a set of n numbers written down in order in the form $(x_1, x_2, \ldots, x_n)$. In a plane the coordinates of a point form a pair or a duple, and in a three-dimensional space the coordinates of a point form a triple. In a plane or three-dimensional space the coordinates of a point depend on the way the coordinate system is introduced. But, once the coordinate system is chosen, each point is identified by its coordinates. There is a one-to-one correspondence between points and duples (or triples).

To generalize this idea we refer to each n-tuple as a "point" but we make no attempt to visualize a geometric picture of the set of points for values of n larger than 3. In the next section we shall introduce an algebraic structure in the set of n-tuples. For each n, the set of all n-tuples, together with the algebraic structure to be introduced, constitutes what we call an *n-dimensional coordinate space*. For a system of equations in n unknowns, then, the n-tuples that satisfy the system are a subset of the n-dimensional coordinate space, and we refer to this subset as the *solution set* of the system. Each point in the solution is a *particular solution*. A *general solution* is a parametrization of the solution set.

In a plane, the single linear equation

$$2x_1 + 3x_2 = 6 \qquad (1.1)$$

has a straight line as the solution set, as shown in Figure 2–1.1. The points $(3, 0)$ and $(0, 2)$ are on this straight line, and they are particular solutions to the equation (1.1). Considering (1.1) as a system of equations (consisting of only one equation),

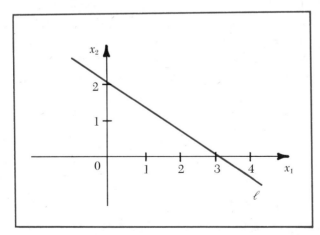

Figure 2–1.1 The solution set in $\mathscr{R}^2$ of the linear equation $2x_1 + 3x_2 = 6$ is a straight line. The points $(3, 0)$ and $(0, 2)$ are on the line and determine the line.

since both x_1 and x_2 have non-zero coefficients, either can be taken as a basic variable. Thus, we can have either

$$x_1 = 3 - \frac{3}{2} x_2 \quad \text{or} \tag{1.2}$$

$$x_2 = 2 - \frac{2}{3} x_1 \tag{1.3}$$

as general solutions to (1.1). In either case, as the parameter takes on various values, the corresponding point takes on various positions on the line. When the parameter runs through its full range of values, the corresponding point sweeps out the entire line.

In a three-dimensional space a single linear equation generally has a plane for its solution set. For example, the equation

$$x_1 + 2x_2 + 3x_3 = 6 \tag{1.4}$$

has (6, 0, 0), (0, 3, 0), and (0, 0, 2) among its particular solutions. The solution set is a plane containing the points (6, 0, 0), (0, 3, 0), and (0, 0, 2), as illustrated in Figure 2-1.2. The points (6, 0, 0), (0, 3, 0), and (0, 0, 2) were chosen because they are particularly easy to obtain. Take 0 as the value of two of the variables and solve for the third. The chosen points are on the coordinate axes and suffice to locate the plane in space.

There are many ways to parametrize the plane. If the method of reduction to basic form is used, three parametrizations can be obtained. Since there is only one equation this amounts to solving for one of the three variables in terms of the other two. The three basic parametric representations of the plane solution set are

$$x_1 = 6 - 2x_2 - 3x_3, \tag{1.5}$$

$$x_2 = 3 - \frac{1}{2} x_1 - \frac{3}{2} x_3, \tag{1.6}$$

$$x_3 = 2 - \frac{1}{3} x_1 - \frac{2}{3} x_2. \tag{1.7}$$

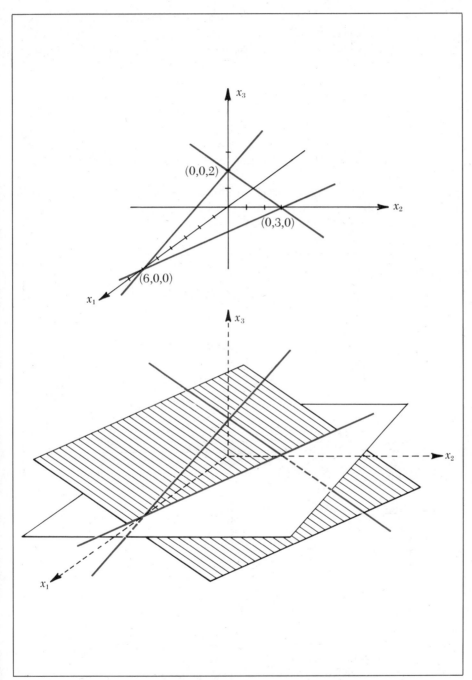

Figure 2-1.2 The solution set in $\mathscr{R}^3$ of the linear equation $x_1 + 2x_2 + 3x_3 = 6$ is a plane. The points (6, 0, 0), (0, 3, 0), and (0, 0, 2) are on the plane and determine it.

In any one of the three parametric representations, specifying the values of the two parameters determines the value of the third variable, and definitely locates a point on the plane. There is, therefore, a one-to-one correspondence between pairs of values of the parameters and points in the plane. In a sense, then, the parameters may be considered to provide coordinates in this plane. This, generally, is the point of view we wish to promote for solutions of systems of linear equations. A general solution of a system of equations in the form of a parametrization amounts to a coordinatization of the solution set.

Although the three parametrizations, (1.5), (1.6), (1.7) are equally useful, a sketch of the situation is somewhat easier to describe if we use (1.7). For a choice of values of (x_1, x_2), the

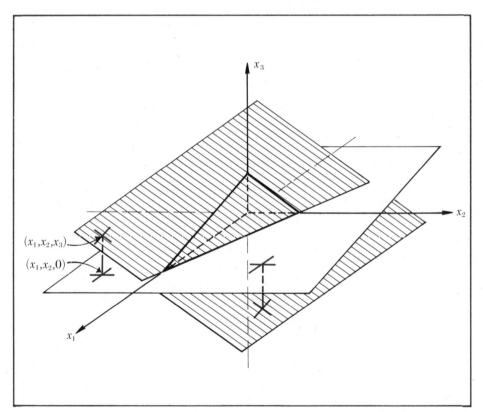

Figure 2-1.3 A parametrization of the solution set of the linear equation $x_1 + 2x_2 + 3x_3 = 6$. For each choice of (x_1, x_2), the point $\left(x_1, x_2, 2 - \frac{1}{3}x_1 - \frac{2}{3}x_2\right)$ is in the solution set. Think of $x_3 = 2 - \frac{1}{3}x_1 - \frac{2}{3}x_2$ as the height of the plane above (or below) the point $(x_1, x_2, 0)$.

corresponding point (x_1, x_2, x_3) is in the plane vertically above or below the point $(x_1, x_2, 0)$.

Generally, in a system of linear equations in two variables, each equation can be represented by a straight line in a plane. The set of points that satisfy all the equations simultaneously is the set of points common to all the lines. For two non-parallel lines this intersection is a single point. For the corresponding system the solution would consist of a single pair of numbers—a unique solution. If the two lines are parallel they are either identical or they have no point of intersection. Then the solution set is either identical to the common line or is empty. Three lines might be located as in Figure 2–1.4a. In this case the intersection is empty and the system of equations has no solution. For a system of three

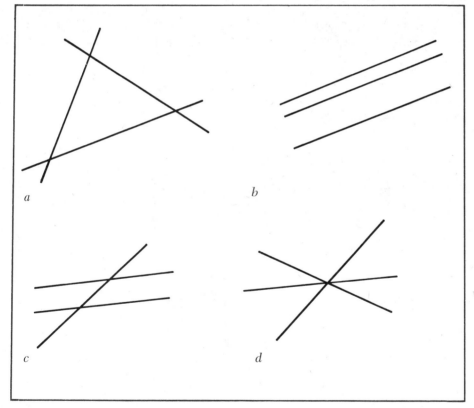

Figure 2–1.4 There are four patterns for the intersections of three distinct lines in a plane. In b, the three lines are parallel. In c, two are parallel and the third is parallel to neither of the first two. In a and d no pair is parallel. Only in d is there a point common to all three lines.

or more equations, if any pair of non-identical lines is parallel, as in Figures 2–1.4b or 2–1.4c, or any three lines form a pattern like that in Figure 2–1.4a, the system will have no solution. It is also possible for a system of many equations to be represented by lines all passing through a common point, as in Figure 2–1.4d. Whether a system of equations has a solution

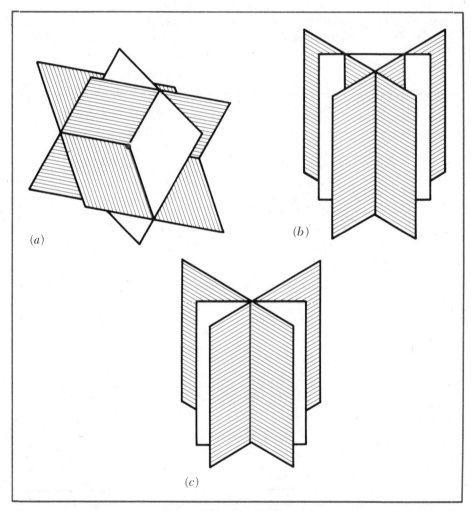

Figure 2–1.5 Patterns for three distinct planes, no two of which are parallel. There are three intersection patterns. In (a) they have a common point. In (b) they have no point in common. In (c) they have a common line.

depends not so much on the number of equations as it does on how they are related to each other.

In three dimensions the situation is only slightly more complicated. Two non-parallel planes intersect in a line. So the solution set for two equations is either a line (if the planes are not parallel) or a plane (if the planes are identical) or empty (if the planes are parallel but not identical). For three planes, no two of which are parallel, the possibilities are as in Figure 2–1.5. In Figure 2–1.5a the solution is unique, a single point; in 2–1.5b there is no solution; and in 2–1.5c the solution set is a line.

These geometric considerations are quite helpful in giving one an idea of what the solution of a system of linear equations looks like. The picture is sufficiently suggestive even for higher dimensional problems. But while these pictures are suggestive to the intuition they are not useful for numerical results. We must transfer this geometric picture into algebraic terms and depend on the algebra for more complicated problems.

EXERCISES 2–1

1. The following linear equations in two unknowns can be represented by lines in a plane. Draw a coordinate plane and sketch these lines.

    A. $2x_1 - 3x_2 = 1$
    B. $5x_1 - 7x_2 = 3$
    C. $2x_1 + x_2 = 13$
    D. $6x_1 - 9x_2 = 2$

2. Determine whether the lines representing A and B intersect. Find the point of intersection, if there is one.

3. Determine whether the lines representing A and C intersect. Find the point of intersection if there is one.

4. Determine whether the lines representing A and D intersect. Find the point of intersection if there is one.

5. Determine whether the lines representing A, B, and C have a common point.

    For the following systems of equations, determine

whether the lines represented by the equations conform to the configuration represented by Figures 2–1.4a, b, c, or d.

6. $2x_1 - 3x_2 = 1$
   $2x_1 - 3x_2 = 2$
   $2x_1 + x_2 = 13.$

7. $2x_1 - 3x_2 = 1$
   $2x_1 - 3x_2 = 2$
   $2x_1 - 3x_2 = 3.$

8. $2x_1 - 3x_2 = 1$
   $2x_1 + x_2 = 5$
   $3x_1 + 4x_2 = 10.$

For the following systems of equations, determine whether the planes represented by the equation conform to the configuration represented by Figures 2–1.5a, b, c, or some other configuration.

9. $x_1 + 2x_2 + 3x_3 = 14$
   $2x_1 - x_2 + x_3 = 3$
   $x_1 + x_2 + x_3 = 6.$

10. $x_1 + 2x_2 + 3x_3 = 14$
    $2x_1 - x_2 + x_3 = 3$
    $-x_1 + 3x_2 + 2x_3 = 11,$

11. $x_1 + 2x_2 + 3x_3 = 14$
    $2x_1 - x_2 + x_3 = 3$
    $3x_1 - 4x_2 - x_3 = -5.$

12. $x_1 + 2x_2 + 3x_3 = 14$
    $2x_1 - x_2 + x_3 = 3$
    $2x_1 + 4x_2 + 6x_3 = 9.$

2–2 VECTORS

We assume the student to be at least acquainted with vectors as they are usually introduced in elementary mathematics courses such as analytic geometry and calculus. However, we shall use those elementary ideas only for descriptive and motivational purposes. Previous acquaintance with vectors is not required logically, but it helps if one can associate the ideas we introduce with some familiar concepts.

A vector is ordinarily described as a "little arrow" positioned in space with a well-defined direction. For our purposes we shall consider vectors located with their tails at a fixed point, which we take as the origin. Then the head of the arrow falls at a point in that space. Thus, there is a one-to-one correspondence between points and vectors: to every point there corresponds a vector—the vector from the origin to that point—and to every vector there corresponds a point—the point situated at the head of the vector. The kinds of vectors we are talking about (with tails fixed to the origin) are often called *position vectors*. Each point identifies the position of a point located at the head of the vector.

Let us now introduce a coordinate system, with its origin located at the common tail-point of our position vectors. Since there is a one-to-one correspondence between coordinates and points, and a one-to-one correspondence between points and vectors, there is a one-to-one correspondence between coordinates and vectors. In this sense we shall refer to these coordinates as coordinates of the vectors.

In elementary vector analysis addition of vectors and multiplication of vectors by scalars are defined. This gives the set of vectors an algebraic structure. We wish to introduce a corresponding algebraic structure into the set of coordinates. In vector analysis, the sum of two vectors is defined by the

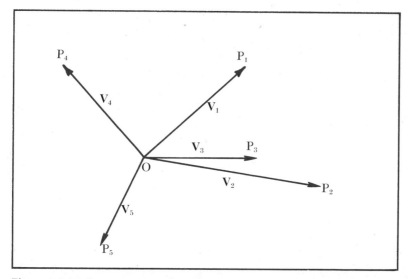

Figure 2–2.1 Once a point is selected as an origin, each point determines a unique position vector which has its tail at the origin and its head at the point.

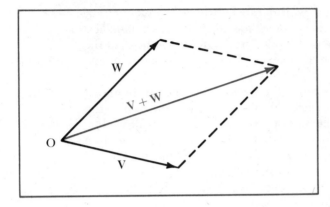

Figure 2-2.2 The parallelogram rule for addition of vectors. To add **V** and **W**, locate **V** and **W** with their tails at the origin. Complete the parallelogram with **V** and **W** as adjacent sides. The sum **V** + **W** is the diagonal with its tail at the origin.

parallelogram rule: *if **V** and **W** are position vectors, their sum, **V** + **W**, is defined to be the vector lying in the diagonal of a parallelogram with **V** and **W** as adjacent sides.* (See Figure 2-2.2.)

If a is a scalar and **V** is a vector, *the product of a and **V**, a**V**, is defined to be a vector in the same direction as **V** (if a is positive) whose length is the length of **V** multiplied by a.* If a is negative, a**V** is in the opposite direction with length equal to the length of **V** multiplied by $-a$. (See Figure 2-2.3.) The zero vector is a vector of zero length, hence the head and tail of the zero vector coincide and the zero vector has no direction.

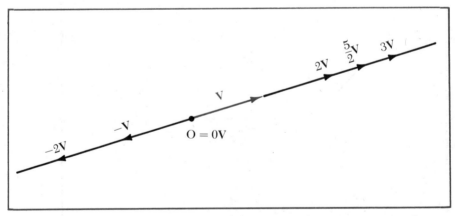

Figure 2-2.3 A scalar multiple of a vector. If **V** is a vector in a real vector space and a is a real scalar, a**V** is a vector in the same direction as **V** and a times as long as **V** if $a > 0$, and a**V** is in the opposite direction as **V** and $-a$ times as long if $a < 0$.

In elementary vector analysis it is shown that if **V** is represented by the coordinates (a_1, a_2, a_3) and **W** is represented by the coordinates (b_1, b_2, b_3), then $\mathbf{V} + \mathbf{W}$ is represented by the coordinates $(a_1 + b_1, a_2 + b_2, a_3 + b_3)$, and $a\mathbf{V}$ is represented by the coordinates (aa_1, aa_2, aa_3). We will not prove these statements here, and we will not depend on the assumption that the student has seen proofs of these statements elsewhere. It will suffice for our purposes to define the addition of vectors and multiplication by scalars directly in terms of the coordinates. The addition of vectors in terms of the parallelogram rule, and scalar multiplication in terms of length are important for the interpretation of what is going on, but proofs will not depend on these interpretations.

Definition *Let $(a_1, a_2, \ldots, a_n)$ and $(b_1, b_2, \ldots, b_n)$ be n-tuples of numbers and let a be a number (or scalar). We define **vector addition** and **scalar multiplication** by the rules*

$$(a_1, a_2, \ldots, a_n) + (b_1, b_2, \ldots, b_n) = (a_1 + b_1, a_2 + b_2, \ldots, a_n + b_n), \quad (2.1)$$

$$a(a_1, a_2, \ldots, a_n) = (aa_1, aa_2, \ldots, aa_n). \quad (2.2)$$

The operations on the left are defined in terms of the operations given on the right. If by "numbers" we mean real numbers, the set of n-tuples with these operations defined is denoted by $\mathscr{R}^n$, and it is called the real n-dimensional coordinate space.

It will make matters more convenient if we introduce some abbreviations. Let A stand for the n-tuple $(a_1, a_2, \ldots, a_n)$. As often as possible, if we use a lower case letter of the alphabet to denote the elements of an n-tuple, we will use the corresponding capital letter to denote the n-tuple as a whole.

In what follows we often have need to write down a one-column matrix. One-column matrices are awkward to display on a printed page, or even in written work. It turns out to be convenient to identify an n-tuple and a one-column matrix. That is, we make the **convention**

$$(a_1, a_2, \ldots, a_n) = \begin{bmatrix} a_1 \\ a_2 \\ \vdots \\ a_n \end{bmatrix} \quad (2.3)$$

Although this is not a standard notational convention, n-tuples and one-column matrices are so often used for the same purpose that the identification causes no problem.

Theorem 2.1 *The n-tuples of an n-dimensional coordinate space, with the definitions of vector addition and scalar multiplication given by formulas (2.1) and (2.2), satisfy the following conditions.*

1) For $A, B \in \mathscr{R}^n$, $A + B \in \mathscr{R}^n$.
2) For $A, B, C \in \mathscr{R}^n$, $(A + B) + C = A + (B + C)$.
3) For $0 = (0, 0, \ldots, 0)$ and all $A \in \mathscr{R}^n$, we have $0 + A = A$.
4) For every $A \in \mathscr{R}^n$, there is an n-tuple, called $-A$, such that $(-A) + A = 0$.
5) For $A, B \in \mathscr{R}^n$, $A + B = B + A$.
6) For $A \in \mathscr{R}^n$ and a a scalar, $aA \in \mathscr{R}^n$.
7) For a and b scalars and $A \in \mathscr{R}^n$, $a(bA) = (ab)A$.
8) For a and b scalars and $A \in \mathscr{R}^n$, $(a + b)A = aA + bA$.
9) For a a scalar and $A, B \in \mathscr{R}^n$, $a(A + B) = aA + aB$.
10) For $A \in \mathscr{R}^n$, $1 \cdot A = A$.

Proof. 1) and 6) are merely matters of definition. The other eight conditions can all be verified by direct calculation in terms of the definitions. These calculations we leave for the reader to carry out for himself. □

n-tuples certainly have interesting properties other than those listed in Theorem 2.1, and those listed may seem pretty trivial. The point is that these same rules have been found to apply to a large number of other collections of objects. These rules form the basis for the following definition.

Definition *A vector space over a field of scalars $\mathscr{F}$ is a non-empty set $\mathscr{V}$ of objects together with operations of addition and multiplication by scalars satisfying the following conditions.*

1) For $\alpha, \beta \in \mathscr{V}$, $\alpha + \beta \in \mathscr{V}$.
2) For $\alpha, \beta, \gamma \in \mathscr{V}$, $(\alpha + \beta) + \gamma = \alpha + (\beta + \gamma)$.
3) There is an element $0 \in \mathscr{V}$ such that for all $\alpha \in \mathscr{V}$ we have $0 + \alpha = \alpha$.
4) For each $\alpha \in \mathscr{V}$ there is an element $-\alpha \in \mathscr{V}$ such that $(-\alpha) + \alpha = 0$.
5) For $\alpha, \beta \in \mathscr{V}$, $\alpha + \beta = \beta + \alpha$.

6) For $a \in \mathscr{F}$ and $\alpha \in \mathscr{V}$, $a\alpha \in \mathscr{V}$.
7) For $a, b \in \mathscr{F}$ and $\alpha \in \mathscr{V}$, $a(b\alpha) = (ab)\alpha$.
8) For $a, b \in \mathscr{F}$ and $\alpha \in \mathscr{V}$, $(a+b)\alpha = a\alpha + b\alpha$.
9) For $a \in \mathscr{F}$ and $\alpha, \beta \in \mathscr{V}$, $a(\alpha+\beta) = a\alpha + a\beta$.
10) For $\alpha \in \mathscr{V}$, $1\alpha = \alpha$.

What we have done here is quite customary in the development of mathematics. We start with a definite concrete model, in this case the two- and three-dimensional vectors. We work with them enough to observe the properties that are needed for what we wish to do, in this case the properties of Theorem 2.1. We use these properties to define an abstraction of the model we started with. In this case, as is usually true, there are vastly more systems satisfying the defining conditions than we were considering at the start. Furthermore, any characteristic or theorem, applying to the concrete model, which depends only on these defining properties applies equally well to other systems which have these same properties.

What is gained and what is lost in doing this sort of thing? Simplicity is usually gained by stripping away those features of the original model that are difficult to ignore, but that do not contribute in an essential way to an understanding of the model. For example, since the n-tuples are numbers certain kinds of calculations are possible. In the abstraction these calculations are not available. Many theorems can be proved by means of these calculations. With the calculations unavailable, ingenuity has been required to discover new proofs. Happily, the new proofs are usually simpler and more conceptually relevant to the ideas involved. The defining conditions are the keys to applying the theory in new places. In a new situation it is merely necessary to show that the defining conditions hold. As soon as that is done the entire theory can be applied.

What is lost is usually that the results in the theory are somewhat less specific. This loss depends on what has been ignored in choosing the defining conditions. We will later define the dimension of a vector space, and then restrict our attention to vector spaces with finite dimension. For finite dimensional vector spaces there is no loss at all.

EXERCISES 2-2

1. Let $\mathscr{P}_n$ be the set of all polynomials with real coefficients of degree less than n together with the zero polynomial

(which has no degree). The sum of two polynomials is defined in the usual way (add corresponding coefficients), and multiplying a polynomial by a real number is defined in the usual way (multiply all coefficients by the real number). Show that with these definitions, the set $\mathscr{P}_n$ is a vector space over the real numbers.

2. Let $\mathscr{P}$ be the set of all polynomials with real coefficients (including the zero polynomial). With addition and scalar multiplication defined as in Exercise 1, show that $\mathscr{P}$ is a vector space over the real numbers.

3. Let $\mathscr{V}$ be the set of all real valued functions defined on any non-empty set $\mathscr{S}$. For f and g any two functions in $\mathscr{V}$, define $f + g$ by the rule

$$(f + g)(x) = f(x) + g(x)$$

for all $x \in \mathscr{S}$. Similarly, for any real number a, define af by the rule

$$(af)(x) = af(x).$$

Show that with these definitions, the set $\mathscr{V}$ is a vector space over the real numbers.

4. Let $\mathscr{C}$ be the set of continuous real valued functions defined on the real numbers. With addition and scalar multiplication defined as in Exercise 3, show that $\mathscr{C}$ is a vector space over the real numbers.

5. Let $\mathscr{D}$ be the set of differentiable real valued functions defined on the real numbers. With addition and scalar multiplication defined as in Exercise 3, show that $\mathscr{D}$ is a vector space over the real numbers.

6. Let $\mathscr{E}$ be the set of real valued functions defined on the real numbers which take on only non-negative values. With addition and scalar multiplication defined as in Exercise 3, show that $\mathscr{E}$ is *not* a vector space over the real numbers.

7. Let $\mathscr{R}$, the field of real numbers, occupy both the roles of $\mathscr{F}$ and $\mathscr{V}$ in the definition of a vector space. Use ordinary addition and multiplication as the operations of vector addition and scalar multiplication. In particular, we look at the product ab as the product of a from $\mathscr{R}$ considered as the scalar field and b from $\mathscr{R}$ considered as the

set of vectors. Show that, in this light, $\mathscr{R}$ is a vector space over itself.

8. Let $\mathscr{R}$, the field of real numbers, occupy the role of $\mathscr{V}$ and let $\mathscr{Q}$, the field of rational numbers, occupy the role of $\mathscr{F}$ in the definition of a vector space. Let vector addition be the ordinary addition of real numbers, and let scalar multiplication be the ordinary multiplication of a rational number by a real number. Show that $\mathscr{R}$ is a vector space over $\mathscr{Q}$.

9. For a fixed n, let $\mathscr{L}_n$ be the set of all n-tuples of integers. Let addition and scalar multiplication be defined as in (2.1) and (2.2), where $\mathscr{R}$ is taken for the scalar field. Show that $\mathscr{L}_n$ is *not* a vector space.

10. For a fixed n, let $\mathscr{L}_n$ be the set of all n-tuples of integers. Let $\mathscr{Z}$ denote the set of integers. Let addition and scalar multiplication be defined as in (2.1) and (2.2), where the scalars are taken from $\mathscr{Z}$. Show that $\mathscr{L}_n$ is not a vector space.

11. Let $\mathscr{M}$ be the set of all $m \times n$ matrices with elements from a field $\mathscr{F}$. Show that $\mathscr{M}$ is a vector space over $\mathscr{F}$, where matrix addition is taken for vector addition and multiplication of a matrix by a scalar is taken for scalar multiplication.

12. Let $\mathscr{V}$ consist of a single object 0. That is, $\mathscr{V} = \{0\}$. Define $0 + 0 = 0$ and $a0 = 0$ for all $a \in \mathscr{F}$. Show that with these definitions $\mathscr{V}$ is a vector space.

13. Show that $0\alpha = 0$ for all $\alpha \in \mathscr{V}$. (Hint: See Appendix 3.)

14. Show that $a0 = 0$ for all $a \in \mathscr{F}$. (Hint: See Appendix 3.)

15. Show that if $\alpha \in \mathscr{V}$ and $\alpha \neq 0$, then $a\alpha = 0$ for $a \in \mathscr{F}$ implies $a = 0$.

16. Show that if $\mathscr{V}$ contains more than one element, it contains at least as many elements as are contained in the field $\mathscr{F}$.

2-3 SOLUTION SETS AS SETS OF VECTORS

Let us re-examine the examples discussed in Section 2–1 in terms of coordinate spaces and vectors. The parametrization (1.3) leads to the computation,

$$(x_1, x_2) = \left(x_1, 2 - \frac{2}{3} x_1\right)$$

$$= (0, 2) + \left(x_1, -\frac{2}{3} x_1\right) \quad (3.1)$$

$$= (0, 2) + x_1 \left(1, -\frac{2}{3}\right).$$

Figure 2–3.1 shows the solution set $\mathscr{L}$ as a straight line in the coordinate plane. $(1, -2/3)$ represents a vector parallel to the line $\mathscr{L}$. For each value of x_1, $x_1(1, -2/3)$ is a scalar multiple of $(1, -2/3)$ and is the position vector of a point on a line $\mathscr{L}'$ parallel to $\mathscr{L}$. As x_1 runs through the real numbers, $x_1(1, -2/3)$ runs through the position vectors (with heads) on $\mathscr{L}'$. Adding $(0, 2)$ to the position vectors on $\mathscr{L}'$ translates $\mathscr{L}'$ parallel to itself to $\mathscr{L}$.

Let us examine the parametrization given by (1.7) from the same point of view. From (1.7) we obtain

$$(x_1, x_2, x_3) = \left(x_1, x_2, 2 - \frac{1}{3} x_1 - \frac{2}{3} x_2\right)$$

$$= (0, 0, 2) + \left(x_1, 0, -\frac{1}{3} x_1\right) + \left(0, x_2, -\frac{2}{3} x_2\right) \quad (3.2)$$

$$= (0, 0, 2) + x_1 \left(1, 0, -\frac{1}{3}\right) + x_2 \left(0, 1, -\frac{2}{3}\right).$$

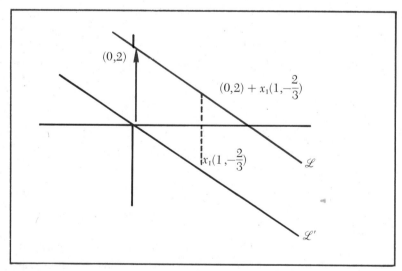

Figure 2–3.1   The solution set of a linear equation $2x_1 + 3x_2 = 6$ in the plane $\mathscr{R}^2$. The line $\mathscr{L}'$ contains the origin $(0, 0)$ and $(1, -2/3)$. Every point on $\mathscr{L}'$ is moved to a point on $\mathscr{L}$ by adding $(0, 2)$. Thus, the solution set $\mathscr{L}$ is parallel to the line $\mathscr{L}'$.

For each value of x_1, $x_1(1, 0, -1/3)$ is a scalar multiple of $(1, 0, -1/3)$ and the end point of $x_1(1, 0, -1/3)$ lies on a line through the origin and the point $(1, 0, -1/3)$. Similarly, the end point of $x_2(0, 1, -2/3)$ lies on a line through the origin and the point $(0, 1, -2/3)$. Thus, as x_1 and x_2 run through the real numbers, $x_1(1, 0, -1/3) + x_2(0, 1, -2/3)$ runs through (position vectors of) points on a plane containing the origin and the points $(1, 0, -1/3)$ and $(0, 1, -2/3)$. Call this plane $\mathscr{L}'$. Adding $(0, 0, 2)$ to the position vectors in $\mathscr{L}'$ translates the plane $\mathscr{L}'$ parallel to itself to a new position, as illustrated in Figure 2-3.2.

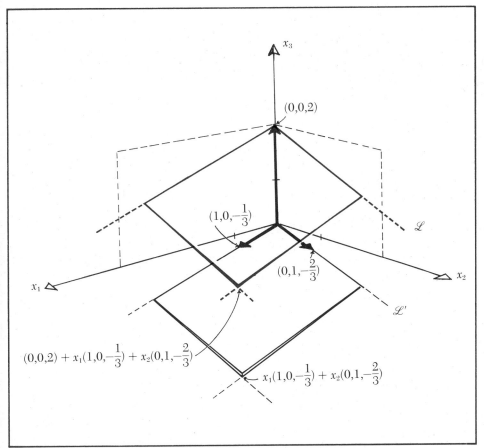

Figure 2-3.2 The solution set of a linear equation $x_1 + 2x_2 + 3x_3 = 6$ in $\mathscr{R}^3$. The plane $\mathscr{L}'$ contains the origin and the points $(1, 0, -1/3)$ and $(0, 1, -2/3)$. Every point on $\mathscr{L}'$ is moved to a point on $\mathscr{L}$ by adding $(0, 0, 2)$. Thus, the solution set $\mathscr{L}$ is parallel to he plane $\mathscr{L}'$.

Consider what happens if we change only the constant term in equation (1.4). Suppose, for example, we have

$$x_1 + 2x_2 + 3x_3 = 9, \tag{3.3}$$

which results in the parametric solution

$$x_3 = 3 - \frac{1}{3} x_1 - \frac{2}{3} x_2. \tag{3.4}$$

The vector form of this parametric expression is

$$(x_1, x_2, x_3) = (0, 0, 3) + x_1 \left(1, 0, -\frac{1}{3}\right) + x_2 \left(0, 1, -\frac{2}{3}\right). \tag{3.5}$$

What we wish to emphasize here is that this solution set is also a plane parallel to $\mathscr{L}'$, and the plane $\mathscr{L}'$ is still generated by vectors that can be represented in the form

$$x_1 \left(1, 0, -\frac{1}{3}\right) + x_2 \left(0, 1, -\frac{2}{3}\right). \tag{3.6}$$

In fact, if only the constant term is changed in equation (1.4), for the various values of this constant we get a family of parallel planes. All the resulting parametrizations will contain the vectors in (3.6) added to a particular vector that depends on the constant term. The vectors in expression (3.6) do not depend on the constant term.

The vectors in (3.6) generate the plane $\mathscr{L}'$, and this plane is also the solution set when the constant term on the right side of equation (1.4) is taken to be 0. This suggests that the problem of solving a system of linear equations can be profitably separated into two problems. One problem is to find the solution set when the constant terms are set equal to zero, and the other is to find a particular solution. In the example just discussed, these two steps correspond to finding the plane $\mathscr{L}'$ and then translating the plane parallel to itself to the correct position. The first of these two problems motivates the discussion in the rest of this chapter.

EXERCISES 2-3

1. In (3.1), take $x_1 = 3$ to obtain (3, 0). Show that (3, 0) is a particular solution of equation (1.1).

2. Show that
$$x_1 = 3 + 3t$$
$$x_2 = -2t$$
(3.7)

is another parametrization of the solution set for equation (1.1).

3. Consider the equation $2x_1 + 3x_2 = 5$. Show that
$$(x_1, x_2) = \left(0, \frac{5}{3}\right) + x_1\left(1, -\frac{2}{3}\right)$$
(3.8)

is a parametrization of the solution set. Compare (3.8) with the parametrization (3.1) and interpret the comparison geometrically.

4. Take $x_1 = 2$, $x_2 = 3$ in (3.6), and obtain $\left(2, 3, -\frac{2}{3}\right)$ as a particular solution of (1.4).

5. Show that
$$(x_1, x_2, x_3) = (3, 3, -1) + s(1, 1, -1) + t(2, -1, 0)$$
(3.9)

is another parametrization of the solution set for (1.4).

6. Show that
$$(x_1, x_2, x_3) = (2, 3, 1) + s(1, 1, -1) + t(2, -1, 0)$$
(3.10)

is a parametrization of the solution set of the equation $x_1 + 2x_2 + 3x_3 = 11$.

7. Show that the solution sets represented by the parametrizations (3.9) and (3.10) have no points in common. (They are parallel planes.)

8. Show that $(x_1, x_2, x_3) = (4, 5, 0) + x_3(-1, -1, 1)$ is a parametrization in vector form of the solution set for Exercise 10 of Section 2-1.

9. Given the parametrization
$$(x_1, x_2, x_3, x_4) = (2, -1, 3, 0) + x_4(2, 3, -4, 1),$$

find a system of linear equations for which this parametrization gives the solution set.

10. Given the parametrization
$$(x_1, x_2, x_3, x_4, x_5) = (1, 2, 0, 4, 0) + x_3(6, 7, 1, 9, 0)$$
$$+ x_5(9, 7, 0, 3, 1),$$

find a system of linear equations for which this parametrization gives the solution set.

2-4 SUBSPACES AND HOMOGENEOUS EQUATIONS

Definition *Let $\mathscr{V}$ be a vector space over a field $\mathscr{F}$ of scalars. A **subspace** of $\mathscr{V}$ is a non-empty subset $\mathscr{S}$ of $\mathscr{V}$ that is also a vector space over $\mathscr{F}$. The operations of vector addition and multiplication by scalars in $\mathscr{S}$ must be the same as those defined in $\mathscr{V}$. With these two operations $\mathscr{S}$ must satisfy the ten conditions given in the definition of a vector space.*

Of the ten conditions that $\mathscr{S}$ must satisfy, several are true in any subset. Those that are true are the universal statements, specifically 2), 5), 7), 8), 9), and 10) as given in the definition of a vector space. While the other four statements are not true for every subset of $\mathscr{V}$, 1) and 6) are the only ones that must be checked for non-empty subsets, because from these will follow 3) and 4). Expressed for a subset $\mathscr{S}$, 1) says, "For α and $\beta \in \mathscr{S}$, $\alpha + \beta \in \mathscr{S}$," and 6) says, "For $a \in \mathscr{F}$ and $\alpha \in \mathscr{S}$, $a\alpha \in \mathscr{S}$." If $\mathscr{S}$ is non-empty there is at least one vector $\alpha \in \mathscr{S}$. Then $0 = 0 \cdot \alpha \in \mathscr{S}$ because of condition 6). Also, for any $\alpha \in \mathscr{S}$, $-\alpha = (-1)\alpha \in \mathscr{S}$ because of condition 6). Thus, a subset of a vector space is a subspace if conditions 1) and 6) are satisfied for that subset.

The two conditions that a subset $\mathscr{S}$ must satisfy to be a subspace are so important that they are worth expressing in other forms and other words. If for a subset $\mathscr{S}$ the condition, "For α and $\beta \in \mathscr{S}$, $\alpha + \beta \in \mathscr{S}$," is satisfied, we say that $\mathscr{S}$ is *closed under vector addition*. If for a subset $\mathscr{S}$ the condition, "For $a \in \mathscr{F}$ and $\alpha \in \mathscr{S}$, $a\alpha \in \mathscr{S}$," is satisfied, we say that $\mathscr{S}$ is *closed under scalar multiplication*.

The conditions for closure under vector addition and closure under scalar multiplication can be combined in a single statement. If α and β are vectors and a and b are scalars, the expression

$$a\alpha + b\beta \tag{4.1}$$

is called a *linear combination* of α and β. For the particular choices $a = b = 1$, (4.1) is a vector sum. For $b = 0$ and an arbitrary a, (4.1) is a scalar multiple. Thus, if $\mathscr{S}$ is closed under linear combinations, it is closed under both vector addition and scalar multiplication. On the other hand, if $\mathscr{S}$ is closed under scalar multiplication it contains $a\alpha$ and $b\beta$, and if it is closed under vector addition it contains $a\alpha + b\beta$. Thus, closure under linear combinations is equivalent to closure under both vector addition and scalar multiplication.

Theorem 4.1 *A non-empty subset $\mathscr{S}$ of a vector space is a subspace if and only if it is closed under both vector addition and scalar multiplication or, equivalently, if and only if it is closed under linear combinations.*

Proof. Closure under vector addition and closure under scalar multiplication are two of the ten conditions a subspace must satisfy. On the other hand, we have shown that for a non-empty subset of a vector space these two conditions are sufficient. □

A linear equation is said to be *homogeneous* if the constant term is 0. Our next objective is to show that the solution set of a system of homogeneous linear equations in n variables is a subspace of an n-dimensional coordinate space.

Consider the single linear homogeneous equation in n variables,

$$a_1 x_1 + a_2 x_2 + \cdots + a_n x_n = 0. \qquad (4.2)$$

If $(x_1, x_2, \ldots, x_n)$ and $(y_1, y_2, \ldots, y_n)$ are solutions of (4.2), then their sum $(x_1 + y_1, x_2 + y_2, \ldots, x_n + y_n)$ is also a solution since

$$\begin{aligned} &a_1(x_1 + y_1) + a_2(x_2 + y_2) + \cdots + a_n(x_n + y_n) \\ &= a_1 x_1 + a_2 x_2 + \cdots + a_n x_n + a_1 y_1 + a_2 y_2 + \cdots + a_n y_n \\ &= 0 + 0 = 0. \qquad (4.3) \end{aligned}$$

Also, for any scalar a, $a(x_1, x_2, \ldots, x_n) = (ax_1, ax_2, \ldots, ax_n)$ is a solution of (4.2) since

$$\begin{aligned} &a_1(ax_1) + a_2(ax_2) + \cdots + a_n(ax_n) \\ &\qquad = a(a_1 x_1 + a_2 x_2 + \cdots + a_n x_n) = 0. \ (4.4) \end{aligned}$$

This shows that the set of solutions of a single linear homogeneous equation is a subspace of $\mathscr{R}^n$. In the same way we could show that the solution set of any system of homogeneous linear equations is a subspace. But it is easier, more informative, and of more use in the long run to prove and use a more general principle.

Theorem 4.2 *The intersection of any collection of subspaces of a vector space is a subspace.*

Proof. Let $\mathscr{V}$ be a vector space and let $\{\mathscr{S}_i\}$ be any collection of subspaces. Let $\cap_i \mathscr{S}_i$ denote the intersection. The intersection is not empty since the zero vector is in every subspace. Let α and β be in the intersection and let a and b be any scalars. For each subspace $\mathscr{S}_i$ in the collection, $\alpha \in \mathscr{S}_i$ and $\beta \in \mathscr{S}_i$. Since $\mathscr{S}_i$ is a subspace, $a\alpha + b\beta \in \mathscr{S}_i$. Since this is true for each $\mathscr{S}_i$, $a\alpha + b\beta$ is in the intersection. □

Now continue our consideration of a system of linear homogeneous equations. Let $\mathscr{S}_i$ be the solution subspace of the i-th equation in the system. The solution set of the entire system is, by definition, the set of n-tuples that satisfy all the equations of the system. Thus, it is precisely the set of vectors in each and every $\mathscr{S}_i$. This is the intersection of the $\mathscr{S}_i$, which is a subspace according to Theorem 4.2.

In three dimensions the solution set of a single linear homogeneous equation is (generally) a plane containing the origin. The intersection of two distinct planes will be a line through the origin. (The planes cannot be parallel since the planes under consideration all contain the origin.) The intersection of three distinct planes must be a single point, the origin, as in Figure 2–1.5a, or a straight line through the origin, as in Figure 2–1.5c. The situation in Figure 2–1.5b cannot occur since the origin is contained in every plane under consideration.

Such visualizations become increasingly complex in systems of equations involving more variables. But we wish to show that the general situation is quite similar to that described in the previous paragraph. Specifically, we wish to show that the solution set of a single linear homogeneous equation in n variables is (generally) a subspace of dimension $n-1$; that it is essentially an $(n-1)$-dimensional coordinate space. And as further equations are considered, for each new equation the intersection remains the same or the intersection is smaller with a decrease in the dimension of exactly 1. But to make this kind of discussion meaningful we have to be more precise about what we mean by "dimension." The next few sections have this as their objective.

EXERCISES 2-4

Some of the exercises in Section 2-2 are easier to work in the context of this section. The vector space in Exercise 3 of Section 2-2 is a *function space*. All that is required is that $\mathscr{S}$ be non-empty and the functions have values in the field over which the function space is to be defined. The spaces described in Exercises 4 and 5 of Section 2-2 can be considered to be subspaces of an appropriate function space. These function spaces are vector spaces because of Exercise 3 of Section 2-2. Then only conditions 1) and 6) need be verified to show that the sets described in Exercises 4 and 5 are subspaces, and hence vector spaces. Use this method in the following two exercises.

1. Let $\mathscr{F}$ be the set of all real valued functions f of a real variable that satisfy the differential equation $\dfrac{d^2f}{dx^2} + f = 0$. Show that $\mathscr{F}$ is a vector space over the real numbers.

2. Let $\mathscr{G}$ be the set of all real valued functions f of a real variable that vanish for $x = 1$ (i.e. $f(1) = 0$). Show that $\mathscr{G}$ is a vector space over the real numbers.

3. Consider the equations
$$x_1 + 2x_2 + 3x_3 = 0 \qquad (4.5)$$
$$2x_1 - x_2 + x_3 = 0. \qquad (4.6)$$
Show that $(x_1, x_2, x_3) = x_2(-2, 1, 0) + x_3(-3, 0, 1)$ is a general solution of (4.5) and $(x_1, x_2, x_3) = x_2 \left(\dfrac{1}{2}, 1, 0\right) + x_3 \left(-\dfrac{1}{2}, 0, 1\right)$ is a general solution of (4.6). Geometrically, each is a plane. Take $x_3 = -x_2 = s$ in both parametrizations to show that $(x_1, x_2, x_3) = s(-1, -1, 1)$ is contained in both general solutions. That is, it is in the intersection. Use a reduction to basic form for the system $\{(4.5), (4.6)\}$ to obtain this parametrization directly.

4. Show that $(x_1, x_2, x_3) = (x_2 + x_3)(-2, 1, 0) + x_3(-1, -1, 1)$ is a general solution of (4.5), and $(x_1, x_2, x_3) = (x_2 + x_3)\left(\dfrac{1}{2}, 1, 0\right) + x_3(-1, -1, 1)$ is a general solution of (4.6).

5. Show that the set of all real valued functions defined on

the real numbers that are both continuous and vanish at $x = 1$ is a vector space over the real numbers.

6. $(x_1, x_2, x_3) = x_1(1, 0, -4) + x_2(0, 1, -5)$ is a parametric representation of the solution set for the homogeneous equation $4x_1 + 5x_2 + x_3 = 0$. Show, by verifying the closure properties for the solution set directly in terms of the parametric representation, that the solution set is a subspace.

Use a reduction to basic form to obtain parametric representations for the solution subspaces for the following systems of homogeneous equations.

7. $7x_1 + 8x_2 + 4x_3 = 0$

8. $7x_1 + 8x_2 + 4x_3 = 0$
   $4x_1 + 5x_2 + x_3 = 0$

9. $2x_1 + 3x_2 + 3x_3 + 2x_4 = 0$
   $3x_1 + 2x_2 - 2x_3 + x_4 = 0$

10. $2x_1 + x_2 - 3x_3 - x_4 = 0$
    $3x_1 + 2x_2 + x_3 + 9x_4 = 0$
    $x_1 - 2x_2 + 2x_3 + 9x_4 = 0$

11. $2x_1 + x_2 - 3x_3 - x_4 = 0$
    $3x_1 + 2x_2 - 4x_3 + 9x_4 = 0$
    $x_1 + 2x_2 \phantom{{}-4x_3} + 7x_4 = 0$

12. $x_1 + 2x_2 + 3x_3 + 8x_4 = 0$
    $2x_1 - x_2 - 4x_3 + x_4 = 0$
    $5x_1 + 3x_2 + x_3 + 19x_4 = 0$

13. Let $\mathscr{R}^n$ be the n-dimensional real coordinate space. Show that each of the following subsets of $\mathscr{R}^n$ is not a subspace of $\mathscr{R}^n$.

    a) The set of all $(x_1, x_2, \ldots, x_n)$ for which $x_1 + x_2 + \cdots + x_n = 1$.
    b) The set of all $(x_1, x_2, \ldots, x_n)$ for which $x_1 \geq 0$.
    c) The set of all $(x_1, x_2, \ldots, x_n)$ for which $|x_1| \leq 100$.
    d) The set of all $(x_1, x_2, \ldots, x_n)$ for which all x_i are rational numbers.
    e) The set of all $(x_1, x_2, \ldots, x_n)$ for which at least one $x_i = 0$.

2-5 LINEAR COMBINATIONS

In a vector space, vector addition and scalar multiplication are defined for pairs of elements. Conditions 2), 5), 7), 8), and 9) allow us to add and multiply in more complicated expressions without being specific or concerned about the order in which the operations are performed. For example, the various expressions $(\alpha+\beta)+\gamma$, $\alpha+(\beta+\gamma)$, $(\alpha+\gamma)+\beta$, $\alpha+(\gamma+\beta)$, etc., are all equal and can be written as $\alpha+\beta+\gamma$ without ambiguity. A similar remark applies to the meaningfulness of an expression like $a\alpha+b\beta+c\gamma$.

Definition   *Let $\{\alpha_1, \alpha_2, \ldots, \alpha_r\}$ be a collection of vectors and let $\{a_1, a_2, \ldots, a_r\}$ be a corresponding collection of scalars. The expression*

$$a_1\alpha_1 + a_2\alpha_2 + \cdots + a_r\alpha_r \tag{5.1}$$

*is called a **linear combination** of the vectors in the set $\{\alpha_1, \alpha_2, \ldots, \alpha_r\}$.* By the remarks of the previous paragraph, the vector designated by (5.1) is unambiguously specified regardless of the order in which the indicated operations are performed. If $\mathscr{S}$ is any subset (finite or infinite) of a vector space, a vector is a linear combination of the vectors in $\mathscr{S}$ if it is a linear combination of a finite number of vectors selected from $\mathscr{S}$.

Consider the single linear homogeneous equation

$$x_1 + 2x_2 + 3x_3 + 4x_4 = 0, \tag{5.2}$$

with the parametric solution

$$x_1 = -2x_2 - 3x_3 - 4x_4. \tag{5.3}$$

In the spirit of the discussion of Section 2–3, the parametric solution can be used to express the 4-tuple representing a solution in the form

$$\begin{aligned}(x_1, x_2, x_3, x_4) &= (-2x_2 - 3x_3 - 4x_4, x_2, x_3, x_4) \\ &= (-2x_2, x_2, 0, 0) + (-3x_3, 0, x_3, 0) + (-4x_4, 0, 0, x_4) \\ &= x_2(-2, 1, 0, 0) + x_3(-3, 0, 1, 0) + x_4(-4, 0, 0, 1). \end{aligned} \tag{5.4}$$

The point of the expression in (5.4) is that it gives the solution 4-tuples as linear combinations of the 4-tuples in the set $\{(-2,1,0,0), (-3,0,1,0), (-4,0,0,1)\}$. Furthermore, since there is no restriction on the choices for the values of the parameters, all solutions are obtained by taking all possible linear combinations. We consider (5.4) as equivalent to the parametric solution — it is just the parametric solution expressed in vector form.

In a more general setting consider the following system of homogeneous linear equations in basic form.

$$
\begin{aligned}
x_1 \phantom{{}+x_2} &+ a_{1,m+1}x_{m+1} + \cdots + a_{1n}x_n = 0 \\
\phantom{x_1 +{}} x_2 &+ a_{2,m+1}x_{m+1} + \cdots + a_{2n}x_n = 0 \\
&\vdots \\
x_m &+ a_{m,m+1}x_{m+1} + \cdots + a_{mn}x_n = 0.
\end{aligned}
\qquad (5.5)
$$

The solution in parametric form is

$$
\begin{aligned}
x_1 &= -a_{1,m+1}x_{m+1} - \cdots - a_{1n}x_n \\
x_2 &= -a_{2,m+1}x_{m+1} - \cdots - a_{2n}x_n \\
&\vdots \\
x_m &= -a_{m,m+1}x_{m+1} - \cdots - a_{mn}x_n.
\end{aligned}
\qquad (5.6)
$$

Following the form of the calculation we did in the particular example above, we can write down an expression in vector form containing the information in (5.6).

$$
\begin{aligned}
(x_1, x_2, \ldots, x_n) &= x_{m+1}(-a_{1,m+1}, -a_{2,m+1}, \ldots, -a_{m,m+1}, 1, 0, \ldots, 0) \\
&+ x_{m+2}(-a_{1,m+2}, -a_{2,m+2}, \ldots, -a_{m,m+2}, 0, 1, \ldots, 0) \\
&+ \cdots \\
&+ x_n(-a_{1n}, -a_{2n}, \ldots, -a_{mn}, 0, 0, \ldots, 1).
\end{aligned}
\qquad (5.7)
$$

If we denote the n-tuple $(x_1, x_2, \ldots, x_n)$ by the letter X, and the n-tuple $(-a_{1,m+j}, -a_{2,m+j}, \ldots, -a_{m,m+j}, 0, \ldots, 1, \ldots, 0)$ by the letter A_j, then formula (5.7) can be written in the more compact form

$$
X = x_{m+1}A_1 + x_{m+2}A_2 + \cdots + x_n A_{n-m}. \qquad (5.8)
$$

X is a linear combination of the n-tuples in the set $\{A_1 A_2, \ldots,$

$A_{n-m}\}$. The scalars $x_{m+1}, x_{m+2}, \ldots, x_n$ are the parameters in these linear combinations, and we refer to (5.8) as the *parametric solution in vector form*. We say that $\{A_1, A_2, \ldots, A_{n-m}\}$ *generates* the solution set.

If we start with a set of homogeneous linear equations, solving the system amounts to finding a set of generators for the subspace of solutions. We shall show before long that every subspace has a finite set of generators, and that every subspace is the solution set of a system of homogeneous linear equations. In this sense, a subspace can be characterized either as the solution set of a system of homogeneous linear equations or as a set of linear combinations of a finite set of generators. The numerical methods we developed in Chapter 1 allow us to pass from either kind of characterization to the other.

If $\mathscr{S}$ is a subset of a vector space $\mathscr{V}$, the set of all linear combinations of vectors in $\mathscr{S}$ is denoted by $\langle \mathscr{S} \rangle$. The set $\langle \mathscr{S} \rangle$ is called the set *spanned* or *generated* by $\mathscr{S}$ or the *span* of $\mathscr{S}$. For example, the solution set for the equation (5.2) can be written in the form $\langle (-2, 1, 0, 0), (-3, 0, 1, 0), (-4, 0, 0, 1) \rangle$. Similarly, the solution set for the system (5.5) is $\langle A_1, A_2, \ldots, A_{n-m} \rangle$.

In order to permit some theorems, like the following one, to be stated without adding a special phrase to cover the case where $\mathscr{S}$ is the empty set, we define $\langle \varnothing \rangle = \{0\}$. That is, $\langle \varnothing \rangle$ is the subspace consisting of only the zero vector.

Theorem 5.1 *For any subset $\mathscr{S}$, $\langle \mathscr{S} \rangle$ is a subspace.*

Proof. Let $\alpha, \beta \in \langle \mathscr{S} \rangle$. This means there exist finite subsets $\{\alpha_1, \ldots, \alpha_r\} \subset \mathscr{S}$ and $\{\beta_1, \ldots, \beta_t\} \subset \mathscr{S}$ such that

$$\alpha = a_1\alpha_1 + a_2\alpha_2 + \cdots + a_r\alpha_r, \quad (5.9)$$
$$\beta = b_1\beta_1 + b_2\beta_2 + \cdots + b_t\beta_t.$$

Then for any scalars a and b, we have

$$a\alpha + b\beta = aa_1\alpha_1 + \cdots + aa_r\alpha_r + bb_1\beta_1 + \cdots + bb_t\beta_t. \quad (5.10)$$

This shows that $a\alpha + b\beta \in \langle \mathscr{S} \rangle$ and, since $\langle \mathscr{S} \rangle$ is closed under linear combinations, $\langle \mathscr{S} \rangle$ is a subspace. ☐

Theorem 5.2 *For any subset $\mathscr{S}$ and any subspace $\mathscr{T}$*

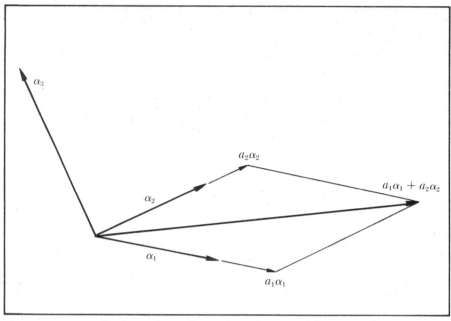

Figure 2–5.1 This figure shows three vectors $\{\alpha_1, \alpha_2, \alpha_3\}$ lying in a plane. The linear combinations of any two of them will generate the plane containing them. This plane is the subspace spanned by $\{\alpha_1, \alpha_2, \alpha_3\}$.

containing $\mathscr{S}$, $\langle \mathscr{S} \rangle \subset \mathscr{T}$. That is, $\langle \mathscr{S} \rangle$ is the smallest subspace containing $\mathscr{S}$.

Proof. It is clear enough that $\mathscr{S} \subset \langle \mathscr{S} \rangle$. If $\mathscr{T}$ is any subspace containing $\mathscr{S}$, $\mathscr{T}$ contains all linear combinations of elements in $\mathscr{T}$ and, hence, all linear combinations of elements in $\mathscr{S}$. Thus $\langle \mathscr{S} \rangle \subset \mathscr{T}$. □

Corollary 5.3 If $\mathscr{S}$ is a subspace, $\langle \mathscr{S} \rangle = \mathscr{S}$. □

Corollary 5.4 For any subset $\mathscr{S}$ of a vector space, $\langle \langle \mathscr{S} \rangle \rangle = \langle \mathscr{S} \rangle$.

Proof. Since $\langle \mathscr{S} \rangle$ is a subspace, $\langle \langle \mathscr{S} \rangle \rangle = \langle \mathscr{S} \rangle$. □

If $\mathscr{S} = \{A_1, A_2, \ldots, A_n\}$ is a finite set of m-tuples, the problem of determining whether a particular m-tuple B is or is not an element of $\langle \mathscr{S} \rangle$ is a problem that has already been considered. Let $A_j = (a_{1j}, a_{2j}, \ldots, a_{mj})$ and let $B = (b_1,$

$b_2, \ldots, b_m$). Then the problem of finding scalars $x_1, x_2, \ldots, x_n$ such that

$$x_1 A_1 + x_2 A_2 + \cdots + x_n A_n = B, \qquad (5.11)$$

is exactly equivalent to solving the system

$$\begin{aligned} a_{11}x_1 + a_{12}x_2 + \cdots + a_{1n}x_n &= b_1 \\ a_{21}x_1 + a_{22}x_2 + \cdots + a_{2n}x_n &= b_2 \\ &\vdots \\ a_{m1}x_1 + a_{m2}x_2 + \cdots + a_{mn}x_n &= b_m. \end{aligned} \qquad (5.12)$$

If the system (5.12) has a solution, the solution expresses B as a linear combination in the form (5.11), and $B \in \langle \mathcal{S} \rangle$. If the system (5.12) does not have a solution then $B \notin \langle \mathcal{S} \rangle$.

Theorem 5.5 *If $\mathcal{S} \subset \mathcal{T}$ then $\langle \mathcal{S} \rangle \subset \langle \mathcal{T} \rangle$.*

Proof. Since $\mathcal{S} \subset \mathcal{T} \subset \langle \mathcal{T} \rangle$, $\langle \mathcal{T} \rangle$ is a subspace containing $\mathcal{S}$. Since $\langle \mathcal{S} \rangle$ is the smallest subspace containing $\mathcal{S}$, $\langle \mathcal{S} \rangle \subset \langle \mathcal{T} \rangle$. □

For a numerical example, consider the following system of homogeneous linear equations.

$$\begin{aligned} x_1 + 2x_2 + 4x_3 - x_4 + 2x_5 &= 0 \\ 2x_1 + 5x_2 + 9x_3 - 3x_4 + 5x_5 &= 0 \\ 3x_1 + 5x_2 + 11x_3 - x_4 + 4x_5 &= 0 \\ -x_1 - x_2 - 3x_3 + 2x_4 - 3x_5 &= 0. \end{aligned} \qquad (5.13)$$

The coefficient matrix can be reduced to basic form in several different ways. One of the possible reduced basic forms is the following,

$$\begin{bmatrix} 1 & 0 & 2 & 0 & 1 \\ 0 & 1 & 1 & 0 & 0 \\ 0 & 0 & 0 & 1 & -1 \\ 0 & 0 & 0 & 0 & 0 \end{bmatrix} \qquad (5.14)$$

Our aim is to be able to write down a set of generators for the solution subspace directly from the basic form. But to see what to look for we translate (5.14) into equation form,

$$\begin{aligned} x_1 \quad &+ 2x_3 \quad\quad\ + x_5 = 0 \\ x_2 &+\ x_3 \quad\quad\quad\quad\ \ = 0 \\ &\quad\quad\quad\quad\ x_4 - x_5 = 0. \end{aligned} \qquad (5.15)$$

The corresponding parametric form is

$$\begin{aligned} x_1 &= -2x_3 - x_5 \\ x_2 &= -\ x_3 \\ x_3 &= \quad x_3 \\ x_4 &= \quad\quad\quad\ x_5 \\ x_5 &= \quad\quad\quad\ x_5 \,. \end{aligned} \qquad (5.16)$$

Finally, the vector parametric form is

$$(x_1, x_2, x_3, x_4, x_5) = x_3(-2, -1, 1, 0, 0) + x_5(-1, 0, 0, 1, 1). \quad (5.17)$$

The point is, how do we see that $\{(-2, -1, 1, 0, 0), (-1, 0, 0, 1, 1)\}$ is the set of generators directly from the reduced matrix (5.14)? From (5.17) we can see that one generator is obtained by taking $x_3 = 1$, $x_5 = 0$, and the other generator is obtained by taking $x_3 = 0$, $x_5 = 1$. This is just as easily obtained from the system (5.15), which can be visualized from looking at (5.14). From (5.14) we can tell that x_3 and x_5 are parameters. This is enough to tell us that one generator is of the form (, , 1, , 0) and the other is of the form (, , 0, , 1). Visualizing (5.15) we can fill in the remaining three positions for each generator.

To specify the solution set for the system (5.13) it is quite sufficient to give the set of generators for the solution set. All other information is implicit. Because of the 1 in the third position of the first generator and the 0 in the third place of the second, we can tell that x_3 is the first parameter. Similarly, we can tell that x_5 is the second parameter. (Notice that a similar remark applies to x_4. That simply means that whatever the parameter is, both x_4 and x_5 have the same value as that parameter.)

Now let $\mathscr{S}$ be the subspace spanned by $\{(1, 1, 1, 1), (2, 3, 1, 3), (0, -1, 1, -1), (-1, -2, 1, 0), (1, 1, 3, 5)\}$. Is $(5, 8, 0, 4)$ an element of $\mathscr{S}$? The system of equations for this problem, of the form given in (5.12), is

$$\begin{aligned} x_1 + 2x_2 \quad\quad\ -\ x_4 &+\ x_5 = 5 \\ x_1 + 3x_2 - x_3 - 2x_4 &+\ x_5 = 8 \\ x_1 +\ x_2 + x_3 +\ x_4 &+ 3x_5 = 0 \\ x_1 + 3x_2 - x_3 \quad\quad\ \ &+ 5x_5 = 4. \end{aligned} \qquad (5.18)$$

The corresponding augmented matrix, reduced to a basic form, is

$$\begin{bmatrix} 1 & 0 & 2 & 0 & -1 & 1 \\ 0 & 1 & -1 & 0 & 2 & 1 \\ 0 & 0 & 0 & 1 & 2 & -2 \\ 0 & 0 & 0 & 0 & 0 & 0 \end{bmatrix}. \qquad (5.19)$$

For this basic form, x_1, x_2, and x_4 are basic variables and x_3 and x_5 are parameters. There is a significant difference between this problem and other systems of linear equations that we have considered. Since we want to determine whether $(5, 8, 0, 4)$ is an element of $\mathscr{S}$, we do not need to know *all* possible representations of $(5, 8, 0, 4)$ as linear combinations of the generators. In fact, from the form of (5.19) alone we can tell that the system (5.18) has at least one solution. That is, $(5, 8, 0, 4)$ is in $\mathscr{S}$. However, it is desirable to produce at least one representation of $(5, 8, 0, 4)$ as a linear combination. For this purpose it is most convenient to give all parameters the value 0. This gives $x_1 = 1$, $x_2 = 1$, and $x_4 = -2$. Since these are just the numbers in the last column of (5.19), the representation of $(5, 8, 0, 4)$ is easily read off from the basic form (5.19). Verify that

$$(1, 1, 1, 1) + (2, 3, 1, 3) - 2(-1, -2, 1, 0) = (5, 8, 0, 4). \qquad (5.20)$$

The sequence of pivot operations leading from (5.18) to (5.19) depends only on the coefficients appearing on the left sides of the equations in (5.18). Therefore, if we are asked to determine whether several vectors are in $\mathscr{S}$, it is just as easy to handle the several vectors simultaneously. For example, suppose we are asked to determine if $(1, 0, 0)$, $(0, 1, 0)$, and $(0, 0, 1)$ are in the subspace $\mathscr{S}$ spanned by $\{(6, -1, -3), (2, 0, -1), (-1, 1, 1)\}$ We set up the matrix

$$\begin{bmatrix} 6 & 2 & -1 & 1 & 0 & 0 \\ -1 & 0 & 1 & 0 & 1 & 0 \\ -3 & -1 & 1 & 0 & 0 & 1 \end{bmatrix}. \qquad (5.21)$$

Here, the first three columns and column four, the first three columns and column five, and the first three columns and column six represent each of the three problems posed. Since only the first three columns correspond to the generators of $\mathscr{S}$, all pivot operations should be confined to the first three columns. When (5.21) is reduced to basic form—actually row-echelon form—in this way we obtain

$$\begin{bmatrix} 1 & 0 & 0 & 1 & -1 & 2 \\ 0 & 1 & 0 & -2 & 3 & -5 \\ 0 & 0 & 1 & 1 & 0 & 2 \end{bmatrix}. \quad (5.22)$$

In (5.22) it is easy to see that the fourth column is a linear combination of the first three columns. The numbers in the fourth column give this relation.

$$(1, -2, 1) = 1(1, 0, 0) - 2(0, 1, 0) + 1(0, 0, 1). \quad (5.23)$$

Since (5.22) was obtained from (5.21) by elementary row operations, any linear relations among the columns were preserved. That is,

$$(1, 0, 0) = 1(6, -1, -3) - 2(2, 0, -1) + 1(-1, 1, 1). \quad (5.24)$$

In the same way, the numbers in the fifth column give

$$(0, 1, 0) = -(6, -1, -3) + 3(2, 0, -1) + 0(-1, 1, 1), \quad (5.25)$$

and the numbers in the sixth column give

$$(0, 0, 1) = 2(6, -1, -3) - 5(2, 0, -1) + 2(-1, 1, 1). \quad (5.26)$$

EXERCISES 2–5

Find a set of generators for the solution subspace for the following systems of homogeneous linear equations.

1. $3x_1 + x_2 - x_4 = 0$
   $ 2x_1 + 3x_4 + x_5 = 0$
   $-x_1 + x_3 = 0$

2. $x_1 + x_2 = 0$
   $x_1 - 2x_2 = 0$

3. $2x_1 - 3x_2 - 8x_3 = 0$
   $-x_1 + 2x_2 + 5x_3 = 0$
   $3x_1 + 4x_2 + 5x_3 = 0$

4. $4x_1 - 8x_2 + 16x_3 = 0$
   $x_1 - 3x_2 + 6x_3 = 0$
   $2x_1 + x_2 - 2x_3 = 0$

5. $2x_1 + 2x_2 + 5x_3 + 6x_4 = 0$
   $x_1 + x_2 - 2x_3 + 2x_4 = 0$

6. $x_1 - 3x_2 + x_3 + x_4 = 0$
   $-x_1 + 3x_2 + 2x_3 - x_4 = 0$
   $2x_1 - 6x_2 - x_3 + 2x_4 = 0$
   $-2x_1 + 6x_2 + x_3 - x_4 = 0$

7. $3x_1 + x_2 + 9x_3 = 0$
   $3x_1 + 2x_2 + 12x_3 = 0$
   $2x_1 + x_2 + 7x_3 = 0$

8. $2x_1 + 6x_3 + x_4 + 2x_5 + 4x_6 = 0$
   $2x_1 + 2x_2 + 10x_3 + x_5 + 3x_6 = 0$
   $x_1 + x_2 + 5x_3 + x_4 + 3x_5 + x_6 = 0$
   $3x_1 + 9x_3 + x_4 + x_5 + 7x_6 = 0$
   $5x_1 + 3x_2 + 21x_3 + 5x_4 + 2x_5 + 18x_6 = 0$

9. Let $\mathscr{S}$ be spanned by $\{(6, -1, -3), (2, 0, -1), (14, -2, -7), (-1, 1, 1), (7, -2, -4)\}$. Determine whether $(-11, 3, 6) \in \mathscr{S}$ and if it is express it as a linear combination of the generators.

10. For the space $\mathscr{S}$ given in Exercise 9, show that three of the generators suffice to span $\mathscr{S}$.

11. Let the subspace $\mathscr{S}$ be spanned by $\{(0, 1, 0, 0), (1, 0, -2, 0), (0, 1, 2, 1)\}$. Determine whether $(1, 1, 1, 2) \in \mathscr{S}$. Determine whether $(1, 1, 2, 2) \in \mathscr{S}$.

12. Let $\mathscr{P}_4$ be the set of polynomials with real coefficients of degree less than four. Let $\mathscr{S}$ be the subspace spanned by the polynomials $\{t^2, t^3 - 2t, t^2 + 2t + 1\}$. Determine whether $t^3 + t^2 + t + 2$ is in $\mathscr{S}$. Determine whether $t^3 + t^2 + 2t + 2$ is in $\mathscr{S}$. (Hint: Find a relation between this exercise and Exercise 11.)

13. For the subspace $\mathscr{S}$ described in Exercise 11, determine a condition that a 4-tuple (x_1, x_2, x_3, x_4) must satisfy in order to be an element of $\mathscr{S}$. (Hint: Write down a matrix with the 4-tuples spanning $\mathscr{S}$ in the first three columns and the 4-tuple (x_1, x_2, x_3, x_4) in the last column. Reduce this matrix to basic form and determine a condition on the x_i that will give this reduced matrix the required form.)

2-6 LINEAR INDEPENDENCE

Suppose the subspace $\mathscr{S}$ is spanned by the set $\mathscr{A} = \{\alpha_1, \ldots, \alpha_r\}$. We raise the following questions. Can we find a smaller set that also spans $\mathscr{S}$? Since the set $\mathscr{A}$ spans $\mathscr{S}$, every element in $\mathscr{S}$ can be expressed as a linear combination of the vectors in $\mathscr{A}$. Are other expressions possible? It turns out that the concept of linear independence is crucial in answering both these questions.

To see what is involved, suppose that one of the vectors in $\mathscr{A}$ is a linear combination of the others. To be specific, suppose

$$\alpha_r = a_1 \alpha_1 + a_2 \alpha_2 + \cdots + a_{r-1} \alpha_{r-1}. \tag{6.1}$$

Since $\mathscr{A}$ spans $\mathscr{S}$, any element $\beta \in \mathscr{S}$ can be represented in the form

$$\beta = b_1 \alpha_1 + b_2 \alpha_2 + \cdots + b_r \alpha_r. \tag{6.2}$$

By substituting the representation (6.1) for α_r in (6.2) we can obtain

$$\beta = (b_1 + b_r a_1) \alpha_1 + (b_2 + b_r a_2) \alpha_2 + \cdots \\ + (b_{r-1} + b_r a_{r-1}) \alpha_{r-1}. \tag{6.3}$$

If $b_r \alpha_r \neq 0$, the representation of β in (6.3) is different from the representation in (6.2). Since the β chosen was an arbitrary element in $\mathscr{S}$ and the right side of (6.3) involves only the vectors $\{\alpha_1, \ldots, \alpha_{r-1}\}$, $\mathscr{S}$ is spanned by $\{\alpha_1, \ldots, \alpha_{r-1}\}$.

Thus, if one vector in $\mathscr{A}$ can be expressed as a linear combination of the others, representations as linear combinations of elements in $\mathscr{S}$ are not generally unique, and the subspace spanned by $\mathscr{A}$ can be spanned by a set smaller than $\mathscr{A}$. The converse is also true. Specifically, suppose the proper subset $\{\alpha_1, \ldots, \alpha_{r-1}\}$ also spans $\mathscr{S}$. Then, since $\alpha_r \in \mathscr{S}$, α_r can be expressed as a linear combination of the elements in $\{\alpha_1, \ldots, \alpha_{r-1}\}$. Also, suppose there is an element β in $\mathscr{S}$ that can be represented as a linear combination of the elements in $\mathscr{A}$ in at least two different ways. For example, suppose

$$\beta = b_1 \alpha_1 + \cdots + b_r \alpha_r = c_1 \alpha_1 + \cdots + c_r \alpha_r. \tag{6.4}$$

Then

$$(b_1 - c_1) \alpha_1 + \cdots + (b_r - c_r) \alpha_r = 0. \tag{6.5}$$

Since we have assumed that the two representations in (6.4) are different, at least one of the coefficients in (6.5) is non-zero. If $(b_r - c_r) \neq 0$, then

$$\alpha_r = \frac{-1}{b_r - c_r} \{(b_1 - c_1)\alpha_1 + \cdots + (b_{r-1} - c_{r-1})\alpha_{r-1}\}. \quad (6.6)$$

Definition Let $\mathcal{A} = \{\alpha_1, \ldots, \alpha_r\}$ *be an indexed set of vectors. A **linear relation** among the vectors in this set is an expression of the form*

$$a_1\alpha_1 + a_2\alpha_2 + \cdots + a_r\alpha_r = 0. \quad (6.7)$$

*This linear relation is said to be **trivial** if $a_1 = a_2 = \cdots = a_r = 0$, and it is **non-trivial** if at least one $a_j \neq 0$. The set $\{\alpha_1, \ldots, \alpha_r\}$ is said to be **linearly dependent** if there exists a non-trivial linear relation among the elements of the set, and it is said to be **linearly independent** if the only linear relation among the elements of the set is trivial.*

We have stated the definition of linear dependence and independence for finite sets, but the definition can easily be extended to infinite sets. An arbitrary set is linearly independent if and only if every finite subset is linearly independent. It is linearly dependent if and only if it contains a finite linearly dependent subset.

It is clear from the definition that if a set $\mathcal{A}$ is linearly dependent, any set containing $\mathcal{A}$ is also linearly dependent. If $\mathcal{A}$ is linearly independent, any subset of $\mathcal{A}$ is linearly independent. These two statements are logically equivalent.

Theorem 6.1 Let $\{\alpha_1, \alpha_2, \ldots, \alpha_r\}$ *be a set of vectors, indexed in some order. This set is linearly dependent if and only if at least one vector is a linear combination of the preceding vectors.* ("Preceding" *refers to the ordering of the indices.*)

Proof. If $\{\alpha_1, \ldots, \alpha_r\}$ is linearly dependent, there is a non-trivial linear relation of the form

$$a_1\alpha_1 + a_2\alpha_2 + \cdots + a_r\alpha_r = 0. \quad (6.8)$$

This relation has at least one non-zero coefficient. Therefore, there is a last non-zero coefficient in (6.8). Let the last non-zero coefficient be a_k. Then we can solve for α_k in the form

$$\alpha_k = \frac{-1}{a_k}(a_1\alpha_1 + \cdots + a_{k-1}\alpha_{k-1}). \tag{6.9}$$

There is one more minor point; a_k may be the only non-zero coefficient in (6.8). This would mean $\alpha_k = 0$. If $k > 1$, this would mean $\alpha_k = 0\alpha_1 + \cdots + 0\alpha_{k-1}$. If $k = 1$ there are no preceding vectors. We can either add the condition $\alpha_1 \neq 0$ to the hypothesis of the theorem or we can agree to accept $\alpha_1 = 0$ as being within the meaning of the conclusion. It doesn't matter which alternative is chosen, but since we have already agreed that $\langle \varnothing \rangle = \{0\}$ we shall choose the latter alternative.

It is clear that if one vector in $\{\alpha_1, \ldots, \alpha_r\}$ is a linear combination of the preceding ones the set is linearly dependent. □

To say that α_k is a linear combination of $\{\alpha_1, \ldots, \alpha_{k-1}\}$ is equivalent to saying that $\alpha_k \in \langle \alpha_1, \ldots, \alpha_{k-1} \rangle$. In case $k = 1$ this condition takes the form $\alpha_1 \in \langle \varnothing \rangle$. This is true if and only if $\alpha_1 = 0$. Thus, Theorem 6.1 can be stated in the equivalent form, "The set $\{\alpha_1, \ldots, \alpha_r\}$ is linearly dependent if and only if for at least one k, $\alpha_k \in \langle \alpha_1, \ldots, \alpha_{k-1} \rangle$." Equivalently, "The set $\{\alpha_1, \ldots, \alpha_r\}$ is linearly independent if and only if for each $k = 1, \ldots, r$ $\alpha_k \notin \langle \alpha_1, \ldots, \alpha_{k-1} \rangle$. (For $k = 1$ we interpret this notation to mean $\alpha_1 \notin \langle \varnothing \rangle$.)"

If $\{A_1, \ldots, A_r\}$ is a set of m-tuples, the numerical calculations required to determine whether this set is linearly independent or not are already familiar. All that is required is a proper interpretation of the work. In attempting to find a linear relation we must solve the system of equations of the form

$$x_1 A_1 + x_2 A_2 + \cdots + x_r A_r = 0. \tag{6.10}$$

Let $A_j = (a_{1j}, a_{2j}, \ldots, a_{mj})$. In terms of m-tuples, the system (6.10) amounts to a system of m linear homogeneous equations in r unknowns. The system is solved by performing a sequence of pivot operations. To obtain a conclusion of the type mentioned in Theorem 6.1 (that one vector is a linear combination of the *preceding* vectors), it is necessary to select the variables for the pivot operations in a particular order. First, try to find a pivot element in the first column, then the second, the third, and so on. The necessary conclusion can then be read off from the resulting basic form. The following example will illustrate how this is done.

Consider the set of 4-tuples $\{A_1 = (1, 2, 1, 3), A_2 = (1, 1, 0, 2), A_3 = (2, 4, 3, 6), A_4 = (1, 3, 4, 4), A_5 = (1, 2, 1, 4), A_6 = (4, 8, 6, 11)\}$. The equation (6.10) is equivalent to the system

$$\begin{aligned} x_1 + x_2 + 2x_3 + x_4 + x_5 + 4x_6 &= 0 \\ 2x_1 + x_2 + 4x_3 + 3x_4 + 2x_5 + 8x_6 &= 0 \\ x_1 \phantom{{}+x_2} + 3x_3 + 4x_4 + x_5 + 6x_6 &= 0 \\ 3x_1 + 2x_2 + 6x_3 + 4x_4 + 4x_5 + 11x_6 &= 0. \end{aligned} \qquad (6.11)$$

The matrix corresponding to this system is

$$\begin{array}{c} A_1\ A_2\ A_3\ A_4\ A_5\ A_6 \\ \begin{bmatrix} 1 & 1 & 2 & 1 & 1 & 4 \\ 2 & 1 & 4 & 3 & 2 & 8 \\ \fbox{1} & 0 & 3 & 4 & 1 & 6 \\ 3 & 2 & 6 & 4 & 4 & 11 \end{bmatrix} \end{array} \qquad (6.12)$$

We now perform the sequence of pivot operations, making the selections from the columns in the same order in which they appear in the matrix (6.12).

$$\begin{bmatrix} 0 & \fbox{1} & -1 & -3 & 0 & -2 \\ 0 & 1 & -2 & -5 & 0 & -4 \\ 1 & 0 & 3 & 4 & 1 & 6 \\ 0 & 2 & -3 & -8 & 1 & -7 \end{bmatrix}$$

$$\begin{bmatrix} 0 & 1 & -1 & -3 & 0 & -2 \\ 0 & 0 & \fbox{-1} & -2 & 0 & -2 \\ 1 & 0 & 3 & 4 & 1 & 6 \\ 0 & 0 & -1 & -2 & 1 & -3 \end{bmatrix}$$

$$\begin{bmatrix} 0 & 1 & 0 & -1 & 0 & 0 \\ 0 & 0 & 1 & 2 & 0 & 2 \\ 1 & 0 & 0 & -2 & 1 & 0 \\ 0 & 0 & 0 & 0 & \fbox{1} & -1 \end{bmatrix}$$

$$\begin{bmatrix} 0 & 1 & 0 & -1 & 0 & 0 \\ 0 & 0 & 1 & 2 & 0 & 2 \\ 1 & 0 & 0 & -2 & 0 & 1 \\ 0 & 0 & 0 & 0 & 1 & -1 \end{bmatrix}.$$

It is desirable to obtain the basic form in row-echelon form, so we take an additional step.

$$\begin{array}{cccccc} A_1 & A_2 & A_3 & A_4 & A_5 & A_6 \end{array}$$
$$\begin{bmatrix} 1 & 0 & 0 & -2 & 0 & 1 \\ 0 & 1 & 0 & -1 & 0 & 0 \\ 0 & 0 & 1 & 2 & 0 & 2 \\ 0 & 0 & 0 & 0 & 1 & -1 \end{bmatrix}. \tag{6.13}$$

Any solution of (6.11) amounts to a linear relation among the columns of (6.12). Since the elementary row operations lead to equivalent systems of equations, they also preserve the relations among the columns. Thus, any linear relation among the columns of (6.13) is equivalent to the same linear relation among the columns of (6.12). Relations among the columns of (6.13) are particularly easy to see.

Even though the columns of the matrix (6.13) are not the 4-tuples of the set we started with we have labeled each column in (6.13) by the name of the 4-tuple from which it was obtained. We read (6.13) in the following way. The numbers in column 4 tell us that

$$A_4 = -2A_1 - A_2 + 2A_3, \tag{6.14}$$

and the numbers in column 6 tell us that

$$A_6 = A_1 + 2A_3 - A_5. \tag{6.15}$$

Writing the relations indicated in (6.14) and (6.15) in full, we see that

$$(1, 3, 4, 4) = -2(1, 2, 1, 3) - (1, 1, 0, 2) + 2(2, 4, 3, 6), \tag{6.16}$$

$$(4, 8, 6, 11) = (1, 2, 1, 3) + 2(2, 4, 3, 6) - (1, 2, 1, 4). \tag{6.17}$$

Also, since columns 1, 2, 3, and 5 of (6.13) are linearly independent (none is a linear combination of the preceding ones), the set $\{A_1, A_2, A_3, A_5\} = \{(1, 2, 1, 3), (1, 1, 0, 2), (2, 4, 3, 6), (1, 2, 1, 4)\}$ is linearly independent. This set is a maximal linearly independent subset of the given set in the sense that it is not properly contained in a larger linearly independent subset.

In mathematics, the term "maximal" and "minimal" are

usually used in a special way. A subset 𝒮 is *maximal* with a property if it is not contained properly in another subset with that property. Thus, a maximal linearly independent subset is not properly contained in a larger linearly independent subset. The definition allows for the possibility of several maximal subsets, and for the possibility that different maximal subsets contain different numbers of elements. In particular, a maximal subset with a given property might not be the largest subset with the given property. Similar remarks apply to a *minimal* subset with a given property.

EXERCISES 2-6

1. For the set $\mathscr{S} = \{(3, 1, 2), (3, 4, 3), (-3, 2, -1), (9, 9, 8), (1, 1, 1), (4, -4, 1)\}$ determine the relations that express one vector as a linear combination of the preceding vectors.

2. For the set $\mathscr{S} = \{(3, 3, 1), (1, 4, 1), (8, 5, 2), (2, 3, 1), (12, -7, 0), (4, 7, 2)\}$ determine the relations that express one vector as a linear combination of the preceding vectors.

3. For the set $\mathscr{S} = \{(1, -2, 1, -3), (0, 1, 2, 1), (-2, 6, 2, 8), (0, 0, 1, -1), (1, 0, 4, 0), (0, 0, 0, 1), (4, -10, -1, -10)\}$ determine the relations that express one vector as a linear combination of the preceding vectors.

4. Consider the set $\{(2, 1, 1, 2), (1, 2, -2, -1), (3, 3, -1, 1), (-1, 3, 0, 1), (1, 4, 1, 3), (3, 4, 2, 2), (3, -3, 1, 2)\}$. Find a maximal linearly independent subset by eliminating every vector which is a linear combination of the preceding vectors.

5. For the set $\mathscr{S} = \{(1, 1, 1, 1), (2, 3, 1, 3), (0, -1, 1, -1), (-1, -2, 1, 0), (1, 1, 3, 5)\}$ determine the relations that express one vector as a linear combination of the preceding vectors.

6. For the set 𝒮 in Exercise 5, determine all linearly independent subsets containing exactly three elements.

7. For the set $\{(1, 2, 1, 3), (1, 1, 0, 2), (2, 4, 3, 6), (1, 3, 4, 4), (1, 2, 1, 4), (4, 8, 4, 11)\}$, determine all linearly independent subsets containing exactly four elements.

8. Show that a set of vectors that includes the zero vector is linearly dependent.

9. Show that if a set consisting of a single vector is linearly dependent, that vector must be the zero vector.

10. $\begin{bmatrix} 1 & 0 & 1 & 0 & 1 & 0 \\ 0 & 1 & 1 & 0 & 0 & 0 \\ 0 & 0 & 0 & 1 & 1 & 0 \\ 0 & 0 & 0 & 0 & 0 & 1 \end{bmatrix}$

This matrix is in row-echelon form. Find all minimal linearly dependent sets of columns in this matrix. Find a maximal linearly independent subset.

2-7 BASES AND DIMENSION

Theorem 7.1 *Let $\mathscr{S}$ be a subspace which we assume is spanned by the finite set $\{\alpha_1, \ldots, \alpha_r\}$. Let $\{\beta_1, \beta_2, \ldots\}$ be a linearly independent set of vectors from $\mathscr{S}$. Then the number of vectors in $\{\beta_1, \beta_2, \ldots\}$ is less than or equal to r.*

Proof. Since $\beta_1 \in \mathscr{S}$, β_1 is a linear combination of the vectors in $\{\alpha_1, \ldots, \alpha_r\}$. Thus the set $\{\beta_1, \alpha_1, \ldots, \alpha_r\}$ is linearly dependent. Because of Theorem 6.1 this means that one of the α_j is a linear combination of the preceding elements in $\{\beta_1, \alpha_1, \ldots, \alpha_r\}$. Since $\{\alpha_1, \ldots, \alpha_r\}$ spans $\mathscr{S}$, so also does $\{\beta_1, \alpha_1, \ldots, \alpha_r\}$. But α_j can be removed and the resulting set will still span $\mathscr{S}$. To avoid cumbersome notation we will reindex the elements in $\{\alpha_1, \alpha_2, \ldots, \alpha_r\}$ so that after the removal we have $\{\beta_1, \alpha_2, \ldots, \alpha_r\}$.

Now the process is repeated. That is, since $\beta_2 \in \mathscr{S}$ and $\{\beta_1, \alpha_2, \ldots, \alpha_r\}$ spans $\mathscr{S}$, $\{\beta_1, \beta_2, \alpha_2, \ldots, \alpha_r\}$ is a linearly dependent set. One of the vectors in this set is a linear combination of the preceding ones. This vector cannot be one of the β_i's since $\{\beta_1, \beta_2\}$ is linearly independent. $\{\beta_1, \beta_2, \alpha_2, \ldots, \alpha_r\}$ spans $\mathscr{S}$, and the dependent vector can be removed to obtain a smaller set that still spans $\mathscr{S}$. We reindex this set so that $\{\beta_1, \beta_2, \alpha_3, \ldots, \alpha_r\}$ spans $\mathscr{S}$.

Again, the set $\{\beta_1, \beta_2, \beta_3, \alpha_3, \ldots, \alpha_r\}$ spans $\mathscr{S}$ and is linearly dependent. One of the α_j's is a linear combination of the others and can be removed.

If the number of elements in the linearly independent

2-7 Bases and Dimension

set $\{\beta_1, \beta_2, \ldots\}$ is greater than r we would arrive at the spanning set $\{\beta_1, \beta_2, \ldots, \beta_r\}$. But this would mean that $\beta_{r+1} \in \mathscr{S}$ is a linear combination of the elements in $\{\beta_1, \beta_2, \ldots, \beta_r\}$, which would contradict the assumed linear independence of $\{\beta_1, \beta_2, \ldots\}$. Thus, the number of elements in the set $\{\beta_1, \beta_2, \ldots\}$ is at most equal to r. □

In less formal language, we have shown that in every finitely spanned subspace the number of elements in any linearly independent set is at most equal to the number of elements in any spanning set.

The argument used to prove Theorem 7.1 is known as the *Steinitz replacement principle*. Let $\mathscr{S}$ be a subspace, $\mathscr{A}$ a subset spanning $\mathscr{S}$, and $\mathscr{B}$ a linearly independent subset of $\mathscr{S}$. The Steinitz replacement principle says that the elements of the linearly independent set can be used to replace some of the elements of the spanning set in a one-for-one replacement. We end up with a new spanning set containing the linearly independent set as a subset.

Theorem 7.2 *If $\mathscr{V}$ is a vector space spanned by a finite subset, every subspace of $\mathscr{V}$ is spanned by a finite linearly independent subset.*

Proof. Suppose $\mathscr{V}$ is spanned by a set with n elements. Let $\mathscr{S}$ be a subspace of $\mathscr{V}$. If $\mathscr{S} = \{0\}$, $\mathscr{S}$ is spanned by $\varnothing$. If $\mathscr{S} \neq \{0\}$, there is a $\beta_1 \in \mathscr{S}$, $\beta_1 \neq 0$. If $\mathscr{S} \neq \langle \beta_1 \rangle$, there is a $\beta_2 \in \mathscr{S}$, $\beta_2 \notin \langle \beta_1 \rangle$. This means $\{\beta_1, \beta_2\}$ is linearly independent. Again, if $\mathscr{S} \neq \langle \beta_1, \beta_2 \rangle$, there is a $\beta_3 \in \mathscr{S}$, $\beta_3 \notin \langle \beta_1, \beta_2 \rangle$. The set $\{\beta_1, \beta_2, \beta_3\}$ is linearly independent. This cannot continue indefinitely since, by Theorem 7.1, we cannot obtain a linearly independent set with more than n elements. Thus, we must eventually obtain a linearly independent set $\{\beta_1, \ldots, \beta_k\}$ such that $\mathscr{S} = \langle \beta_1, \ldots, \beta_k \rangle$ and $k \leq n$. □

Corollary 7.3 *If $\mathscr{V}$ is spanned by a finite subset it is spanned by a finite linearly independent subset.*

Proof. $\mathscr{V}$ is one of its own subspaces. □

Definition. *Let $\mathscr{S}$ be a subspace of a vector space $\mathscr{V}$. A linearly independent subset of $\mathscr{S}$ that spans $\mathscr{S}$ is called a basis of $\mathscr{S}$.* (The plural is *bases*.) Corollary 7.3 shows that if a vector space is spanned by a finite set it has a finite basis, and Theorem 7.2 shows that every subspace has a finite basis. Henceforth, we shall restrict our attention to vector spaces with a finite basis.

Theorem 7.4   For each subspace $\mathscr{S}$, any two bases of $\mathscr{S}$ have the same number of elements.

Proof.   Let $\{\alpha_1, \ldots, \alpha_r\}$ and $\{\beta_1, \ldots, \beta_s\}$ be bases of $\mathscr{S}$. Since the first spans $\mathscr{S}$ and the second is linearly independent, $s \leq r$ by Theorem 7.1. And since the first is linearly independent and the second spans $\mathscr{S}$, $r \leq s$. Thus, $r = s$.   □

Definition   Since all bases of a given subspace have the same number of elements, this number is a property associated with the subspace. This number is called the **dimension** of the subspace. We denote the dimension of $\mathscr{S}$ by dim $\mathscr{S}$. In particular, $\mathscr{V}$ is one of the subspaces and has a dimension. Accordingly, our agreement to limit our discussion to vector spaces with a finite basis means that we are considering only **finite dimensional vector spaces**.

It is often desirable to choose a basis carefully in order to make a calculation easier or to make a conclusion more evident. For example, in analytic geometry for a problem involving an ellipse, a pair of coordinate axes coinciding with the major and minor axes yields a simple equation for the ellipse. In linear algebra the following theorem is most useful for this purpose.

Theorem 7.5   Let $\{\alpha_1, \ldots, \alpha_r\}$ be a linearly independent subset of a subspace $\mathscr{S}$. There exists a basis $\{\alpha_1, \ldots, \alpha_r, \alpha_{r+1}, \ldots, \alpha_m\}$ of $\mathscr{S}$ with $\{\alpha_1, \ldots, \alpha_r\}$ as a subset.

Proof.   Let $\{\beta_1, \ldots, \beta_m\}$ be any basis for $\mathscr{S}$ and consider the set $\{\alpha_1, \ldots, \alpha_r, \beta_1, \ldots, \beta_m\}$. This set spans $\mathscr{S}$. If this set is linearly dependent (as it will certainly be if $r \geq 1$) then one of the vectors is a linear combination of the preceding vectors. This one will not be one of the α_i because of the linear independence of $\{\alpha_1, \ldots, \alpha_r\}$. The dependent β_j can be removed to obtain a smaller set spanning $\mathscr{S}$. If the resulting set is also linearly dependent, another of the β_j will be a linear combination of the preceding vectors and can be removed to obtain a smaller set spanning $\mathscr{S}$. We continue in this way, discarding elements as long as we have a linearly dependent spanning set. At no stage will we discard one of the α_i. Since our original spanning set was finite this process must terminate with a basis containing $\{\alpha_1, \ldots, \alpha_r\}$. □

This theorem will be used many times in this text and we will refer to it by saying that we "extend a linearly independent set to a basis." In many cases the linearly independent set we start with will consist of a single non-zero vector that has been chosen to have a certain required property.

Theorem 7.6 *Let $\mathscr{S}$ be a subspace of $\mathscr{V}$. There is a basis of $\mathscr{V}$ that includes as a subset a basis of $\mathscr{S}$.*

Proof. Let $\{\alpha_1, \ldots, \alpha_r\}$ be a basis of $\mathscr{S}$. Since this set is linearly independent, it can be extended to a basis of $\mathscr{V}$. □

Theorem 7.7 *If $\mathscr{S}$ and $\mathscr{T}$ are subspaces of $\mathscr{V}$ such that $\mathscr{S} \subset \mathscr{T}$, then $\dim \mathscr{S} \leq \dim \mathscr{T}$. If $\mathscr{S} \subset \mathscr{T}$ and $\dim \mathscr{S} = \dim \mathscr{T}$, then $\mathscr{S} = \mathscr{T}$.*

Proof. Let $\{\alpha_1, \ldots, \alpha_r\}$ be a basis of $\mathscr{S}$. Since $\{\alpha_1, \ldots, \alpha_r\} \subset \mathscr{T}$, this linearly independent set can be extended to a basis $\{\alpha_1, \ldots, \alpha_s\}$ of $\mathscr{T}$. Clearly $r \leq s$. If $r = s$, then $\mathscr{S} = \langle \alpha_1, \ldots, \alpha_r \rangle = \mathscr{T}$. □

Theorem 7.8 *Let $\mathscr{S}$ be a subspace of $\mathscr{V}$ of dimension m. If $\mathscr{A} = \{\alpha_1, \ldots, \alpha_m\}$ is a linearly independent subset of $\mathscr{S}$, $\mathscr{A}$ is a basis of $\mathscr{S}$.*

Proof. By Theorem 7.5, $\mathscr{A}$ can be extended to a basis of $\mathscr{S}$. By Theorem 7.4, this basis will have m elements, the dimension of $\mathscr{S}$. But since $\mathscr{A}$ already has m elements the extended set can be no larger. □

Theorem 7.9 *Let $\mathscr{S}$ be a subspace of $\mathscr{V}$ of dimension m. If $\mathscr{A} = \{\alpha_1, \ldots, \alpha_m\}$ is a subset of $\mathscr{S}$ that spans $\mathscr{S}$, $\mathscr{A}$ is a basis of $\mathscr{S}$.*

Proof. If $\mathscr{A}$ is linearly dependent, one vector in $\mathscr{A}$ is a linear combination of the others. This vector can be removed from $\mathscr{A}$ and the remaining set will still span $\mathscr{S}$. But by

Theorem 7.1 a set with less than m elements cannot span $\mathscr{S}$. Thus $\mathscr{A}$ is linearly independent and, hence, a basis of $\mathscr{S}$. □

For a subspace of dimension m, a basis must span $\mathscr{S}$, must be linearly independent, and must have m elements. Theorems 7.4, 7.8, and 7.9 show that any two of these three properties imply the third. To determine the dimension of a subspace $\mathscr{S}$ we must show that a certain subset of $\mathscr{S}$ is linearly independent and spans $\mathscr{S}$. Then the number of elements in this subset is the dimension of $\mathscr{S}$. But often we know the dimension of $\mathscr{S}$. Then to show that a subset containing $\dim \mathscr{S}$ elements is a basis, we need only verify that it spans $\mathscr{S}$ or that it is linearly independent.

We have previously defined a coordinate space with the adjective "n-dimensional." Now we have given an independent definition of the concept of dimension. We should show that an n-dimensional coordinate space is really of dimension n. Let $D_j = (\delta_{1j}, \delta_{2j}, \ldots, \delta_{nj}) = (0, 0, \ldots, 1, \ldots, 0, 0)$, where the n-tuple consists of zeros except for a 1 in the j-th position. We wish to show that $\{D_1, \ldots, D_n\}$ is a basis of the n-dimensional coordinate space. Let $(a_1, \ldots, a_n)$ be an arbitrary n-tuple. Then

$$(a_1, a_2, \ldots, a_n) = \sum_{j=1}^{n} a_j D_j, \qquad (7.1)$$

and, hence, $\{D_1, \ldots, D_n\}$ spans $\mathscr{R}^n$. On the other hand, if

$$0 = \sum_{j=1}^{n} x_j D_j \qquad (7.2)$$

is a linear relation, then for each i, $0 = \sum_{j=1}^{n} x_j \delta_{ij} = x_i$. This shows that the linear relation must be trivial and, hence, $\{D_1, \ldots, D_n\}$ is linearly independent. $\{D_1, \ldots, D_n\}$ is called the *natural basis* for $\mathscr{R}^n$. Because of formula (7.1) it is a particularly convenient basis.

Consider the system of equations (5.13). We found there that $\{(-2, -1, 1, 0, 0), (-1, 0, 0, 1, 1)\}$ spanned the subspace of solutions. By the methods discussed in Section 2–6 we can show that this set is linearly independent. Thus, the subspace of solutions of the system (5.13) is of dimension 2.

Generally, the method we have used to solve a system of

linear homogeneous equations leads to a basis of the solution subspace. In Section 2-5, the considerations leading from the system of equations in (5.5) to the linear combinations in (5.8) show that the solution subspace is spanned by $\{A_1, A_2, \ldots, A_{n-m}\}$. If $X - 0$ in equation (5.8), this would require $x_{m+1} = x_{m+2} = \cdots = x_n = 0$ in (5.7). These are the coefficients of the A_i in (5.8). Thus $\{A_1, \ldots, A_{n-m}\}$ is linearly independent. This shows that $\{A_1, \ldots, A_{n-m}\}$ is a basis of the solution subspace. The $x_{m+1}, \ldots, x_n$ in (5.8) are the parameters of the solution set. They are the coefficients of the representations of the elements of the solution subspace with respect to this basis. The fact that $\{A_1, \ldots, A_{n-m}\}$ spans the solution set means the parameters can be assigned arbitrarily and all solutions will be obtained. The fact that $\{A_1, \ldots, A_{n-m}\}$ is linearly independent means that a distinct solution is obtained for each choice of the parameters.

Notice that the dimension of the solution subspace plus the number of non-zero equations in the reduced basic form is equal to the number of variables.

In Section 2-6 we considered the set $\{A_1 = (1, 2, 1, 3), A_2 = (1, 1, 0, 2), A_3 = (2, 4, 3, 6), A_4 = (1, 3, 4, 4), A_5 = (1, 2, 1, 4), A_6 = (4, 8, 6, 11)\}$ and raised the question of expressing any vectors in this set as linear combinations of the preceding vectors. This lead to the matrix (6.12), which can be obtained directly by writing the 4-tuples in the set as columns of a matrix. We obtained

$$\begin{array}{c} A_1\ A_2\ A_3\ A_4\ A_5\ A_6 \\ \begin{bmatrix} 1 & 1 & 2 & 1 & 1 & 4 \\ 2 & 1 & 4 & 3 & 2 & 8 \\ 1 & 0 & 3 & 4 & 1 & 6 \\ 3 & 2 & 6 & 4 & 4 & 11 \end{bmatrix}. \end{array} \quad (6.12)$$

We then reduced this matrix to row-echelon form, and obtained

$$\begin{array}{c} A_1\ A_2\ A_3\ A_4\ A_5\ A_6 \\ \begin{bmatrix} 1 & 0 & 0 & -2 & 0 & 1 \\ 0 & 1 & 0 & -1 & 0 & 0 \\ 0 & 0 & 1 & 2 & 0 & 2 \\ 0 & 0 & 0 & 0 & 1 & -1 \end{bmatrix}. \end{array} \quad (6.13)$$

From (6.13) we can conclude that A_4 is a linear combination of elements of $\{A_1, A_2, A_3\}$ and that A_6 is a linear combination of elements of $\{A_1, A_2, A_3, A_5\}$. Thus, as in the discussion at the beginning of Section 2-6, $\{A_1, A_2, A_3, A_5\}$ and $\{A_1, A_2, A_3, A_4, A_5, A_6\}$ span the same subspace. We can also see from (6.13) that $\{A_1, A_2, A_3, A_5\}$ is linearly independent. Thus, $\{A_1, A_2, A_3, A_5\}$ is a basis of the subspace spanned by the given set. This also tells us that this subspace is of dimension 4 and since it is a subspace of $\mathscr{R}^4$ it must be all of $\mathscr{R}^4$. Thus, $\{A_1, A_2, A_3, A_5\}$ is a basis of $\mathscr{R}^4$.

In Section 2-5, page 73, we raised the question as to whether $(1, 0, 0)$, $(0, 1, 0)$, and $(0, 0, 1)$ were in the subspace spanned by $\{(6, -1, -3), (2, 0, -1), (-1, 1, 1)\}$. We showed in the calculations between (5.21) and (5.22) that they were. Since $\{(1, 0, 0), (0, 1, 0), (0, 0, 1)\}$ is the natural basis of $\mathscr{R}^3$, this amounts to showing that the set $\{(6, -1, -3), (2, 0, -1), (-1, 1, 1)\}$ spans $\mathscr{R}^3$. Since it also has three elements it must be a basis of $\mathscr{R}^3$.

Also in Section 2-5, page 72, we considered the subspace $\mathscr{S}$ spanned by $\{(1, 1, 1, 1), (2, 3, 1, 3), (0, -1, 1, -1), (-1, -2, 1, 0), (1, 1, 3, 5)\}$. $\mathscr{S}$ is a subspace of $\mathscr{R}^4$, so the dimension of $\mathscr{S}$ is at most 4. Since the spanning set contains five elements we know that at least one is a linear combination of the others. As in the example just considered, we write the 4-tuples as columns of a matrix and reduce to row-echelon form. This form is the first five columns of (5.19). We can see from this form that $\{(1, 1, 1, 1), (2, 3, 1, 3), (-1, -2, 1, 0)\}$ is a basis of $\mathscr{S}$, which is therefore of dimension 3 and a proper subspace of $\mathscr{R}^4$.

Theorem 7.5 asserts that a linearly independent set can be extended to a basis. The proof of the theorem, carried out by using elementary row operations, provides an effective method for making the extension. For example, consider the linearly independent set $\{(-2, -1, 1, 0, 0), (-1, 0, 0, 1, 1)\}$. We wish to extend this set to a basis of $\mathscr{R}^5$. For the basis of $\mathscr{R}^5$ that is used in the proof we might as well take the natural basis. Thus we are led to set up the matrix

$$\begin{bmatrix} -2 & -1 & 1 & 0 & 0 & 0 & 0 \\ -1 & 0 & 0 & 1 & 0 & 0 & 0 \\ 1 & 0 & 0 & 0 & 1 & 0 & 0 \\ 0 & 1 & 0 & 0 & 0 & 1 & 0 \\ 0 & 1 & 0 & 0 & 0 & 0 & 1 \end{bmatrix}. \tag{7.3}$$

A sequence of pivot operations, followed by a rearrangement of the rows, leads to the row-echelon form

$$\begin{bmatrix} 1 & 0 & 0 & 0 & 1 & 0 & 0 \\ 0 & 1 & 0 & 0 & 0 & 0 & 1 \\ 0 & 0 & 1 & 0 & 2 & 0 & 1 \\ 0 & 0 & 0 & 1 & 1 & 0 & 0 \\ 0 & 0 & 0 & 0 & 0 & 1 & -1 \end{bmatrix}. \qquad (7.4)$$

From (7.4) we see that columns 1, 2, 3, 4, 6 are linearly independent, and hence the same columns of (7.3) are linearly independent. Thus $\{(-2, -1, 1, 0, 0), (-1, 0, 0, 1, 1), (1, 0, 0, 0, 0), (0, 1, 0, 0, 0), (0, 0, 0, 1, 0)\}$ is a basis of $\mathscr{R}^5$ that contains $\{(-2, -1, 1, 0, 0), (-1, 0, 0, 1, 1)\}$ as a subset.

EXERCISES 2-7

For each of the following systems of homogeneous linear equations find a basis for the solution subspace $\mathscr{S}$. Determine the dimension of $\mathscr{S}$ in each case. For each exercise the dimension of the coordinate space under consideration is equal to the number of unknowns.

1. $\begin{aligned}3x_1 + x_2 \phantom{{}+x_3} - x_4 \phantom{{}+x_5} &= 0 \\ 2x_1 \phantom{{}+x_2+x_3} + 3x_4 + x_5 &= 0 \\ -x_1 \phantom{{}+x_2} + x_3 \phantom{{}+x_4+x_5} &= 0\end{aligned}$

2. $\begin{aligned}x_1 + x_2 &= 0 \\ x_1 - 2x_2 &= 0\end{aligned}$

3. $\begin{aligned}2x_1 - 3x_2 - 8x_3 &= 0 \\ -x_1 + 2x_2 + 5x_3 &= 0 \\ 3x_1 + 4x_2 + 5x_3 &= 0\end{aligned}$

4. $\begin{aligned}4x_1 - 8x_2 + 16x_3 &= 0 \\ x_1 - 3x_2 + 6x_3 &= 0 \\ 2x_1 + x_2 - 2x_3 &= 0\end{aligned}$

5. $\begin{aligned}2x_1 + 2x_2 + 5x_3 + 6x_4 &= 0 \\ x_1 + x_2 - 2x_3 + 2x_4 &= 0\end{aligned}$

6. $\begin{aligned}x_1 - 3x_2 + x_3 + x_4 &= 0 \\ -x_1 + 3x_2 + 2x_3 - x_4 &= 0 \\ 2x_1 - 6x_2 - x_3 + 2x_4 &= 0 \\ -2x_1 + 6x_2 + x_3 - x_4 &= 0\end{aligned}$

7. $\begin{aligned}3x_1 + x_2 + 9x_3 &= 0 \\ 3x_1 + 2x_2 + 12x_3 &= 0 \\ 2x_1 + x_2 + 7x_3 &= 0\end{aligned}$

8. $\begin{aligned} 2x_1 \phantom{{}+{}} &\phantom{{}+{}} 6x_3 + x_4 + 2x_5 + 4x_6 = 0 \\ 2x_1 + 2x_2 &+ 10x_3 \phantom{{}+{}1x_4} + x_5 + 3x_6 = 0 \\ x_1 + x_2 &+ 5x_3 + x_4 + 3x_5 + x_6 = 0 \\ 3x_1 \phantom{{}+{}} &+ 9x_3 + x_4 + x_5 + 7x_6 = 0 \\ 5x_1 + 3x_2 &+ 21x_3 + 5x_4 + 2x_5 + 18x_6 = 0 \end{aligned}$

9. Find a basis for $\mathscr{R}^6$ that includes as a subset a basis for the solution subspace $\mathscr{S}$ for the system in Exercise 8.

10. Determine whether $\{(1, 0, 1), (1, 1, 0), (0, 1, 1)\}$ is a basis of $\mathscr{R}^3$.

11. Determine whether $\{(1, 0, -1), (1, -1, 0), (0, 1, -1)\}$ is a basis of $\mathscr{R}^3$.

12. Find *all* linearly independent subsets of $\{(1, 2, 3), (2, 5, 5), (4, 9, 11), (-1, -3, -1), (2, 5, 4)\}$ that contain three elements.

13. Consider, again, Exercise 4 of Section 2–6. Find a maximal linearly independent subset of $\{A_1 = (2, 1, 1, 2), A_2 = (1, 2, -2, -1), A_3 = (3, 3, -1, 1), A_4 = (-1, 3, 0, 1), A_5 = (1, 4, 1, 3), A_6 = (3, 4, 2, 2), A_7 = (3, -3, 1, 2)\}$. When these 4-tuples are written in columns and the resulting matrix is reduced to row-echelon form, we obtain

$$\begin{array}{c} A_1\ A_2\ A_3\ A_4\ A_5\ A_6\ A_7 \\ \begin{bmatrix} 1 & 0 & 1 & 0 & 1 & 0 & 1 \\ 0 & 1 & 1 & 0 & 0 & 0 & 0 \\ 0 & 0 & 0 & 1 & 1 & 0 & 2 \\ 0 & 0 & 0 & 0 & 0 & 1 & -1 \end{bmatrix} \end{array} \quad (7.5)$$

The matrix (7.5) tells us that $\{A_1, A_2, A_4, A_6\}$ is a maximal linearly independent set. Since the subspace spanned by the $\{A_i\}$ is then of dimension 4, every maximal linearly independent subset would have four elements. A 7-element set has 35 4-element subsets. While we could enumerate them all and determine which are linearly independent, it is in the spirit of mathematics to try to find a systematic way to find all maximal linearly independent subsets.

From the row-echelon form (7.5) we see that $\{A_1, A_2, A_3\}$, $\{A_1, A_4, A_5\}$, and $\{A_1, A_4, A_6, A_7\}$ are minimal linearly dependent subsets. These particular subsets are chosen because form (7.5) shows that A_3, A_5, and A_7 are the 4-tuples that are linear combinations of the previous A_i. It is possible to show that all linear relations are linear combinations of these three involving A_3, A_5, and A_7. The coefficients of a linear relation are obtained by regarding the matrix (7.5) as representing a system of homogeneous linear equations. From that point of view we would write $\{(-1,-1,1,0,0,0,0),(-1,0,0,-1,1,0,0),(-1,0,0,-2,0,1,1)\}$ as the generators of the solution subspace of this system of equations. These 7-tuples represent the linear relations

$$\begin{aligned} -A_1 - A_2 + A_3 &= 0 \\ -A_1 \qquad\qquad - A_4 + A_5 &= 0 \\ -A_1 \qquad\qquad - 2A_4 \qquad + A_6 + A_7 &= 0. \end{aligned} \qquad (7.6)$$

In the sense that every solution of the system of homogeneous linear equations represented by the matrix (7.5) is a linear combination of $\{(-1,-1,1,0,0,0,0),(-1,0,0,-1,1,0,0),(-1,0,0,-2,0,1,1)\}$, every linear relation involving the $\{A_i\}$ is a linear combination of the relations given in (7.6).

From the relations in (7.6) it can be seen that if a linear relation involves A_2 or A_3 non-trivially (with a non-zero coefficient) it must involve both non-trivially. Also, if a linear relation involves A_6 or A_7 non-trivially, it must involve both non-trivially. From the row-echelon form (7.5) it can be seen that $\{A_1, A_2, A_3, A_4, A_5\}$ spans a 3-dimensional subspace.

We now have enough information to find all maximal linearly independent subsets of $\{A_1, \ldots, A_7\}$. All maximal linearly independent subsets have four elements. All 4-element subsets of $\{A_1, \ldots, A_5\}$ are linearly dependent. If a 4-element subset contains A_6 or A_7, but not both, it will be linearly dependent if and only if the other three elements form a linearly dependent subset. The 4-element subsets that contain both A_6 and A_7 contain two others among the first five. The subsets that are linearly dependent can be identified and deleted from the list of 35 4-element subsets. Carry out these deletions. (You should find that 12 of the 35 4-element subsets are linearly dependent.)

14. Let $\mathscr{M}_{m,n}$ be the set of $m \times n$ matrices with elements from a field $\mathscr{F}$. We have shown in Exercise 11 of Section 2-2 that $\mathscr{M}_{m,n}$ is a vector space over $\mathscr{F}$. Let E_{ij} be the $m \times n$ matrix with zero in all positions except for a 1 in row i column j. Show that the set $\{E_{ij}\}$ of all such matrices is a basis of $\mathscr{M}_{m,n}$. Show that the dimension of $\mathscr{M}_{m,n}$ as a vector space over $\mathscr{F}$ is mn.

2-8 BASES AND EQUATIONS

If we look at a single homogeneous equation in n unknowns of the form

$$a_1 x_1 + a_2 x_2 + \cdots + a_n x_n = 0, \tag{8.1}$$

one thing that should be striking is the symmetry between the roles of the coefficients $(a_1, a_2, \ldots, a_n)$ and the unknowns $(x_1, x_2, \ldots, x_n)$. We are accustomed to thinking of the coefficients as given and being required to find suitable values for the x_i. But there is no reason why we should not regard the x_i as given and solve for the a_i. We have been considering the a_i as the coefficients in an equation and the x_i as the coordinates of an n-tuple. The problem, which is symmetric to the one we have been considering, is to have a set of n-tuples given and to be asked to find all equations for which the given n-tuples are solutions. It should be clear from the symmetry of equation (8.1) that the methods for solving this kind of problem are exactly the same as those we have been using.

These ideas are illustrated by the following example. Let $\mathscr{S}$ be a subspace of $\mathscr{R}^6$ spanned by $\{A_1 = (1, -1, 2, 1, 1, 4)$, $A_2 = (-1, 1, -1, 1, -1, -2)$, $A_3 = (2, -1, 4, 3, 2, 8)$, $A_4 = (1, 0, 3, 4, 1, 6)$, $A_5 = (3, -2, 6, 4, 4, 11)$. The coefficients of an equation that has these five 6-tuples as solutions must satisfy the following conditions.

$$\begin{aligned}
a_1 - a_2 + 2a_3 + a_4 + a_5 + 4a_6 &= 0 \\
-a_1 + a_2 - a_3 + a_4 - a_5 - 2a_6 &= 0 \\
2a_1 - a_2 + 4a_3 + 3a_4 + 2a_5 + 8a_6 &= 0 \\
a_1 \phantom{{}-a_2} + 3a_3 + 4a_4 + a_5 + 6a_6 &= 0 \\
3a_1 - 2a_2 + 6a_3 + 4a_4 + 4a_5 + 11a_6 &= 0.
\end{aligned} \tag{8.2}$$

This is a system of equations of a familiar form; the only difference is that a_i's are used for the unknowns instead of x_i's. The coefficient matrix for this system is

$$\begin{bmatrix} 1 & -1 & 2 & 1 & 1 & 4 \\ -1 & 1 & -1 & 1 & -1 & -2 \\ 2 & -1 & 4 & 3 & 2 & 8 \\ 1 & 0 & 3 & 4 & 1 & 6 \\ 3 & -2 & 6 & 4 & 4 & 11 \end{bmatrix}. \tag{8.3}$$

We would now perform row operations to reduce this matrix to basic form, and obtain a general solution for the system (8.2). However, to preserve the convention of writing the n-tuples representing vectors in columns, we prefer to write the coefficient matrix of the system (8.2) in the form

$$\begin{bmatrix} 1 & -1 & 2 & 1 & 3 \\ -1 & 1 & -1 & 0 & -2 \\ 2 & -1 & 4 & 3 & 6 \\ 1 & 1 & 3 & 4 & 4 \\ 1 & -1 & 2 & 1 & 4 \\ 4 & -2 & 8 & 6 & 11 \end{bmatrix} \qquad (8.4)$$

Using this matrix the system of equations in (8.2) can be written in the form

$$[a_1 \ a_2 \ a_3 \ a_4 \ a_5 \ a_6] \begin{bmatrix} 1 & -1 & 2 & 1 & 3 \\ -1 & 1 & -1 & 0 & -2 \\ 2 & -1 & 4 & 2 & 6 \\ 1 & 1 & 3 & 4 & 4 \\ 1 & -1 & 2 & 1 & 4 \\ 4 & -2 & 8 & 6 & 11 \end{bmatrix} = 0. \qquad (8.5)$$

In equation (8.5) we preserve our convention of writing the coefficients of an equation in a row and the representation of a vector in a column.

The matrix (8.4) is obtained from (8.3) by interchanging rows and columns. We say that the matrix (8.4) is the *transpose* of the matrix (8.3). If A denotes a matrix, we denote its transpose by A^T. Note that $(A^T)^T = A$.

Since we would solve the system (8.2) by performing elementary row operations on the matrix (8.3), we can achieve the same result by performing elementary column operations on the matrix (8.4). When (8.4) is reduced to basic form (column-echelon form) by column operations, we obtain

$$\begin{bmatrix} 1 & 0 & 0 & 0 & 0 \\ 0 & 1 & 0 & 0 & 0 \\ 0 & 0 & 1 & 0 & 0 \\ -2 & 1 & 2 & 0 & 0 \\ 0 & 0 & 0 & 1 & 0 \\ 1 & 0 & 2 & -1 & 0 \end{bmatrix} \qquad (8.6)$$

In (8.6) the basic elements are designated in color. Using the same ideas we use to solve a system of linear equations, we see that

$$\begin{bmatrix} 2 & -1 & -2 & 1 & 0 & 0 \\ -1 & 0 & -2 & 0 & 1 & 1 \end{bmatrix}, \tag{8.7}$$

are solutions to the system (8.2). We write these solutions as row matrices since they represent equations. These equations are

$$\begin{aligned} 2x_1 - x_2 - 2x_3 + x_4 &= 0 \\ -x_1 - 2x_3 + x_5 + x_6 &= 0. \end{aligned} \tag{8.8}$$

It is instructive to complete the circle by solving the system of equations in (8.8). The coefficient matrix of this system is

$$\begin{bmatrix} 2 & -1 & -2 & 1 & 0 & 0 \\ -1 & 0 & -2 & 0 & 1 & 1 \end{bmatrix}, \tag{8.9}$$

and it is already in reduced basic form. Columns 1, 2, 3 and 5 correspond to the parameters of the solution set and $\{(1, 0, 0, -2, 0, 1), (0, 1, 0, 1, 0, 0), (0, 0, 1, 2, 0, 2), (0, 0, 0, 0, 1, -1)\}$ is a basis of the solution subspace. These are precisely the non-zero columns of the matrix (8.6).

In the matrix (8.9) the coefficients of the basic variables are printed in color. To be sure, the fifth element in the second row would serve as a coefficient of a basic variable as well as the sixth element. In the reduced form (8.6), the coefficients of the basic variables have been colored, and they appear in rows 1, 2, 3 and 5. In (8.9) they are in columns 4 and 6. The rows of (8.6) that contain basic elements correspond to the columns of (8.9) that do not contain basic elements, and the columns of (8.9) that contain basic elements correspond to the rows of (8.6) that do not contain basic elements. In other words, the basic variables of one system correspond to the parameters of the other.

Write the matrices (8.6) and (8.9) as the factors of a product,

$$\begin{bmatrix} 2 & -1 & -2 & 1 & 0 & 0 \\ -1 & 0 & -2 & 0 & 1 & 1 \end{bmatrix} \begin{bmatrix} 1 & 0 & 0 & 0 \\ 0 & 1 & 0 & 0 \\ 0 & 0 & 1 & 0 \\ -2 & 1 & 2 & 0 \\ 0 & 0 & 0 & 1 \\ 1 & 0 & 2 & -1 \end{bmatrix}. \tag{8.10}$$

In the product (8.10), we calculate terms of the form $a_1x_1 + a_2x_2 + \cdots + a_nx_n$ where $[a_1 \ a_2 \ \ldots \ a_n]$ is a row of the first factor and $(x_1, x_2, \ldots, x_n)$ is a column from the second factor. These terms are all zero. We say that a row of the first factor *annihilates* a column of the second factor in this case. In (8.10) each row of the first factor annihilates every column of the second factor and each column of the second factor annihilates every row of the first factor.

The rows of the first matrix factor in (8.10) represent equations, and the columns of the second factor represent vectors. The vectors span the subspace of all vectors annihilated by the given equations. The relation between the two matrices in (8.10) is so symmetrical that it suggests a reformulation of our viewpoint to exploit this symmetry.

The sum of two homogeneous linear equations is a homogeneous linear equation, and the scalar multiple of a homogeneous linear equation is a homogeneous linear equation. In fact, it is not difficult to show that the set of all linear homogeneous equations in n unknowns, with these operations of addition and scalar multiplication, forms a vector space of dimension n. Any given system of homogeneous linear equations generates a subspace of equations. The elementary row operations allow us to form linear combinations of the given equations. The equations in a basic system are linearly independent, and span the same subspace of equations as the original system. The equations in the basic system, therefore, are a basis for this subspace.

Starting with a system of equations $\mathscr{E}$, we can obtain the corresponding solution subspace $\mathscr{S}(\mathscr{E})$, the subspace annihilated by $\mathscr{E}$. Starting with a set of vectors $\mathscr{S}$, we can obtain the subspace $\mathscr{E}(\mathscr{S})$ of equations annihilated by $\mathscr{S}$. If $\mathscr{E}$ is a subspace of equations, $\mathscr{E}(\mathscr{S}(\mathscr{E})) = \mathscr{E}$. If $\mathscr{S}$ is a subspace of vectors, $\mathscr{S}(\mathscr{E}(\mathscr{S})) = \mathscr{S}$. Thus, there is a one-to-one correspondence between subspaces of the vector space and subspaces of the space of equations: one being the annihilator of the other.

In (8.10), the columns of the second factor constitute a basis of the subspace annihilated by the rows of the first factor, and the rows of the first factor constitute a basis of the subspace annihilated by the columns of the second factor. In this case the subspace $\mathscr{E}$ of equations has a dimension equal to the number of basic elements appearing in the first matrix. The dimension of the subspace $\mathscr{S}$ of vectors has a dimension equal to the number of basic elements appearing in the second matrix. It is clear from the way the basic elements from the two factors are matched that the dimension of $\mathscr{E}$ plus the dimension of $\mathscr{S}$ is equal to the dimension n of the vector space. In this particular case, the dimension of $\mathscr{E}$ is 2, the dimension of $\mathscr{S}$ is 4, and the dimension of $\mathscr{V}$ is 6.

We have seen in Theorem 7.2 that every subspace $\mathscr{S}$ of $\mathscr{V}$ has a finite basis. Now we have shown how we can get a subspace $\mathscr{E}$ of equations for which $\mathscr{S}$ is the solution set. This shows that every subspace of $\mathscr{S}$ is the solution set of a system of equations. Thus each subspace of $\mathscr{V}$ can be characterized in two different ways; by giving a spanning set of vectors or by giving a defining system of equations. The numerical work of solving a system of equations amounts to passing from one characterization to the other. For finite dimensional vector spaces the same methods apply for going in either direction.

EXERCISES 2-8

For each of the following exercises, find a basis for the set of all equations for which the given vectors are solutions, and from these equations find a basis for the solution set.

1. $\{(1, 2, 1, 2), (6, 1, 2, 5)\}$

2. $\{(1, 2, 3, 2, 1), (1, -1, 2, -1, 1), (2, 1, -1, 1, 2), (1, -1, 1, -1, 1)\}$

3. $\{(1, -1, 2, 1, 3), (-1, 1, -1, 0, 2), (2, -1, 4, 3, 6), (1, 1, 3, 4, 4), (1, -1, 2, 1, 4), (4, -2, 8, 6, 11)\}$

For each of the following exercises, a matrix in basic form is given. Find a matrix in basic form which is complementary to it, in the sense of this section.

4. $\begin{bmatrix} 1 & 0 & -2 & 0 & 3 \\ 0 & 1 & 1 & 0 & 2 \\ 0 & 0 & 0 & 1 & 1 \end{bmatrix}$

5. $\begin{bmatrix} 1 & 2 & 1 & -2 & 1 & 0 \\ 2 & 1 & 0 & -1 & 2 & 1 \end{bmatrix}$

6. $\begin{bmatrix} 1 & 0 & 1 & 3 & 2 & 0 \\ -2 & 1 & 0 & -2 & 1 & 0 \\ 0 & 0 & 0 & 1 & 1 & 1 \end{bmatrix}$

7. $[1 \ 1 \ 1 \ 1 \ 1]$

For each of the following systems of equations, find a linearly independent set of equations with the same solution set.

8. $2x_1 - 3x_2 - 8x_3 = 0$
$-x_1 + 2x_2 + 5x_3 = 0$
$3x_1 + 4x_2 + 5x_3 = 0$

9. $4x_1 - 8x_2 + 16x_3 = 0$
$x_1 - 3x_2 + 6x_3 = 0$
$2x_1 + x_2 - 2x_3 = 0$

10. $2x_1 \quad\quad + 6x_3 + x_4 + 2x_5 + 4x_6 = 0$
$2x_1 + 2x_2 + 10x_3 \quad\quad + x_5 + 3x_6 = 0$
$x_1 + x_2 + 5x_3 + x_4 + 3x_5 + x_6 = 0$
$3x_1 \quad\quad + 9x_3 + x_4 + x_5 + 7x_6 = 0$
$5x_1 + 3x_2 + 21x_3 + 5x_4 + 2x_5 + 18x_6 = 0$

Summary

A *coordinate space* is constructed by defining the addition of *n*-tuples and the multiplication of an *n*-tuple by a number. The coordinate space serves two purposes. It is the space in which all practical calculations take place. It is also the model used to define an abstract *vector space*. A *subspace* of a vector space is a subset which is itself a vector space.

We define *linear combinations* and *linear independence*. We show that for any given set of vectors, the set of all linear combinations of these vectors is a subspace. It is the subspace *spanned* or generated by the given subset.

A set of vectors contained in a subspace is a *basis* of the subspace if it is linearly independent and spans the subspace. The principal theorem is that (for subspaces spanned by a finite set) the number of elements in a basis depends only on the subspace and not on the way the basis is chosen. The number of elements in a basis is the *dimension* of the subspace.

In a coordinate space, the solution set of a system of linear *homogeneous* equations is a subspace. Conversely, every subspace is the solution set of a system of linear homogeneous equations.

Solving a system of linear homogeneous equations involves finding a basis for the solution subspace of the system.

It amounts to passing from a description of the subspace as a solution set to a description in terms of a spanning set (a parametric representation).

The numerical technique used to solve a system of linear homogeneous equations can also be used to pass from a description of a subspace in terms of a spanning set to a description as the solution set of a system of linear homogeneous equations.

3
BASES AND COORDINATE SYSTEMS

In this chapter we use the concept of a basis to show that every abstract vector space spanned by a finite set is essentially equivalent to a coordinate space. This equivalence is exploited by using the coordinate space to allow numerical calculations, which cannot be carried out abstractly.

Establishing a correspondence between an abstract vector space and a coordinate space is called coordinatization. We say that we introduce coordinates. However, the correspondence between an abstract vector space and a coordinate space can be made in many ways. The numerical work can be considerably simplified if this correspondence is chosen carefully. We must show what is involved in making the choices and how to change from one coordinate system to another.

3-1 INTRODUCING COORDINATES

Let $\mathscr{V}$ be a vector space spanned (as we consistently assume) by a finite set. As shown in Section 2-7, $\mathscr{V}$ has a basis with a finite number of elements. Let $\{\alpha_1, \alpha_2, \ldots, \alpha_n\} = \mathscr{A}$ be a basis of $\mathscr{V}$. For each $\beta \in \mathscr{V}$, β has a representation as a linear combination in the form

$$\beta = b_1\alpha_1 + b_2\alpha_2 + \cdots + b_n\alpha_n \qquad (1.1)$$

since $\mathscr{A}$ spans $\mathscr{V}$. The representation of β as a linear combination in the form (1.1) is unique, as shown in Section 2–6. We say that the resulting n-tuple $(b_1, b_2, \ldots, b_n)$ represents β. Thus, for a given basis, every $\beta \in \mathscr{V}$ has a unique representation by an n-tuple.

Conversely, given any n-tuple $(b_1, b_2, \ldots, b_n)$ the expression

$$b_1 \alpha_1 + b_2 \alpha_2 + \cdots + b_n \alpha_n \qquad (1.2)$$

is a vector in $\mathscr{V}$. Thus there is a one-to-one correspondence between vectors in $\mathscr{V}$ and n-tuples which represent them.

If $\gamma = c_1 \alpha_1 + c_2 \alpha_2 + \cdots + c_n \alpha_n$ is another vector in $\mathscr{V}$, then $\beta + \gamma$ is represented by the linear combination

$$\begin{aligned}\beta + \gamma &= (b_1 \alpha_1 + b_2 \alpha_2 + \cdots + b_n \alpha_n) + (c_1 \alpha_1 + c_2 \alpha_2 + \cdots + c_n \alpha_n) \\ &= (b_1 + c_1) \alpha_1 + (b_2 + c_2) \alpha_2 + \cdots + (b_n + c_n) \alpha_n. \end{aligned} \qquad (1.3)$$

This means $\beta + \gamma$ is represented by the n-tuple

$$\begin{aligned}(b_1 + c_1, b_2 + c_2, \ldots, b_n + c_n) = &(b_1, b_2, \ldots, b_n) \\ &+ (c_1, c_2, \ldots, c_n). \end{aligned} \qquad (1.4)$$

Thus we see that addition of vectors corresponds to addition of the representing n-tuples. In a similar way $b\beta$ is represented by

$$(bb_1, bb_2, \ldots, bb_n) = b(b_1, b_2, \ldots, b_n). \qquad (1.5)$$

That is, scalar multiplication in $\mathscr{V}$ and scalar multiplication in the representing coordinate space correspond. In this sense, the coordinate space *represents* the vector space. Once the correspondence between vectors and n-tuples has been made, the indicated calculations can be performed with the n-tuples.

The n-dimensional coordinate space satisfies the axioms for a vector space and is, therefore, a vector space in its own right. But we prefer to encourage a point of view that regards the abstract vector space as more fundamental and the coordinate space as a mere representation. Much the same idea occurs in the distinction between numbers and numerals. Twenty-seven is a number, but 27 and XXVII are numerals— one Arabic and the other Roman. In grade school arithmetic we learn to calculate, and the rules are expressed in terms of numerals. For example, consider the rules involving carrying in multiplication. The distinction between concept and representing symbol is blurred when we pronounce a number and write a representing symbol. Such casual blurring seldom causes trouble because we consistently use the same symbol for each number. But anyone who has ever tried to represent numbers in a base other than base ten should realize that this distinction is real and necessary.

Suppose that another basis $\{\alpha_1', \alpha_2', \ldots, \alpha_n'\} = \mathscr{A}'$ of $\mathscr{V}$ is selected. Using this basis, β has a representation as a linear combination of the elements of $\mathscr{A}'$ similar to (1.1). When this is done each vector will be represented by two n-tuples; one for each choice of a basis. In such a situation we will refer to $(b_1, b_2, \ldots, b_n)$ as the representation of β *with respect to the basis* $\mathscr{A}$ to distinguish it from other possible representations.

Let $\mathscr{D} = \{D_1, D_2, \ldots, D_n\}$ be the natural basis of the n-dimensional coordinate space $\mathscr{R}^n$. Let us find the representation of the n-tuple $(a_1, a_2, \ldots, a_n)$ with respect to this basis. Since

$$\sum_{j=1}^{n} x_j D_j = \sum_{j=1}^{n} x_j (\delta_{1j}, \delta_{2j}, \ldots, \delta_{nj})$$

$$= \left(\sum_{j=1}^{n} x_j \delta_{1j}, \sum_{j=1}^{n} x_j \delta_{2j}, \ldots, \sum_{j=1}^{n} x_j \delta_{nj} \right)$$

$$= (x_1, x_2, \ldots, x_n), \qquad (1.6)$$

the equation

$$\sum_{j=1}^{n} x_j D_j = (a_1, a_2, \ldots, a_n) \qquad (1.7)$$

is satisfied only for $x_j = a_j$ for all j. This shows that each n-tuple is represented by itself with respect to the natural basis. This is certainly comforting. But suppose another basis is chosen, for example, $\{(1, 1, 1), (0, 1, 1), (0, 0, 1)\}$. Then $D_1 = (1, 0, 0) = (1, 1, 1) - (0, 1, 1)$ so that D_1 is represented by the triple $(1, -1, 0)$ with respect to the basis $\{(1, 1, 1), (0, 1, 1), (0, 0, 1)\}$.

Choosing a basis arbitrarily or changing from one to another is the rule rather than the exception. Therefore, it seems desirable to avoid the confusion of representing one n-tuple by another. This is the main reason we prefer to reduce coordinate spaces to a supporting role. Therefore, we shall consistently think of an n-tuple as representing a vector rather than as being a vector itself. But this poses other problems. This means we should speak of "the vector represented by the n-tuple $(a_1, \ldots, a_n)$." It isn't long before such terminology becomes a burden. In conversation such a lengthy expression becomes shortened to "the vector $(a_1, \ldots, a_n)$." This is the same convention we use when we speak of "the number 27." It is more important that we understand clearly what we mean than it is that we adhere to any particular notation or form of expression.

Qualitative problems can be posed in the abstract, but quantitative problems cannot be posed entirely in the abstract. Such problems are usually posed in terms of n-tuples and matrices. For example, suppose we are given two subspaces $\mathscr{S}_1$ and $\mathscr{S}_2$, and we are asked to find $\mathscr{S}_1 \cap \mathscr{S}_2$. This is not a meaningful problem, only a definition, unless the relation between $\mathscr{S}_1$ and $\mathscr{S}_2$ is specified. The relation can be implied by giving bases for $\mathscr{S}_1$ and $\mathscr{S}_2$, or by giving systems of equations that characterize these subspaces. For example, suppose we are given

$$\begin{aligned}\mathscr{S}_1 &= \langle A_1 = (1, 2, 3, -1), A_2 = (2, 1, 6, 3), A_3 \\ &= (0, 1, 0, 2) \rangle, \\ \mathscr{S}_2 &= \langle B_1 = (2, -1, 0, 2), B_2 = (3, 1, 1, -1), B_3 \\ &= (1, 2, 1, -2) \rangle.\end{aligned} \quad (1.8)$$

How is this problem to be interpreted? Without being told, we are expected to understand that there is a 4-dimensional vector space $\mathscr{V}$ with an unspecified basis $\mathscr{A} = \{\alpha_1, \alpha_2, \alpha_3, \alpha_4\}$. The 4-tuple $(1, 2, 3, -1)$ represents the vector $\alpha_1 + 2\alpha_2 + 3\alpha_3 - \alpha_4$, and similarly for the other 4-tuples.

To find the intersection of $\mathscr{S}_1$ and $\mathscr{S}_2$, consider the following matrix.

$$\begin{array}{c} A_1 A_2 A_3 B_1 B_2 B_3 \\ \begin{bmatrix} 1 & 2 & 0 & 2 & 3 & 1 \\ 2 & 1 & 1 & -1 & 1 & 2 \\ 3 & 6 & 0 & 0 & 1 & 1 \\ -1 & 3 & 2 & 2 & -1 & -2 \end{bmatrix}. \end{array} \quad (1.9)$$

Reduce the matrix (1.9) to basic form

$$\begin{array}{c} A_1 A_2 A_3 B_1 B_2 B_3 \\ \begin{bmatrix} 16 & 43 & 0 & 1 & 0 & 0 \\ -5 & -13 & 1 & 0 & 0 & 0 \\ -17 & -45 & 0 & 0 & 1 & 0 \\ 20 & 51 & 0 & 0 & 0 & 1 \end{bmatrix}. \end{array} \quad (1.10)$$

From (1.10) we read off the linear relations

$$A_1 = -5A_3 + 16B_1 - 17B_2 + 20B_3$$
$$A_2 = -13A_3 + 43B_1 - 45B_2 + 51B_3. \quad (1.11)$$

These relations are equivalent to

$$A_1 + 5A_3 = 16B_1 - 17B_2 + 20B_3$$
$$A_2 + 13A_3 = 43B_1 - 45B_2 + 51B_3. \quad (1.12)$$

In each of the relations shown in (1.12), the left side represents a vector in $\mathscr{S}_1$ and the right side represents a vector in $\mathscr{S}_2$. Thus, each represents a vector in $\mathscr{S}_1 \cap \mathscr{S}_2$. Since

$$A_1 + 5A_3 = (1, 7, 3, 9)$$
$$A_2 + 13A_3 = (2, 14, 6, 29), \quad (1.13)$$

we see that $\{(1, 7, 3, 9), (2, 14, 6, 29)\}$ represent vectors in $\mathscr{S}_1 \cap \mathscr{S}_2$. Every vector in $\mathscr{S}_1 \cap \mathscr{S}_2$ leads to an expression like one of those in (1.12), which in turn leads to one like those in (1.11). Thus, every vector in $\mathscr{S}_1 \cap \mathscr{S}_2$ leads to a relation among the columns of (1.9). Conversely, every relation among the columns of (1.9) leads to a vector in $\mathscr{S}_1 \cap \mathscr{S}_2$. Therefore, since all relations among the columns of (1.9) are linear combinations of those given in (1.11), all vectors in $\mathscr{S}_1 \cap \mathscr{S}_2$ are linear combinations of $\{(1, 7, 3, 9), (2, 14, 6, 29)\}$. It is not hard to see that this set is linearly independent. Thus, dim $\mathscr{S}_1 \cap \mathscr{S}_2 = 2$. An acceptable answer to the problem given is to say $\mathscr{S}_1 \cap \mathscr{S}_2 = \langle (1, 7, 3, 9), (2, 14, 6, 29) \rangle$. This means that $\mathscr{S}_1 \cap \mathscr{S}_2$ is spanned by $\{\alpha_1 + 7\alpha_2 + 3\alpha_3 + 9\alpha_4, 2\alpha_1 + 14\alpha_2 + 6\alpha_3 + 29\alpha_4\}$. In neither the original problem, the answer, nor even in the intermediate work is it necessary to refer to the underlying vector space or its basis. All numerical work would be in the form of n-tuples and matrices, and that is all that need be shown. But one must understand the vector space interpretations in order to guide the work properly. How, otherwise, could one get from the matrix (1.10) to the 4-tuples in (1.13)?

When, at a football game, the announcer says, "28 has just gone in for 23," everyone understands what he means. Everyone knows he means, "The player with the numeral 28 has just replaced the player with the numeral 23." Despite what the announcer says, no one will add 28 to the team and subtract 23, and conclude that the team has 5 players more than it had before. In this example, numerals are used to identify, or represent, the players and no operations of addition or multiplication would be meaningful. As another example, suppose numbers are used to represent sums of money. In this setting, addition or subtraction of these numbers would be quite meaningful. But it is doubtful that a meaning could be given to the product of $5 and $7.

Even though the arithmetic of numbers is quite rich, in some applications not all operations available are meaningful. The arithmetic of n-tuples and matrices is even richer, and the variety of applications just as wide. So, while there is no objection to using short expressions that seem to confuse a concept with its representing symbol, beware if this sloppy talk is not supported with an understanding of what is intended. Beware that we don't add a halfback to a fullback to get a guard.

When a basis is chosen and the vector space is represented by a coordinate space, we say we have introduced *coordinates*. We have introduced a coordinate system. If β is represented by $(b_1, \ldots, b_n)$, the b_i are called the *coordinates* of β. Because of (1.4) and (1.5) we say that the vector space operations of vector addition and scalar multiplication are carried out *coordinate-wise* in the representation space. In the representation of β as a linear combination in (1.1), the terms $b_i \alpha_i$ are called the *components* of β. Components are vectors and coordinates are scalars. This distinction, too, is often blurred in casual conversation, but confusion on this point seldom causes difficulty.

The correspondence that we have established between an abstract vector space and a coordinate space is called an *isomorphism*. In algebra this term is used to denote a one-to-one correspondence that preserves the algebraic structure. In this book we are interested in vector spaces, and their algebraic structure involves the operations of vector addition and scalar multiplication. Thus, an isomorphism of vector spaces must be a one-to-one correspondence that preserves vector addition and scalar multiplication. Specifically, let $\mathcal{U}$ and $\mathcal{V}$ be two vector spaces over a field of scalars $\mathcal{F}$. Let f be a one-to-one function mapping $\mathcal{U}$ onto $\mathcal{V}$. If f is to be a vector space isomorphism of $\mathcal{U}$ onto $\mathcal{V}$, we must have

$$\begin{aligned} f(\alpha + \beta) &= f(\alpha) + f(\beta), \text{ and} \\ f(a\alpha) &= a \cdot f(\alpha) \end{aligned} \qquad (1.14)$$

for all vectors α and β in $\mathcal{U}$ and all scalars a. Two isomorphic vector spaces may be distinguishable by an appropriate notation device, but as far as their internal structure is concerned they are indistinguishable. As far as any mathematical statements, limited to their properties as vector spaces, are concerned they are the same.

We have shown in this section that every vector space of finite dimension is isomorphic to a coordinate space. Furthermore, when the field of scalars and the dimension are specified, there is only one coordinate space. Thus, as far as internal properties are concerned, for a given field and dimension, there is really only one abstract vector space. This means, also, that a study of finite dimensional vector spaces could be confined to a study of coordinate spaces. This is, as a matter of fact, characteristic of the early development of linear algebra. But this is not advisable for a number of reasons. The possible (and likely) confusion mentioned above is one reason. Also, many of the arguments and proofs are much simpler when given free of reference to coordinates. An example is the proof of Theorem 7.1 in Chapter 2.

The idea of a finite dimensional coordinate space can be generalized to coordinate spaces with a countable number of coordinates. But, generally, Fourier series and other orthogonal functions, Fourier transforms and Laplace transforms are representative of the tools used to handle problems formulated in such spaces. Proofs based in coordinate spaces must be abandoned and we must start over again. But many coordinate-free arguments carry over or generalize quite easily. Hence, we prefer coordinate-free arguments as much as possible.

EXERCISES 3–1

A real plane is 2-dimensional, and Figure 3–1.1 shows two linearly independent vectors $\{\alpha_1, \alpha_2\}$ attached to a common

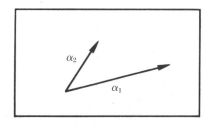

Figure 3–1.1 In a real 2-dimensional vector space, a plane, any two non-colinear vectors can be selected as a basis.

106 • Bases and Coordinate Systems

point we take for an origin. These two vectors (or any other two linearly independent vectors) can be selected as a basis. In Figure 3–1.2 we show a coordinate grid determined by $\{\alpha_1, \alpha_2\}$.

In Figure 3–1.2, the vector β is $-\alpha_1 + 2\alpha_2$ and the end point of that vector would have coordinates $(-1, 2)$ in the coordinate grid shown. The fact that we are accustomed to seeing a coordinate grid in a square or rectangular pattern should not prejudice us to think that is the only way a coordinate grid can be thought of.

1. In Figure 3–1.2, what are the coordinates of the vector γ?

2. Is $\{\beta, \gamma\}$ a linearly independent set?

3. If $\{\beta, \gamma\}$ were used for a basis, what would be the coordinates of α_1? of α_2?

4. Let $\mathscr{V}$ be a 3-dimensional vector space over $\mathscr{R}$, and let $\mathscr{A} = \{\alpha_1, \alpha_2, \alpha_3\}$ be a basis of $\mathscr{V}$. If $\xi = 3\alpha_1 + 5\alpha_2 - 7\alpha_3$, what triple represents ξ with respect to the basis $\mathscr{A}$? What

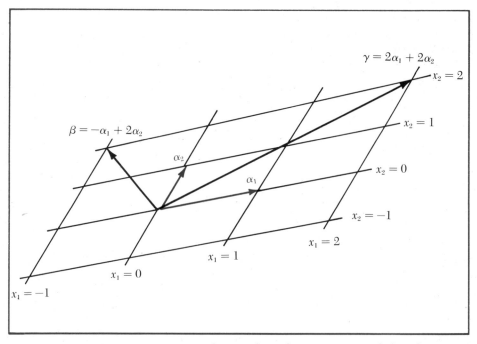

Figure 3–1.2 Any two linear independent vectors in a real plane determine a coordinate grid. The grid lines are shown in this figure.

is the α_1 coordinate of ξ? What is the α_1 component of ξ? What triple represents 3ξ?

Let $\eta = 5\alpha_1 + 3\alpha_3$. What triple represents η? What triple represents $\xi + \eta$? What triple represents $2\xi + 3\eta$?

5. Let $\mathscr{V}$ be a 3-dimensional vector space, and let $\mathscr{A} = \{\alpha_1, \alpha_2, \alpha_3\}$ be a basis of $\mathscr{V}$. Let $\mathscr{B} = \{\beta_1, \beta_2, \beta_3\}$ be a set of vectors in $\mathscr{V}$ where β_1 is represented by (3, 3, 1) with respect to the basis $\mathscr{A}$, β_2 is represented by (1, 4, 1), and β_3 is represented by (2, 3, 1). Show that $\mathscr{B}$ is linearly independent.

Since $\mathscr{B}$ is a linearly independent set with three elements in $\mathscr{V}$, it is a basis of $\mathscr{V}$. Express α_1 as a linear combination of the vectors in $\mathscr{B}$. That is, find a representation of α_1 with respect to the basis $\mathscr{B}$.

Let γ be a vector in $\mathscr{V}$ whose representation with respect to $\mathscr{B}$ is (1, 2, 3). Find the representation of γ with respect to $\mathscr{A}$.

6. Let $\mathscr{S}_1$ and $\mathscr{S}_2$ be the subspaces described in (1.8). Find a system of equations for $\mathscr{S}_1$. Find a system of equations for $\mathscr{S}_2$. Find a system of equations for $\mathscr{S}_1 \cap \mathscr{S}_2$.

Use the equations for $\mathscr{S}_1 \cap \mathscr{S}_2$ to find a basis for $\mathscr{S}_1 \cap \mathscr{S}_2$.

3-2 CHANGE OF BASIS

Coordinates make it possible to deal with specific numerical problems. But coordinates depend on the basis, and the basis can be chosen quite arbitrarily. So it is necessary to consider how the coordinates depend on the choice of basis. How are the coordinates changed when a different basis is chosen?

Even though a basis may be given there are several reasons why it might be desirable to choose a new one. If the desired result is a calculation, it is sometimes much easier to perform the calculation with one basis than with another. Often the desired result is a qualitative conclusion, and the conclusion may be relatively evident when a basis is chosen properly. On a deeper level, we regard the abstract vector space as fundamental and the coordinate system as arbitrary. Thus, meaningful statements should be independent of the choice of a coordinate system. Conversely, one has the conviction that a statement that is independent of the coordinate system ought to have a meaningful interpretation.

Let $\mathscr{A} = \{\alpha_1, \ldots, \alpha_n\}$ and $\mathscr{B} = \{\beta_1, \ldots, \beta_n\}$ be two bases for the vector space $\mathscr{V}$. Think of $\mathscr{A}$ as the given, or old, basis and think of $\mathscr{B}$ as the new basis. Since $\mathscr{A}$ is a basis, each $\beta_j \in \mathscr{B}$ has a representation of the form

$$\beta_j = \sum_{i=1}^{n} p_{ij} \alpha_i. \tag{2.1}$$

The matrix

$$P = [p_{ij}] \tag{2.2}$$

is called the *matrix of transition* from the basis $\mathscr{A}$ to the basis $\mathscr{B}$. Notice that the second index of p_{ij} is the same as the index of β_j. Therefore, the coordinates of β_j appear in column j of P. Furthermore, since $\mathscr{B}$ is a linearly independent set, the columns of P, as n-tuples, are linearly independent. The matrix P is square of order n.

Let ξ be an arbitrary vector in $\mathscr{V}$. Let $(x_1, \ldots, x_n)$ be the representation of ξ with respect to the basis $\mathscr{A}$, and let $(y_1, \ldots, y_n)$ be the representation of ξ with respect to $\mathscr{B}$. How are these two representations related? We can use (2.1) to pass from one representation to the other. Specifically,

$$\begin{aligned}
\xi &= \sum_{j=1}^{n} y_j \beta_j \\
&= \sum_{j=1}^{n} y_j \left(\sum_{i=1}^{n} p_{ij} \alpha_i \right) \\
&= \sum_{i=1}^{n} \left(\sum_{j=1}^{n} y_j p_{ij} \right) \alpha_i \\
&= \sum_{i=1}^{n} x_i \alpha_i.
\end{aligned} \tag{2.3}$$

Since the representation of ξ with respect to the basis $\mathscr{A}$ is unique, we have

$$x_i = \sum_{j=1}^{n} p_{ij} y_j \tag{2.4}$$

for each i. Let X and Y stand for the one-column matrices

$$X = \begin{bmatrix} x_1 \\ x_2 \\ \vdots \\ x_n \end{bmatrix}, \quad Y = \begin{bmatrix} y_1 \\ y_2 \\ \vdots \\ y_n \end{bmatrix}. \quad (2.5)$$

With this notation the n equations of (2.4) can be written as a single equation

$$X = PY \quad (2.6)$$

in matrix form.

There is a way of looking at these formulas that systematizes them and makes them easier to recall. For a given basis $\mathscr{A} = \{\alpha_1, \ldots, \alpha_n\}$, there is a one-to-one correspondence between n-tuples $X = (x_1, \ldots, x_n)$ and vectors $\xi = x_1\alpha_1 + \cdots + x_n\alpha_n$. We can write this correspondence in the following form.

$$[\alpha_1 \ldots \alpha_n] \begin{bmatrix} x_1 \\ \vdots \\ x_n \end{bmatrix} = x_1\alpha_1 + \cdots + x_n\alpha_n = \xi. \quad (2.7)$$

This formula is an abuse of our notational conventions since a matrix is supposed to have field elements as elements. However, the formula is useful and that is enough justification. We can represent formula (2.7) symbolically in the form

$$\mathscr{A}X = \xi. \quad (2.8)$$

The change of coordinates formula (2.1) can be written in the form

$$[\beta_1 \ldots \beta_n] = [\alpha_1 \ldots \alpha_n] P, \quad (2.9)$$

which we can write symbolically in the form

$$\mathscr{B} = \mathscr{A}P. \quad (2.10)$$

Since Y is the representation of the vector ξ with respect to the basis $\mathscr{B}$, we have

Bases and Coordinate Systems

$$\mathscr{B}Y = \xi \qquad (2.11)$$

as the form of formula (2.8) for this basis. Combining formulas (2.10) and (2.11) we have

$$\xi = \mathscr{B}Y = (\mathscr{A}P)Y = \mathscr{A}(PY). \qquad (2.12)$$

Comparing (2.12) with formula (2.8) we obtain

$$\mathscr{A}X = \xi = \mathscr{A}(PY). \qquad (2.13)$$

This suggests that

$$X = PY. \qquad (2.6)$$

If we were using this line of argument to prove formula (2.6) we would have to justify the equality in (2.6) from the equality in (2.13). The argument that does this was given when formula (2.6) was first established. But, in any case, formula (2.6) has already been proved and we are merely offering a compact and systematic way of looking at it.

There is also a way of looking at these formulas by means of diagrams. The diagram in Figure 3–2.1 represents formula (2.8) in the following sense. An n-tuple X in $\mathscr{R}^n$ is associated with a vector $\xi = \mathscr{A}X$ in $\mathscr{V}$. The basis $\mathscr{A}$ determines the particular association through formula (2.8).

In Figure 3–2.2 two of the arrows represent coordinatizations, as in Figure 3–2.1. The other arrow represents the change of coordinates formula (2.6) that associates Y with $X = PY$. Both X and Y are n-tuples in $\mathscr{R}^n$. Formula (2.10) is interpreted as saying that the correspondence $\mathscr{B}$, the diagonal arrow between $\mathscr{R}^n$ and $\mathscr{V}$, is the same as the correspondence

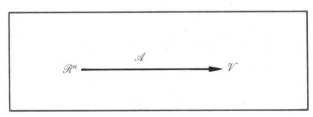

Figure 3–2.1 A coordinatization. If $\mathscr{A} = \{\alpha_1, \ldots, \alpha_n\}$ is a basis of $\mathscr{V}$, to $X = (x_1, \ldots, x_n) \in \mathscr{R}^n$, there corresponds $\mathscr{A}X = \xi = x_1\alpha_1 + \cdots + x_n\alpha_n \in \mathscr{V}$. The correspondence depends on the choice of the basis $\mathscr{A}$.

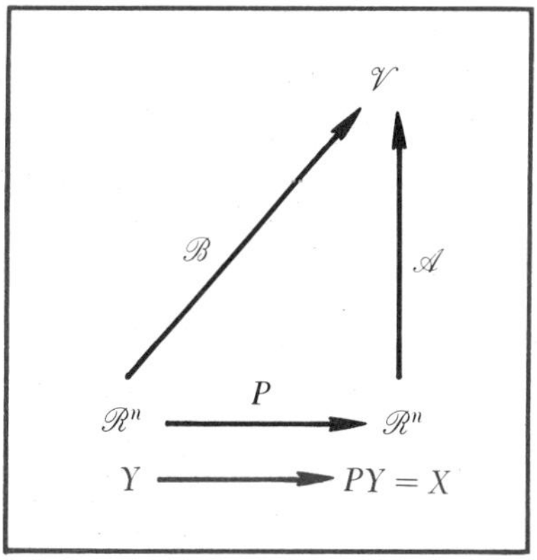

Figure 3-2.2 A change of coordinates. The arrows $\mathscr{A}$ and $\mathscr{B}$ represent coordinatizations with bases $\mathscr{A}$ and $\mathscr{B}$, respectively. P is a matrix of transition. The diagram represents the equation $\mathscr{B} = \mathscr{A}P$. The coordinate maps are $\xi = \mathscr{B}Y$ and $\xi = \mathscr{A}X$. The equality $\mathscr{B} = \mathscr{A}P$ implies that $\mathscr{B}Y = (\mathscr{A}P)Y = \mathscr{A}(PY) = \mathscr{A}X$. Therefore, $PY = X$ is the formula for a change of coordinates.

$\mathscr{A}P$, the horizontal arrow P followed by the vertical arrow $\mathscr{A}$. We say the triangular diagram in Figure 3-2.2 *commutes*. This means that the correspondence represented by the arrow $\mathscr{B}$ is the same as the combined mappings indicated by P and $\mathscr{A}$. That is, $\mathscr{B} = \mathscr{A}P$.

Formula (2.4), or (2.6), makes it possible and convenient to compute all the x_i when all the y_j are known. How do we compute the y_j when the x_i are known? To do this we must have the elements of the basis $\mathscr{A}$ expressed in terms of the basis $\mathscr{B}$. That is,

$$\alpha_j = \sum_{i=1}^{n} q_{ij}\beta_i. \tag{2.14}$$

Then, duplicating the argument that leads to (2.6) we can obtain

$$Y = QX. \tag{2.15}$$

The problem is that we can expect to be given P or Q but not both. How do we find Q if P is given?

With respect to the basis $\mathscr{A}$, β_j is represented by the n-tuple $P_j = (p_{1j}, p_{2j}, \ldots, p_{nj})$ as given in (2.1), and α_j is represented by the n-tuple $D_j = (\delta_{1j}, \delta_{2j}, \ldots, \delta_{nj})$. Our problem is to express the α_j as linear combinations of the β_i, and this is equivalent to expressing the D_j as linear combinations of the P_i. This is precisely the type of problem considered in detail in Section 2-6, where we showed how to express a vector in a linearly dependent set as a linear combination of the preceding vectors. Consider the set of coordinate vectors $\{P_1, P_2, \ldots, P_n, D_1, D_2, \ldots, D_n\}$. Then we arrange these n-tuples into $2n$ columns and perform a sequence of pivot operations to reduce the resulting matrix to row-echelon form. Since the set $\{P_1, P_2, \ldots, P_n\}$ is linearly independent, no P_i will be expressed as a linear combination of the preceding ones. Thus, the steps described in Section 2–6 will lead from

$$\begin{array}{cccccccc} \beta_1 & \beta_2 & \cdots & \beta_n & \alpha_1 & \alpha_2 & \cdots & \alpha_n \\ \begin{bmatrix} p_{11} & p_{12} & \cdots & p_{1n} & 1 & 0 & \cdots & 0 \\ p_{21} & p_{22} & \cdots & p_{2n} & 0 & 1 & \cdots & 0 \\ \cdot & \cdot & & \cdot & \cdot & & & \cdot \\ \cdot & \cdot & & \cdot & \cdot & & & \cdot \\ \cdot & \cdot & & \cdot & \cdot & & & \cdot \\ p_{n1} & p_{n2} & \cdots & p_{nn} & 0 & 0 & \cdots & 1 \end{bmatrix} \end{array} \quad (2.16)$$

to the reduced form

$$\begin{array}{cccccccc} \beta_1 & \beta_2 & \cdots & \beta_n & \alpha_1 & \alpha_2 & \cdots & \alpha_n \\ \begin{bmatrix} 1 & 0 & \cdots & 0 & q_{11} & q_{12} & \cdots & q_{1n} \\ 0 & 1 & \cdots & 0 & q_{21} & q_{22} & \cdots & q_{2n} \\ \cdot & \cdot & & \cdot & \cdot & & & \cdot \\ \cdot & \cdot & & \cdot & \cdot & & & \cdot \\ \cdot & \cdot & & \cdot & \cdot & & & \cdot \\ 0 & 0 & \cdots & 1 & q_{n1} & q_{n2} & \cdots & q_{nn} \end{bmatrix} \end{array} \quad (2.17)$$

This give us the relations written out explicitly in (2.14).

By substituting (2.15) in (2.6) we get

$$X = PY = P(QX) = (PQ)X. \quad (2.18)$$

In particular, formula (2.18) holds for $X = D_j, j = 1, 2, \ldots, n$. $D_j = (PQ)D_j$, in terms of coordinates, means

$$\delta_{ij} = \sum_{l=1}^{n} \left[\sum_{k=1}^{n} p_{ik} q_{kl} \right] \delta_{lj}$$

$$= \sum_{k=1}^{n} p_{ik} \left[\sum_{l=1}^{n} q_{kl} \delta_{lj} \right]$$

$$= \sum_{k=1}^{n} p_{ik} q_{kj}. \quad (2.19)$$

3-2 Change of Basis

In terms of matrices, (2.19) takes the form

$$I = PQ. \tag{2.20}$$

Q is called the *inverse matrix* for P, and is designated by P^{-1}. The reduction to basic form leading from (2.16) to (2.17) is the recommended method for finding the inverse of a matrix. Both Jordan and Gaussian elimination require n^3 multiplicative steps.

(2.19) and (2.20) can also be established directly from the definitions. Substitute (2.1) in (2.14) and obtain

$$\alpha_j = \sum_{k=1}^{n} q_{kj} \beta_k$$

$$= \sum_{k=1}^{n} q_{kj} \left(\sum_{i=1}^{n} p_{ik} \alpha_i \right)$$

$$= \sum_{i=1}^{n} \left(\sum_{k=1}^{n} p_{ik} q_{kj} \right) \alpha_i. \tag{2.21}$$

Since the representation of α_j as a linear combination of the basis vectors is unique, we have

$$\sum_{k=1}^{n} p_{ik} q_{kj} = \delta_{ij}. \tag{2.19}$$

Since Q and P are square matrices of the same order, both QP and PQ can be computed. From (2.20) we have $PQ = I$. What is QP? There are several ways this question could be answered. First, we could duplicate the argument leading to (2.19). Second, we could appeal to the symmetry between the roles of P and Q, and conclude that $QP = I$ from $PQ = I$ on the basis of symmetry. But we wish to show more about the existence of an inverse and its properties.

Let P be any square matrix of order n for which the columns, as coordinate vectors, are linearly independent. Taking any basis $\mathcal{A} = \{\alpha_1, \ldots, \alpha_n\}$ we define $\beta_j = \sum_{i=1}^{n} p_{ij} \alpha_i$ for $j = 1, \ldots, n$. The set $\mathcal{B} = \{\beta_1, \ldots, \beta_n\}$ is linearly independent since any linear relation among the β_j implies a corresponding linear relation among the columns of P. Thus, P is a matrix of transition. The matrix Q of transition from $\mathcal{B}$ to $\mathcal{A}$

has the property that $PQ = I$. The only observation we wish to emphasize is that for any square matrix P with linearly independent columns, there is a square matrix Q of the same order such that $PQ = I$. In other words, P has an inverse on the right, called a *right inverse*. Now, the columns of Q are also linearly independent. Thus, there is a square matrix P' of the same order such that $QP' = I$. Now Q has a left inverse P and a right inverse P'. By the argument given in Section 1-1, Q has a unique inverse, that is, $P' = P$. Also, P and Q are inverses of each other. (In the next chapter we shall show, using a simpler argument based on the concept of rank, that if a square matrix has an inverse on one side, it has an inverse on the other side and the inverse is unique.)

EXERCISES 3-2

1. In a vector space $\mathscr{V}$ of dimension 3, $\{(6, -1, -3), (2, 0, -1), (-1, 1, 1)\}$ is chosen as a new basis. Remember, this notation implies there is a basis $\mathscr{A} = \{\alpha_1, \alpha_2, \alpha_3\}$ and the new basis is $\mathscr{B} = \{\beta_1, \beta_2, \beta_3\}$, where $\beta_1 = 6\alpha_1 - \alpha_2 - 3\alpha_3$, $\beta_2 = 2\alpha_1 - \alpha_3$, $\beta_3 = -\alpha_1 + \alpha_2 + \alpha_3$. Find the matrix of transition P from $\mathscr{A}$ to $\mathscr{B}$. Find the matrix of transition Q from $\mathscr{B}$ to $\mathscr{A}$.

2. In a vector space $\mathscr{V}$ of dimension 3, a given vector α is represented by $(2, -1, 3)$. Remember, this implies there is a basis $\mathscr{A} = \{\alpha_1, \alpha_2, \alpha_3\}$, and with respect to this basis, $\alpha = 2\alpha_1 - \alpha_2 + 3\alpha_3$. If we are given a matrix of transition

$$P = \begin{bmatrix} 6 & 2 & -1 \\ -1 & 0 & 1 \\ -3 & -1 & 1 \end{bmatrix}$$

to a new basis, find the new coordinates for α.

3. Let

$$P = \begin{bmatrix} 2 & 2 & 1 \\ 3 & 5 & 2 \\ 4 & 3 & 2 \end{bmatrix}$$

be a matrix of transition from one basis to another. Express the elements of the new basis as linear combinations of the vectors in the old basis. Express the elements of the old basis as linear combinations of the vectors in the new basis.

4. Find the inverse of each of the following matrices and check your answer to see that it actually is an inverse.

   a) $\begin{bmatrix} 5 & 8 \\ 3 & 5 \end{bmatrix}$

   b) $\begin{bmatrix} 2 & -5 & 5 \\ 2 & -3 & 8 \\ 3 & -8 & 7 \end{bmatrix}$

   c) $\begin{bmatrix} 2 & 5 & 8 \\ 4 & 5 & 13 \\ 1 & -6 & -1 \end{bmatrix}$

5. Let
$$\begin{bmatrix} 3 & 2 & 0 \\ 1 & 0 & 1 \\ 4 & 1 & -1 \end{bmatrix}$$
be a matrix of transition from a basis $\mathscr{A}$ to a basis $\mathscr{B}$. Let β be represented by (3, 1, 4) with respect to $\mathscr{A}$. What is the representation of β with respect to $\mathscr{B}$?

6. The inverse of A is
$$\begin{bmatrix} 1 & 3 \\ 2 & 7 \end{bmatrix}$$
Find A.

7. Let $\mathscr{A} = \{(1, 1, 0), (1, 0, 1), (0, 1, 1)\}$ and $\mathscr{B} = \{(1, 0, 3), (2, 1, 1), (0, -1, 1)\}$ be bases of a three-dimensional vector space. Find the matrix of transition from $\mathscr{A}$ to $\mathscr{B}$, and from $\mathscr{B}$ to $\mathscr{A}$.

8. Show that the sequence of pivot operations leading from the matrix (2.16) to (2.17) generally requires n^3 multiplicative steps.

9. Let ξ be a vector in $\mathscr{V}$, a 3-dimensional vector space, represented by the triple (4, 1, −2) with respect to the basis $\mathscr{B} = \{\beta_1, \beta_2, \beta_3\}$. Use Formula (2.11) to express ξ as a linear combination of the vectors in $\mathscr{B}$.

Let $P = \begin{bmatrix} 3 & 1 & 2 \\ 3 & 4 & 3 \\ 1 & 1 & 1 \end{bmatrix}$ be the matrix of transition from the basis $\mathscr{A} = \{\alpha_1, \alpha_2, \alpha_3\}$ to the basis $\mathscr{B}$. Use Formula (2.10) to express the β_i as linear combinations of the vectors in the basis $\mathscr{A}$.

Use Formula (2.13) to express ξ as a linear combination of the vectors in $\mathscr{A}$. What triple represents ξ with respect to the basis $\mathscr{A}$?

Find the inverse Q of the matrix of transition P. Use Formula (2.13) and the representation of ξ with respect to the basis $\mathscr{A}$ to find the representation of ξ with respect to the basis $\mathscr{B}$. (We should obtain the representation $(4, 1, -2)$ that we started with.)

10. Repeat Exercise 9 for the vector ξ represented by $(3, -2, 1)$ with respect to $\mathscr{B}$ and the matrix of transition
$$P = \begin{bmatrix} 3 & 3 & 1 \\ 1 & 4 & 1 \\ 2 & 3 & 1 \end{bmatrix} \text{ from } \mathscr{A} \text{ to } \mathscr{B}.$$

3-3 ELEMENTARY OPERATIONS AND CHANGE OF COORDINATES

One might reasonably guess, from the discussion of the previous section, that a pivot operation corresponds to a change of coordinates. Indeed it does. But also every elementary operation corresponds to a change of coordinates. We will establish these facts and show in detail how this works out.

Let $\mathscr{A} = \{\alpha_1, \alpha_2, \ldots, \alpha_m\}$ be a given basis for the m-dimensional vector space $\mathscr{V}$. Suppose $\{\beta_1, \beta_2, \ldots, \beta_n\}$ is any collection of vectors in $\mathscr{V}$. We assume neither that they are linearly independent nor that they span $\mathscr{V}$, although these possibilities are permitted. Let

$$\beta_j = \sum_{i=1}^{m} a_{ij} \alpha_i. \tag{3.1}$$

Then $A = [a_{ij}]$ is an $m \times n$ matrix in which column j is an m-tuple representing β_j with respect to the basis $\mathscr{A}$. We wish to make certain kinds of changes in the basis and determine how these changes affect the representations of the form of (3.1). In particular, we wish to describe the changes that correspond to elementary operations in terms of the matrix A.

3-3 Elementary Operations and Change of Coordinates • 117

Let $\mathscr{C} = \{\gamma_1, \ldots, \gamma_m\}$ be a new basis which is obtained from $\mathscr{A}$ by replacing only α_r by $c^{-1}\alpha_r$, where $c \neq 0$. That is,

$$\begin{aligned} \gamma_i &= \alpha_i \quad \text{for} \quad i \neq r, \\ \gamma_r &= c^{-1}\alpha_r. \end{aligned} \qquad (3.2)$$

Then

$$\beta_j = \sum_{i=1}^{m} a_{ij}\alpha_i = \sum_{\substack{i=1 \\ i \neq r}}^{m} a_{ij}\gamma_i + ca_{rj}\gamma_r. \qquad (3.3)$$

Since $(a_{1j}, a_{2j}, \ldots, a_{mj})$ is a typical column in A, all columns would be changed by having the r-th element multiplied by c. This corresponds to multiplying the r-th row by c. This is an elementary operation of type I.

Now let $\mathscr{D} = \{\delta_1, \ldots, \delta_m\}$ be a new basis obtained from $\mathscr{A}$ by replacing α_r by $\alpha_r - c\alpha_k$. That is,

$$\begin{aligned} \delta_i &= \alpha_i \quad \text{for} \quad i \neq r, \\ \delta_r &= \alpha_r - c\alpha_k. \end{aligned} \qquad (3.4)$$

Then

$$\beta_j = \sum_{i=1}^{m} a_{ij}\alpha_i$$

$$= \sum_{\substack{i=1 \\ i \neq r,k}}^{m} a_{ij}\alpha_i + a_{rj}(\alpha_r - c\alpha_k) + (a_{kj} + ca_{rj})\alpha_k$$

$$= \sum_{\substack{i=1 \\ i \neq k}}^{m} a_{ij}\delta_i + (a_{kj} + ca_{rj})\delta_k. \qquad (3.5)$$

Since this is a typical column, all columns in A would be changed by adding c times the number in row r to the number in row k. This corresponds to an elementary operation of type II.

Now let $\mathscr{E} = \{\epsilon_1, \ldots, \epsilon_m\}$ be a new basis obtained from $\mathscr{A}$ by interchanging α_r and α_k. That is,

$$\begin{aligned} \epsilon_i &= \alpha_i \quad \text{for} \quad i \neq r, k, \\ \epsilon_r &= \alpha_k, \\ \epsilon_k &= \alpha_r. \end{aligned} \qquad (3.6)$$

Then

$$\beta_j = \sum_{i=1}^{m} a_{ij}\alpha_i$$

$$= \sum_{\substack{i=1 \\ i \neq r,k}} a_{ij}\alpha_i + a_{rj}\alpha_r + a_{kj}\alpha_k \qquad (3.7)$$

$$= \sum_{\substack{i=1 \\ i \neq r,k}}^{m} a_{ij}\epsilon_i + a_{rj}\epsilon_k + a_{kj}\epsilon_r.$$

Since this is a typical column, all columns in A would be changed by interchanging row r and row k. This is an elementary operation of type III.

Starting again with the basis $\mathscr{A}$, let us replace α_r by $\beta = \sum_{i=1}^{m} b_i \alpha_i$. It is necessary to have $b_r \neq 0$. Otherwise, β would be a linear combination of $\{\alpha_1, \ldots, \alpha_{r-1}, \alpha_{r+1}, \ldots, \alpha_m\}$ and the proposed new basis would be linearly dependent. We can then solve for α_r to obtain

$$\alpha_r = \frac{1}{b_r}\left(\beta - \sum_{\substack{i=1 \\ i \neq r}}^{m} b_i \alpha_i\right). \qquad (3.8)$$

The representation of β_j now becomes

$$\beta_j = \sum_{i=1}^{m} a_{ij}\alpha_i$$

$$= \sum_{\substack{i=1 \\ i \neq r}}^{m} a_{ij}\alpha_i + a_{rj}\alpha_r$$

$$= \sum_{\substack{i=1 \\ i \neq r}}^{m} a_{ij}\alpha_i + \frac{a_{rj}}{b_r}\left(\beta - \sum_{\substack{i=1 \\ i \neq r}}^{m} b_i \alpha_i\right) \qquad (3.9)$$

$$= \sum_{\substack{i=1 \\ i \neq r}}^{m} \left(a_{ij} - b_i \frac{a_{rj}}{b_r}\right)\alpha_i + \frac{a_{rj}}{b_r}\beta.$$

Expressed in terms of the matrix A, this means we divide row

3-3 Elementary Operations and Change of Coordinates • 119

r by b_r and add $-b_i$ times this new row r to row i. This can be accomplished by a sequence of elementary row operations. Starting with any given basis we can change to any other basis by making a sequence of replacements of the type described here. Thus, for any change of basis, the representations of the vectors in $\{\beta_1, \ldots, \beta_n\}$ represented in the columns of a matrix A can be changed correspondingly by a sequence of elementary row operations.

If, in the discussion of the previous paragraph, we take β to be β_k, (3.8) takes the form

$$\alpha_r = \frac{1}{a_{rk}} \left(\beta_k - \sum_{\substack{i=1 \\ i \neq r}}^{m} a_{ik} \alpha_i \right), \qquad (3.10)$$

and (3.9) becomes

$$\beta_j = \sum_{\substack{i=1 \\ i \neq r}}^{m} \left(a_{ij} - a_{ik} \frac{a_{rj}}{a_{rk}} \right) \alpha_i + \frac{a_{rj}}{a_{rk}} \beta_k. \qquad (3.11)$$

The notation has been chosen so that it can be compared directly with corresponding steps of the pivot operation in the notation of formulas (4.8) and (4.9) in Chapter 1. Thus, when we pivot on the element a_{rk}, the abstract concept behind this step is that we replace the basis element α_r by β_k, the vector represented by column k.

The elementary operations are all reversible, or invertible. An operation of type I multiplies a row by a scalar $c \neq 0$. From this resulting matrix we can recover the original matrix by multiplying the same row by c^{-1}. Let R_r and R_k denote the r-th and k-th rows of a matrix. When we add c times row r to row k, the new r-th and k-th rows are R_r and $R_k + cR_r$. Now add $-c$ times R_r to $R_k + cR_r$ and we will obtain the previous matrix. Clearly, if we interchange two rows of a matrix, the previous matrix can be recovered by interchanging those rows again. A sequence of elementary operations can be inverted by inverting each operation separately in the reverse order.

A pivot operation is a sequence of elementary operations. Therefore, it can be inverted by a sequence of elementary operations. But the inverse sequence of elementary operations may not be a pivot operation. So, generally, a pivot operation cannot be inverted by a pivot operation. When a pivot operation has been performed, the column containing the pivot element becomes a basic column and it represents an element of the new basis. A pivot operation can be used to recover the previous matrix if and only if the basis element that was removed is represented by one of the columns. In this respect the basic form is of most interest. When a basic form is obtained, there are as many basic columns as there are linearly independent columns in the matrix. A subsequent pivot operation, which will introduce a new basic column, must necessarily remove some other basic column. Thus, for a matrix in basic form every pivot operation is invertible by a pivot operation.

At this point several observations about the basic form are possible. For a basic form, the number of basic columns is equal to the dimension of the subspace spanned by the columns (actually, the subspace spanned by vectors represented by the columns). Thus, the number of basic columns obtained is the same no matter which basic form is obtained. (Notice, here, the conceptual problem in making this conclusion in terms of the dimension of a subspace of the coordinate space. After a pivot operation the columns are changed and span a different subspace of the coordinate space.) Furthermore, for each selection of a set of basic columns in a basic form, these columns form a basis for the subspace generated by the columns. Since the other columns are linear combinations of the basic columns, the other columns are unique. Thus, there is only basic form corresponding to each selection of a complete set of basic columns. Hence, starting with a given matrix there are as many basic forms obtainable as there are choices for a maximal set of linearly independent columns. (See Exercise 13 of Section 2–7.)

To illustrate what is involved in a sequence of pivot operations, let us go through the steps from (2.16) to (2.17) in a numerical example in "slow motion." Let $\mathscr{A} = \{\alpha_1, \alpha_2, \alpha_3\}$ be a given basis, and let $\mathscr{B} = \{\beta_1, \beta_2, \beta_3\}$ be a proposed new basis, where

$$\begin{aligned} \beta_1 &= \alpha_1 + \alpha_2 \\ \beta_2 &= 2\alpha_1 + \alpha_2 + 2\alpha_3 \\ \beta_3 &= 3\alpha_1 - \alpha_2 + 7\alpha_3. \end{aligned} \qquad (3.12)$$

3-3 Elementary Operations and Change of Coordinates • 121

In this specific case (2.16) takes the form

$$\begin{array}{c} \beta_1\ \beta_2\ \beta_3\ \ \alpha_1\ \alpha_2\ \alpha_3 \\ \begin{bmatrix} \textcircled{1} & 2 & 3 & 1 & 0 & 0 \\ 1 & 1 & -1 & 0 & 1 & 0 \\ 0 & 2 & 7 & 0 & 0 & 1 \end{bmatrix}\begin{array}{l}\alpha_1\\ \alpha_2\\ \alpha_3.\end{array} \end{array} \qquad (3.13)$$

(Here we have written the basis vectors to the right of the matrix.) The first pivot element selected is the 1 in the first column, first row. After this pivot operation we have

$$\begin{array}{c} \beta_1\ \beta_2\ \beta_3\ \ \alpha_1\ \alpha_2\ \alpha_3 \\ \begin{bmatrix} 1 & 2 & 3 & 1 & 0 & 0 \\ 0 & \textcircled{-1} & -4 & -1 & 1 & 0 \\ 0 & 2 & 7 & 0 & 0 & 1 \end{bmatrix}\begin{array}{l}\beta_1\\ \alpha_2\\ \alpha_3.\end{array} \end{array} \qquad (3.14)$$

The first pivot amounted to replacing α_1 by β_1, and this replacement has been indicated to the right of matrix (3.14). The new basis is $\{\beta_1, \alpha_2, \alpha_3\}$. Since $\beta_1 = \alpha_1 + \alpha_2$, we have

$$\begin{aligned} \alpha_1 &= \beta_1 - \alpha_2 \\ \beta_2 &= 2(\beta_1 - \alpha_2) + \alpha_2 + 2\alpha_3 \\ &= 2\beta_1 - \alpha_2 + 2\alpha_3 \\ \beta_3 &= 3(\beta_1 - \alpha_2) - \alpha_2 + 7\alpha_3 \\ &= 3\beta_1 - 4\alpha_2 + 7\alpha_3. \end{aligned} \qquad (3.15)$$

Compare these representations with the 4-th, 2-nd, and 3-rd columns of (3.14). Now pivot on the -1 encircled in (3.14).

$$\begin{array}{c} \beta_1\ \beta_2\ \beta_3\ \ \alpha_1\ \ \alpha_2\ \alpha_3 \\ \begin{bmatrix} 1 & 0 & -5 & -1 & 2 & 0 \\ 0 & 1 & 4 & 1 & -1 & 0 \\ 0 & 0 & \textcircled{-1} & -2 & 2 & 1 \end{bmatrix}\begin{array}{l}\beta_1\\ \beta_2\\ \alpha_3\end{array} \end{array} \qquad (3.16)$$

The second pivot amounted to replacing α_2 by β_2. The new basis is $\{\beta_1, \beta_2, \alpha_3\}$. From (3.15) we have

$$\begin{aligned} \alpha_2 &= 2\beta_1 - \beta_2 + 2\alpha_3 \\ \alpha_1 &= \beta_1 - (2\beta_1 - \beta_2 + 2\alpha_3) \\ &= -\beta_1 + \beta_2 - 2\alpha_3 \\ \beta_3 &= 3\beta_1 - 4(2\beta_1 - \beta_2 + 2\alpha_3) \\ &= -5\beta_1 + 4\beta_2 - \alpha_3. \end{aligned} \qquad (3.17)$$

Again, compare these representations with columns 5, 4, and 3 in (3.16). Finally, pivot on the -1 encircled in (3.16).

$$\begin{array}{cccccc} \beta_1 & \beta_2 & \beta_3 & \alpha_1 & \alpha_2 & \alpha_3 \\ \end{array}$$
$$\begin{bmatrix} 1 & 0 & 0 & 9 & -8 & -5 \\ 0 & 1 & 0 & -7 & 7 & 4 \\ 0 & 0 & 1 & 2 & -2 & -1 \end{bmatrix} \begin{array}{c} \beta_1 \\ \beta_2 \\ \beta_3 \end{array} \qquad (3.18)$$

The new basis is $\{\beta_1, \beta_2, \beta_3\}$. From (3.17) we have

$$\begin{aligned} \alpha_3 &= -5\beta_1 + 4\beta_2 - \beta_3 \\ \alpha_2 &= 2\beta_1 - \beta_2 + 2(-5\beta_1 + 4\beta_2 - \beta_3) \\ &= -8\beta_1 + 7\beta_2 - 2\beta_3 \\ \alpha_1 &= -\beta_1 + \beta_2 - 2(-5\beta_1 + 4\beta_2 - \beta_3) \\ &= 9\beta_1 - 7\beta_2 + 2\beta_3. \end{aligned} \qquad (3.19)$$

Compare these representations with columns 6, 5, and 4 in (3.18).

When we write an n-tuple representing a vector in a column of a matrix and perform an elementary row operation, the new column represents the same vector in a new coordinate system. When we perform a pivot operation it means we are making a particular kind of coordinate change. It corresponds to removing whatever basis element is indexed by the row containing the pivot element and replacing it with the vector that is represented by that column.

EXERCISES 3-3

1. Let $\mathscr{A} = \{\alpha_1, \alpha_2, \alpha_3\}$ be a basis of $\mathscr{V}$. Let $\{\beta_1, \beta_2, \beta_3, \beta_4\}$ be a set of vectors in $\mathscr{V}$ that are represented with respect to the basis $\mathscr{A}$ by the columns of the following matrix,

$$\begin{array}{cccc} \beta_1 & \beta_2 & \beta_3 & \beta_4 \end{array}$$
$$\begin{bmatrix} 1 & 4 & 7 & 10 \\ 2 & 5 & 8 & 11 \\ 3 & 6 & 9 & 12 \end{bmatrix} \begin{array}{c} \alpha_1 \\ \alpha_2 \\ \alpha_3 \end{array}$$

It is decided to replace α_2 by β_3 to form a new basis. Encircle the proper pivot element so that the resulting pivot operation will give the corresponding change in coordinates.

2. In a situation similar to that described in Exercise 1, the corresponding matrix is

$$\begin{array}{cccc} \beta_1 & \beta_2 & \beta_3 & \beta_4 \end{array}$$
$$\begin{bmatrix} 1 & 2 & 0 & -3 \\ 0 & 1 & -2 & 0 \\ -1 & 0 & 1 & 1 \end{bmatrix} \begin{array}{c} \alpha_1 \\ \alpha_2 \\ \alpha_3 \end{array}$$

3-3 Elementary Operations and Change of Coordinates • 123

Which of the β_i can be used to replace α_2 to obtain a new basis?

3. Let $\mathscr{A} = \{\alpha_1, \alpha_2, \alpha_3\}$ be a basis of $\mathscr{V}$. Let $\{\beta_1, \beta_2, \beta_3, \beta_4, \beta_5, \beta_6\}$ be a set of vectors in $\mathscr{V}$ that are represented with respect to the basis $\mathscr{A}$ by the columns of the following matrix.

$$\begin{array}{c} \phantom{\begin{bmatrix}}\beta_1 \ \ \beta_2 \ \ \beta_3 \ \ \beta_4 \ \ \beta_5 \ \ \beta_6 \\ \begin{bmatrix} 1 & 2 & 1 & 1 & 2 & 3 \\ 0 & 1 & 2 & \boxed{1} & -1 & 2 \\ -1 & 1 & -3 & 2 & -2 & 1 \end{bmatrix} \begin{array}{c} \alpha_1 \\ \alpha_2 \\ \alpha_3 \end{array} \end{array}$$

After a pivot operation on the encircled element above we have

$$\begin{array}{c} \phantom{\begin{bmatrix}}\beta_1 \ \ \beta_2 \ \ \beta_3 \ \ \beta_4 \ \ \beta_5 \ \ \beta_6 \\ \begin{bmatrix} 1 & \boxed{1} & -1 & 0 & 3 & 1 \\ 0 & 1 & 2 & 1 & -1 & 2 \\ -1 & -1 & -7 & 0 & 0 & -3 \end{bmatrix} \end{array}.$$

What is the new basis? What is the representation of β_3 in the new coordinate system? What is the representation of α_2 in the new coordinate system? A further pivot operation is made on the encircled element above, and we obtain

$$\begin{array}{c} \phantom{\begin{bmatrix}}\beta_1 \ \ \beta_2 \ \ \beta_3 \ \ \beta_4 \ \ \beta_5 \ \ \beta_6 \\ \begin{bmatrix} 1 & 1 & -1 & 0 & 3 & 1 \\ \boxed{-1} & 0 & 3 & 1 & -4 & 1 \\ 0 & 0 & -8 & 0 & 3 & -2 \end{bmatrix} \end{array}.$$

Now what is the basis? A further pivot is made on the element encircled above, and we obtain

$$\begin{array}{c} \phantom{\begin{bmatrix}}\beta_1 \ \ \beta_2 \ \ \beta_3 \ \ \beta_4 \ \ \beta_5 \ \ \beta_6 \\ \begin{bmatrix} 0 & 1 & 2 & 1 & -1 & 2 \\ 1 & 0 & -3 & -1 & 4 & -1 \\ 0 & 0 & -8 & 0 & 3 & -2 \end{bmatrix} \end{array}.$$

What is the basis now? What is the representation of β_4? Express β_4 as a linear combination of the basis vectors.

4. Re-do Exercise 7 of Section 3–2 using the technique described in this section. In case there is any doubt as to how to proceed, let us indicate the start. Write

$$\begin{array}{c} \phantom{\begin{bmatrix}}\alpha_1 \ \ \alpha_2 \ \ \alpha_3 \ \ \beta_1 \ \ \beta_2 \ \ \beta_3 \\ \begin{bmatrix} 1 & 1 & 0 & 1 & 2 & 0 \\ 1 & 0 & 1 & 0 & 1 & -1 \\ 0 & 1 & 1 & 3 & 1 & 1 \end{bmatrix} \end{array}.$$

Here the first three columns represent the vectors in $\mathscr{A}$ and the last three represent the vectors in $\mathscr{B}$. Three successive pivots in the first three columns will introduce the α_i into the basis and express the β_i as linear combinations of these basis elements. This will yield the matrix of transition in the last three columns.

5. Find the matrix of transition from the basis $\mathscr{A} = \{(1, 2, 3), (1, -1, 1), (-2, 4, 0)\}$ to the basis $\mathscr{B} = \{(2, 1, 0), (3, 1, -1), (2, -1, 2)\}$. Find the matrix of transition from $\mathscr{B}$ to $\mathscr{A}$.

6. Let $\mathscr{A} = \{\alpha_1, \alpha_2, \alpha_3, \alpha_4\}$ be a basis of the 4-dimensional vector space $\mathscr{V}$. Let β be a vector in $\mathscr{V}$ represented by $(2, 1, 3, 2)$. If a new basis is obtained by replacing α_3 by β, what is the matrix of transition corresponding to this change of basis?

7. With the situation as described in Exercise 6, let the vector α_4 in $\mathscr{A}$ be replaced by the vector γ represented by $(-2, 1, 3, 1)$. What is the matrix of transition corresponding to this change of basis?

8. Let $P = \begin{bmatrix} 1 & 2 & 0 & 0 \\ 0 & -1 & 0 & 0 \\ 0 & 1 & 1 & 0 \\ 0 & 3 & 0 & 1 \end{bmatrix}$ be a matrix of transition from the basis $\mathscr{A} = \{\alpha_1, \alpha_2, \alpha_3, \alpha_4\}$ to a new basis. The change involves the replacement of a single element of $\mathscr{A}$. Which element is replaced, and what vector takes its place?

9. Let $P = \begin{bmatrix} 1 & 0 & 4 & 0 \\ 0 & 1 & -1 & 0 \\ 0 & 0 & 2 & 0 \\ 0 & 0 & 7 & 1 \end{bmatrix}$ be a matrix of transition from the basis $\mathscr{A} = \{\alpha_1, \alpha_2, \alpha_3, \alpha_4\}$ to a new basis. The change involves the replacement of a single element of $\mathscr{A}$. Which element is replaced, and what vector takes its place?

10. The following matrix is in basic form.

$$\begin{array}{cccccccc} \alpha_1 & \alpha_2 & \alpha_3 & \alpha_4 & \alpha_5 & \alpha_6 & \alpha_7 & \alpha_8 \end{array}$$
$$\begin{bmatrix} 2 & 0 & 3 & 0 & 1 & 3 & 1 & 0 \\ 1 & 1 & 0 & 0 & 0 & 2 & 0 & 0 \\ 3 & 0 & 1 & 1 & 0 & 1 & 0 & 0 \\ 1 & 0 & -2 & 0 & -1 & 0 & 0 & 1 \end{bmatrix}$$

3-3 Elementary Operations and Change of Coordinates • 125

What is the basis? Perform the pivot operation on the encircled element. What is the new basis? What is the matrix of transition to the new basis?

After this pivot operation is performed, perform a second pivot operation on the element in row 2 column 3. What is the new basis? What is the matrix of transition corresponding to the second pivot operation?

3-4 ELEMENTARY MATRICES

So far we have described the elementary operations in terms of direct action on the matrices. The same effect can be achieved by matrix multiplication. What we mean is this: Let A be a given $m \times n$ matrix on which we intend to perform an elementary operation. There is a square matrix E of order m such that EA is in the required form that would be obtained as a result of the elementary operation. The matrix E thus represents the elementary operation. A matrix representing an elementary operation is called an *elementary matrix*.

The required matrix E is simple enough to describe. Let I be the identity matrix of order m. Then $A = IA$. Perform the same elementary row operation on the first (left) factor on both sides. If A' and I' are the matrices obtained as a result of this row operation, then $A' = I'A$. (We do not expect this assertion to be obvious, but its correctness can be established independently from the following discussion.) $I' = E$ is the required elementary matrix. Using this assertion, let us determine the elementary matrix for each type of elementary operation.

Let $E_k(c)$ denote the matrix obtained from the identity matrix by multiplying row k by $c \neq 0$. $E_k(c)$ has non-zero elements only in the main diagonal. The numbers in the main diagonal are all 1's except for a c in the k-th position.

$$E_k(c) = \begin{bmatrix} 1 & 0 & \cdots & 0 & 0 & 0 & \cdots & 0 \\ 0 & 1 & \cdots & 0 & 0 & 0 & \cdots & 0 \\ \cdot & \cdot & & \cdot & \cdot & \cdot & & \cdot \\ \cdot & \cdot & & \cdot & \cdot & \cdot & & \cdot \\ \cdot & \cdot & & \cdot & \cdot & \cdot & & \cdot \\ 0 & 0 & \cdots & 1 & 0 & 0 & \cdots & 0 \\ 0 & 0 & \cdots & 0 & c & 0 & \cdots & 0 \\ 0 & 0 & \cdots & 0 & 0 & 1 & \cdots & 0 \\ \cdot & \cdot & & \cdot & \cdot & \cdot & & \cdot \\ \cdot & \cdot & & \cdot & \cdot & \cdot & & \cdot \\ \cdot & \cdot & & \cdot & \cdot & \cdot & & \cdot \\ 0 & 0 & \cdots & 0 & 0 & 0 & \cdots & 1 \end{bmatrix} \text{(row } k\text{)} \quad (4.1)$$

Let $E_{rk}(c)$ denote the matrix obtained from the identity matrix by adding c times row r to row k. $E_{rk}(c)$ has 1's along the main diagonal. There is a c in the row k—column r position. All other elements are zero.

$$E_{rk}(c) = \begin{bmatrix} 1 & \cdots & 0 & \cdots & 0 & \cdots & 0 \\ \cdot & & \cdot & & \cdot & & \cdot \\ \cdot & & \cdot & & \cdot & & \cdot \\ \cdot & & \cdot & & \cdot & & \cdot \\ 0 & \cdots & 1 & \cdots & c & \cdots & 0 \\ \cdot & & \cdot & & \cdot & & \cdot \\ \cdot & & \cdot & & \cdot & & \cdot \\ \cdot & & \cdot & & \cdot & & \cdot \\ 0 & \cdots & 0 & \cdots & 1 & \cdots & 0 \\ \cdot & & \cdot & & \cdot & & \cdot \\ \cdot & & \cdot & & \cdot & & \cdot \\ \cdot & & \cdot & & \cdot & & \cdot \\ 0 & \cdots & 0 & \cdots & 0 & \cdots & 1 \end{bmatrix} \begin{matrix} \\ \\ \\ \\ (\text{row } k) \\ \\ \\ \\ \\ \\ \\ \\ \end{matrix} \quad (4.2)$$

with Column r indicated above.

Let E_{rk} denote the matrix obtained from the identity matrix by interchanging row r and row k. There are 1's in all positions along the main diagonal except in the r-th and k-th positions, where there are zeros. There is a 1 in the row k-column r position and a 1 in the row r-column k position. All other elements are zeros.

$$E_{rk} = \begin{bmatrix} 1 & \cdots & 0 & \cdots & 0 & \cdots & 0 \\ \cdot & & \cdot & & \cdot & & \cdot \\ \cdot & & \cdot & & \cdot & & \cdot \\ \cdot & & \cdot & & \cdot & & \cdot \\ 0 & \cdots & 0 & \cdots & 1 & \cdots & 0 \\ \cdot & & \cdot & & \cdot & & \cdot \\ \cdot & & \cdot & & \cdot & & \cdot \\ \cdot & & \cdot & & \cdot & & \cdot \\ 0 & \cdots & 1 & \cdots & 0 & \cdots & 0 \\ \cdot & & \cdot & & \cdot & & \cdot \\ \cdot & & \cdot & & \cdot & & \cdot \\ \cdot & & \cdot & & \cdot & & \cdot \\ 0 & \cdots & 0 & \cdots & 0 & \cdots & 1 \end{bmatrix} \begin{matrix} \\ \\ \\ \\ (\text{row } r) \\ \\ \\ \\ (\text{row } k) \\ \\ \\ \\ \end{matrix} \quad (4.3)$$

with Column r and Column k indicated above.

A formal proof that these elementary matrices do, in fact,

perform the elementary operations can be constructed along the following lines. Let $I = [\delta_{ij}]$ be the identity matrix. Formulas for the elements of the various elementary matrices can then be written down, and formal calculations can be made. But it is just as convincing to do the indicated calculations in several particular cases, and we will leave this to the reader.

We do not wish to make a "big thing" out of elementary matrices. Our interest in them is actually very slight, and very temporary. We will use elementary matrices to establish certain facts, and then use these facts rather than the elementary matrices. Later, when we discuss determinants, we will use elementary matrices in this way. Right now we wish to discuss the connection between elementary matrices and matrices that have inverses.

A matrix that does not have an inverse is said to be *singular*. A matrix that has an inverse is *non-singular*. The inverse must be a two-sided inverse. That is, if A is non-singular and A^{-1} is its inverse, then $A^{-1}A = AA^{-1} = I$.

Theorem 4.1 *A square matrix is non-singular if and only if its columns, as coordinate vectors, are linearly independent.*

Proof. Let $A = [a_{ij}]$ be a square matrix of order m. Let $\mathscr{A} = \{\alpha_1, \ldots, \alpha_m\}$ be any basis of an m-dimensional vector space and define β_j by

$$\beta_j = \sum_{i=1}^{m} a_{ij}\alpha_i, \quad j = 1, \ldots, m. \tag{4.4}$$

Then the columns of A represent the vectors in $\mathscr{B} = \{\beta_1, \ldots, \beta_m\}$. $\mathscr{B}$ is a linearly independent set if and only if the columns of A are linearly independent. And $\mathscr{B}$ is a basis of $\mathscr{V}$ if and only if it is linearly independent. If $\mathscr{B}$ is a basis, it spans $\mathscr{V}$ and each α_j has a representation in the form

$$\alpha_j = \sum_{i=1}^{m} b_{ij}\beta_i, \quad j = 1, \ldots, m. \tag{4.5}$$

Then $B = [b_{ij}]$ is the inverse of A (already discussed in detail in Section 3–3). If A has an inverse $B = [b_{ij}]$, then

$$\sum_{i=1}^{m} b_{ij}\beta_i = \sum_{i=1}^{m} b_{ij}\left(\sum_{k=1}^{m} a_{ki}\alpha_k\right)$$

$$= \sum_{k=1}^{m}\left(\sum_{i=1}^{m} a_{ki}b_{ij}\right)\alpha_k \qquad (4.6)$$

$$= \sum_{k=1}^{m} \delta_{kj}\alpha_k$$

$$= \alpha_j.$$

Then $\mathscr{B}$ spans $\mathscr{V}$ and is a basis. A comparison of all these equivalent statements proves the theorem. □

Theorem 4.2 *A square matrix is non-singular if and only if it can be written as a product of elementary matrices.*

Proof. Suppose $A = E_1 E_2 \ldots E_s$ where each E_i is an elementary matrix. Each elementary matrix has an inverse which is an elementary matrix (of the same type). This follows from the fact that each elementary operation has an inverse which is also an elementary operation. Furthermore, the inverse of $E_1 E_2 \ldots E_s$ is $E_s^{-1} \ldots E_2^{-1} E_1^{-1}$. This can be verified by computing $(E_1 E_2 \ldots E_s)(E_s^{-1} \ldots E_2^{-1} E_1^{-1})$. In fact, generally $(AB)^{-1} = B^{-1}A^{-1}$. Then $E_s^{-1} \ldots E_2^{-1} E_1^{-1}$ is the inverse of A.

On the other hand, suppose A has an inverse. By Theorem 4.1 the columns of A are linearly independent. When A is reduced to row-echelon form by a sequence of elementary operations, the reduced row-echelon form must be the identity matrix. Let $E_1, E_2, \ldots, E_s$ be elementary matrices representing the corresponding elementary operations used to reduce A to row-echelon form in the order used. Then

$$E_s \ldots E_2 E_1 A = I.$$

$(E_s \ldots E_2 E_1) = B$ is a product of elementary matrices and, therefore, has an inverse. By the uniqueness of the inverse, A is the inverse of B. Thus $AB = BA = I$, and B is also the inverse of A. □

Let $X = (x_1, \ldots, x_m)$ be a one-column matrix representing a vector α with respect to some basis $\mathscr{A}$. Let P be the matrix of transition to some new basis $\mathscr{B}$. Then QX where $Q = P^{-1}$ is the representation of α with respect to the new basis. For the matrix A discussed in the proof of Theorem 4.1, the columns $A_1, A_2, \ldots, A_n$ represent $\beta_1, \beta_2, \ldots, \beta_n$ with respect to the basis $\mathscr{A}$. Then the columns of QA are QA_1, $QA_2, \ldots, QA_n$, and they represent the β_j with respect to the new basis. Q is non-singular (as shown in Section 3–2) and Q can be represented as a product of elementary matrices. But Q is almost never computed by multiplying elementary matrices. Quite often QA is obtained directly from A by performing the elementary operations, just as we have been doing it so far in this text. Sometimes Q is obtained from some other information.

If QA is obtained directly from A by performing the elementary operations, we will obtain neither the corresponding elementary matrices nor the matrix Q. If it is desirable to obtain Q it is still not necessary to determine the elementary matrices and compute the product. One can adjoin the identity matrix to A to form a matrix $[A \ I]$ with m rows and $n+m$ columns. Then we perform the same elementary operations on the identity matrix that we perform on A. In the end we will obtain both QA and $QI = Q$. In effect, this is what we did when we obtained (2.17) from (2.16) in Section 3–2.

Example: Let the matrix

$$A = \begin{bmatrix} 1 & 2 & 1 \\ 2 & 0 & 1 \\ 3 & -1 & 2 \end{bmatrix}$$

be given, and perform the following elementary operations in order on $[A \ I]$.

$\begin{bmatrix} 1 & 2 & 1 & 1 & 0 & 0 \\ 2 & 0 & 1 & 0 & 1 & 0 \\ 3 & -1 & 2 & 0 & 0 & 1 \end{bmatrix}$ add twice row 1 to row 2.

$\begin{bmatrix} 1 & 2 & 1 & 1 & 0 & 0 \\ 4 & 4 & 3 & 2 & 1 & 0 \\ 3 & -1 & 2 & 0 & 0 & 1 \end{bmatrix}$ subtract row 2 from row 3.

$\begin{bmatrix} 1 & 2 & 1 & 1 & 0 & 0 \\ 4 & 4 & 3 & 2 & 1 & 0 \\ -1 & -5 & -1 & -2 & -1 & 1 \end{bmatrix}$ subtract row 3 from both row 1 and row 2.

$\begin{bmatrix} 2 & 7 & 2 & 3 & 1 & -1 \\ 5 & 9 & 4 & 4 & 2 & -1 \\ -1 & -5 & -1 & -2 & -1 & 1 \end{bmatrix}$

The matrix obtained from A by these operations is
$$\begin{bmatrix} 2 & 7 & 2 \\ 5 & 9 & 4 \\ -1 & -5 & -1 \end{bmatrix},$$
and the matrix Q that effects this change is
$$Q = \begin{bmatrix} 3 & 1 & -1 \\ 4 & 2 & -1 \\ -2 & -1 & 1 \end{bmatrix}.$$

Note that
$$QA = \begin{bmatrix} 3 & 1 & -1 \\ 4 & 2 & -1 \\ -2 & -1 & 1 \end{bmatrix} \begin{bmatrix} 1 & 2 & 1 \\ 2 & 0 & 1 \\ 3 & -1 & 2 \end{bmatrix} = \begin{bmatrix} 2 & 7 & 2 \\ 5 & 9 & 4 \\ -1 & -5 & -1 \end{bmatrix}.$$

EXERCISES 3–4

1. Write $P = \begin{bmatrix} 3 & 1 & -1 \\ 4 & 2 & -1 \\ -2 & -1 & 1 \end{bmatrix}$ as a product of elementary matrices.

2. Identify the elementary row operation represented by each of the following elementary matrices.

   a) $\begin{bmatrix} 0 & 0 & 1 \\ 0 & 1 & 0 \\ 1 & 0 & 0 \end{bmatrix}$

   b) $\begin{bmatrix} 1 & 0 & 0 \\ 0 & 3 & 0 \\ 0 & 0 & 1 \end{bmatrix}$

   c) $\begin{bmatrix} 1 & -2 & 0 \\ 0 & 1 & 0 \\ 0 & 0 & 1 \end{bmatrix}$

   d) $\begin{bmatrix} 1 & 0 & 1 \\ 0 & 1 & 0 \\ 0 & 0 & 1 \end{bmatrix}$

3. Check your identification of the elementary row operations in Exercise 2 by carrying out the proposed row operation on

$$A = \begin{bmatrix} 1 & 2 & 2 \\ 3 & 1 & 0 \\ -1 & 2 & 1 \end{bmatrix},$$

and comparing the result with QA, where Q is the corresponding elementary matrix.

4. The matrix $P = \begin{bmatrix} 1 & 2 & 0 & 0 \\ 0 & 3 & 0 & 0 \\ 0 & -1 & 1 & 0 \\ 0 & 7 & 0 & 1 \end{bmatrix}$ represents a pivot operation. Describe the pivot operation as a sequence of elementary row operations. Write down the corresponding elementary matrices and represent P as a product of elementary matrices.

5. Write $P = \begin{bmatrix} 1 & 0 & 0 & -3 \\ 0 & 1 & 0 & 2 \\ 0 & 0 & 1 & 6 \\ 0 & 0 & 0 & -2 \end{bmatrix}$ as a product of elementary matrices.

A matrix in which all elements above the main diagonal are zero is said to be in lower-triangular form. A non-singular matrix in lower-triangular form is particularly easy to represent as a product of elementary matrices. Consider the following products.

$$\begin{bmatrix} 1 & 0 & 0 \\ 0 & 1 & 0 \\ 4 & 0 & 1 \end{bmatrix} \begin{bmatrix} 1 & 0 & 0 \\ 0 & 1 & 0 \\ 0 & 6 & 1 \end{bmatrix} \begin{bmatrix} 1 & 0 & 0 \\ 0 & 1 & 0 \\ 0 & 0 & 7 \end{bmatrix} = \begin{bmatrix} 1 & 0 & 0 \\ 0 & 1 & 0 \\ 4 & 6 & 7 \end{bmatrix} \quad (4.7)$$

$$\begin{bmatrix} 1 & 0 & 0 \\ 3 & 1 & 0 \\ 0 & 0 & 1 \end{bmatrix} \begin{bmatrix} 1 & 0 & 0 \\ 0 & 5 & 0 \\ 0 & 0 & 1 \end{bmatrix} = \begin{bmatrix} 1 & 0 & 0 \\ 3 & 5 & 0 \\ 0 & 0 & 1 \end{bmatrix} \quad (4.8)$$

Combining (4.7) and (4.8), we obtain

$$\begin{bmatrix} 2 & 0 & 0 \\ 0 & 1 & 0 \\ 0 & 0 & 1 \end{bmatrix} \begin{bmatrix} 1 & 0 & 0 \\ 3 & 5 & 0 \\ 0 & 0 & 1 \end{bmatrix} \begin{bmatrix} 1 & 0 & 0 \\ 0 & 1 & 0 \\ 4 & 6 & 7 \end{bmatrix} = \begin{bmatrix} 2 & 0 & 0 \\ 3 & 5 & 0 \\ 4 & 6 & 7 \end{bmatrix}. \quad (4.9)$$

The matrix on the right side of formula (4.9) is in lower-triangular form. The right-hand factor on the left side of (4.9) is the product of elementary matrices given in formula (4.7). That product provides the longest row of the lower-triangular matrix. The next factor provides the second longest row of the lower-triangular matrix.

6. Write the following matrices as products of elementary matrices.

a) $\begin{bmatrix} 3 & 0 \\ 2 & -5 \end{bmatrix}$

b) $\begin{bmatrix} -3 & 0 & 0 \\ 2 & 1 & 0 \\ 5 & -3 & 9 \end{bmatrix}$

c) $\begin{bmatrix} 2 & 0 & 0 \\ 7 & -3 & 0 \\ 0 & 2 & 3 \end{bmatrix}$

d) $\begin{bmatrix} 3 & -4 & 7 \\ 0 & 2 & 6 \\ 0 & 0 & -1 \end{bmatrix}$

7. Show that a non-singular matrix of order n in lower-triangular form can be written as a product of not more than $\frac{n(n+1)}{2}$ elementary matrices.

8. Show that a lower-triangular matrix of order n with 1's in the main diagonal can be written as a product of not more than $\frac{n(n-1)}{2}$ elementary matrices.

9. In the next section we will show that a square matrix can be written as the product of a lower-triangular matrix and an upper-triangular matrix. Assuming this fact, show that a non-singular matrix of order n can be written as a product of not more than n^2 elementary matrices.

3–5 A USEFUL FACTORIZATION OF A MATRIX

Let A be any $m \times n$ matrix. We saw in Section 3–4 that if A' is any matrix that can be obtained from A by a sequence of elementary row operations, then $A' = QA$, where Q is a non-singular square matrix of order m. In fact, Q is a product of elementary matrices which represent the elementary operations used to reduce A to A'. Since Q is non-singular, $A =$

3-5 A Useful Factorization of a Matrix • 133

$Q^{-1}A'$. This is a representation of A as a product. By choosing the elementary operations used and the form of the reduced matrix A' properly, this factorization can be very useful.

The operations we use are those of the forward course of Gaussian elimination. A' is the form obtained at the end of the forward course, and $Q^{-1} = P$ can be written down without any additional calculation. Let us be specific about how this is done. Let

$$A = \begin{bmatrix} a_{11} & a_{12} & \cdots & a_{1n} \\ a_{21} & a_{22} & \cdots & a_{2n} \\ \cdot & \cdot & \cdots & \cdot \\ \cdot & \cdot & \cdots & \cdot \\ \cdot & \cdot & \cdots & \cdot \\ a_{m1} & a_{m2} & \cdots & a_{mn} \end{bmatrix} \qquad (5.1)$$

We will make a sequence of assumptions to simplify the discussion and notation. If these assumptions are not satisfied the form, but not the essential features, of the conclusion must be modified. We will show how this is done. First let us assume that $a_{11} \neq 0$ so that we can pivot on a_{11}. We divide the first row of A by a_{11}, then add $-a_{i1}$ times the new row 1 to row i. When this forward pivot is completed we obtain

$$A_1 = \begin{bmatrix} 1 & a_{12}' & a_{13}' & \cdots & a_{1n}' \\ 0 & a_{22}' & a_{23}' & \cdots & a_{2n}' \\ 0 & a_{32}' & a_{33}' & \cdots & a_{3n}' \\ \cdot & \cdot & \cdot & \cdots & \cdot \\ \cdot & \cdot & \cdot & \cdots & \cdot \\ 0 & a_{m2}' & a_{m3}' & \cdots & a_{mn}' \end{bmatrix} \qquad (5.2)$$

The matrix Q_1, a product of elementary matrices described in Section 3-4, such that $A_1 = Q_1 A$ is

$$Q_1 = \begin{bmatrix} a_{11}^{-1} & 0 & 0 & \cdots & 0 \\ -a_{21}a_{11}^{-1} & 1 & 0 & \cdots & 0 \\ -a_{31}a_{11}^{-1} & 0 & 1 & \cdots & 0 \\ \cdot & \cdot & \cdot & \cdots & \cdot \\ \cdot & \cdot & \cdot & \cdots & \cdot \\ -a_{m1}a_{11}^{-1} & 0 & 0 & \cdots & 1 \end{bmatrix} \qquad (5.3)$$

For the next step, assume $a_{22}' \neq 0$ so that we can pivot on a_{22}'. When this forward pivot is completed we obtain

$$A_2 = \begin{bmatrix} 1 & a_{12}' & a_{13}' & \cdots & a_{1n}' \\ 0 & 1 & a_{23}'' & \cdots & a_{2n}'' \\ 0 & 0 & a_{33}'' & \cdots & a_{3n}'' \\ \cdot & \cdot & \cdot & \cdots & \cdot \\ \cdot & \cdot & \cdot & \cdots & \cdot \\ \cdot & \cdot & \cdot & \cdots & \cdot \\ 0 & 0 & a_{m3}'' & \cdots & a_{mn}'' \end{bmatrix} \tag{5.4}$$

The matrix Q_2, a product of elementary matrices, such that $A_2 = Q_2 A_1$ is

$$Q_2 = \begin{bmatrix} 1 & 0 & 0 & \cdots & 0 \\ 0 & a_{22}'^{-1} & 0 & \cdots & 0 \\ 0 & -a_{32}' a_{22}'^{-1} & 1 & \cdots & 0 \\ \cdot & \cdot & & \cdots & \cdot \\ \cdot & \cdot & & \cdots & \cdot \\ \cdot & \cdot & & \cdots & \cdot \\ 0 & -a_{m2}' a_{22}'^{-1} & \cdot & \cdots & 1 \end{bmatrix} \tag{5.5}$$

Assume $a_{33}'' \neq 0$ in A_2 so that we can continue in the pattern indicated. In fact, suppose we can obtain the successive matrices $A_1, A_2, \ldots, A_{m-1}$ and $Q_1, Q_2, \ldots, Q_{m-1}$ such that $A_1 = Q_1 A$, $A_2 = Q_2 A_1$, $A_3 = Q_3 A_2$, etc. Then

$$A' = A_{m-1} = Q_{m-1} Q_{m-2} \cdots Q_2 Q_1 A = QA, \tag{5.6}$$

where $Q = Q_{m-1} \cdots Q_2 Q_1$. It is not particularly difficult to compute Q. However, we really don't want Q. We want $Q^{-1} = P$ so we can write (5.6) in the form $A = PA'$, and P can be obtained without computing Q. In fact, P can be obtained with no further computations beyond those required to obtain A' from A by the forward course of Gaussian elimination. Since $Q^{-1} = Q_1^{-1} Q_2^{-1} \cdots Q_{m-1}^{-1}$, we can obtain the inverse of Q by computing the inverse of each Q_k. Now

$$Q_1^{-1} = \begin{bmatrix} a_{11} & 0 & 0 & \cdots & 0 \\ a_{21} & 1 & 0 & \cdots & 0 \\ \cdot & \cdot & \cdot & \cdots & \cdot \\ \cdot & \cdot & \cdot & \cdots & \cdot \\ \cdot & \cdot & \cdot & \cdots & \cdot \\ a_{m1} & 0 & 0 & \cdots & 1 \end{bmatrix}, \tag{5.7}$$

3-5 A Useful Factorization of a Matrix

$$Q_2^{-1} = \begin{bmatrix} 1 & 0 & 0 & \cdots & 0 \\ 0 & a_{22}' & 0 & \cdots & 0 \\ 0 & a_{32}' & 1 & \cdots & 0 \\ \cdot & \cdot & \cdot & \cdots & \cdot \\ \cdot & \cdot & \cdot & \cdots & \cdot \\ \cdot & \cdot & \cdot & \cdots & \cdot \\ 0 & a_{m2}' & 0 & \cdots & 1 \end{bmatrix} \qquad (5.8)$$

and

$$Q_1^{-1} Q_2^{-1} = \begin{bmatrix} a_{11} & 0 & 0 & \cdots & 0 \\ a_{21} & a_{22}' & 0 & \cdots & 0 \\ a_{31} & a_{32}' & 1 & \cdots & 0 \\ \cdot & \cdot & \cdot & \cdots & \cdot \\ \cdot & \cdot & \cdot & \cdots & \cdot \\ \cdot & \cdot & \cdot & \cdots & \cdot \\ a_{m1} & a_{m2}' & 0 & \cdots & 1 \end{bmatrix} \qquad (5.9)$$

Continuing, we can see that P contains the elements of column k of A_{k-1} from the diagonal element down. Since these elements are computed during the forward course of Gaussian elimination, they can be recorded as they are computed. In fact, as the forward course reduction progresses, each elementary row operation has as its aim replacing some element by a 1 (a type I operation) or by a 0 (a type II operation). At the same time an element of P can be recorded in the same position. In other words, the forward course produces elements of P at the same rate that it reduces elements of A to predetermined values.

As a particular case, assume that A is a square matrix of order n. Then $A' = QA$ is also a square matrix of order n and we have arranged it so that elements in A' below the main diagonal are zeros. We say that A' is in *upper triangular form*. Also, we have shown directly that P is a matrix in which all elements above the main diagonal are zeros. We say that P is in *lower triangular form*. To symbolize the form of the factorization we have obtained we will write $P = L$ and $A' = U$. Then $A = LU$, where L is a lower triangular matrix and U is an upper triangular matrix.

The factorization of $A = LU$ requires

$$n(n-1) + (n-1)(n-2) + \cdots + 2 = \frac{n(n^2-1)}{3} \quad (5.10)$$

multiplicative steps.

In Section 1-6, we introduced the Gaussian elimination procedure in the context of solving a system of linear equations of the form

$$AX = B. \quad (5.11)$$

The factorization $A = LU$ allows us to write (5.11) in the form $LUX = B$ and decompose it into two successive problems,

$$LY = B, \quad (5.12)$$

$$UX = Y. \quad (5.13)$$

The fact that L is in lower triangular form and U is in upper triangular form allows us to solve both these systems quite easily. To simplify notation, let $L = [l_{ij}]$, where $l_{ij} = 0$ for $j > i$, and let $U = [u_{ij}]$, where $u_{ij} = 0$ for $j < i$ and $u_{ii} = 1$. Then

$$\begin{aligned} y_1 &= l_{11}^{-1} b_1 \\ y_2 &= l_{22}^{-1}(b_2 - l_{21} y_1) \\ y_3 &= l_{33}^{-1}(b_3 - l_{31} y_1 - l_{32} y_2) \\ &\quad \vdots \\ y_n &= l_{nn}^{-1}(b_n - l_{n1} y_1 - l_{n2} y_2 - \cdots - l_{n(n-1)} y_{n-1}). \end{aligned} \quad (5.14)$$

After the solution to (5.12) is obtained in this way we can get the solution to (5.13).

$$\begin{aligned} x_n &= y_n \\ x_{n-1} &= y_{n-1} - u_{(n-1)n} x_n \\ x_{n-2} &= y_{n-2} - u_{(n-2)n} x_n - u_{(n-2)(n-1)} x_{n-1} \\ &\quad \vdots \\ x_1 &= y_1 - u_{1n} x_n - u_{1(n-1)} x_{n-1} - \cdots - u_{12} x_2. \end{aligned} \quad (5.15)$$

The calculation in (5.14) requires

$$1 + 2 + \cdots + n = \frac{n(n+1)}{2} \tag{5.16}$$

multiplicative steps. The calculation in (5.15) requires

$$1 + 2 + \cdots + (n-1) = \frac{n(n-1)}{2} \tag{5.17}$$

multiplicative steps. Thus, once the factorization $A = LU$ has been obtained, solving the system (5.11) requires $\frac{n(n+1)}{2} + \frac{n(n-1)}{2} = n^2$ multiplicative steps. Including the steps required to factor A, the number of multiplicative steps required to solve the system (5.11) is

$$\frac{n(n^2-1)}{3} + n^2 = \frac{n(n^2 + 3n - 1)}{3}. \tag{5.18}$$

Notice that this is precisely formula (6.14) of Chapter 1, the number of multiplicative steps required for Gaussian elimination as described in Section 1–6.

The connection between Gaussian elimination and solving a system of equations by obtaining a lower-upper triangular factorization is closer than the mere observation that they involve the same number of arithmetic steps. They involve exactly the same arithmetic steps. To see this, suppose we look at the Gaussian elimination method applied to the augmented matrix $[A \ B]$. Assuming $a_{11} \neq 0$, we pivot on a_{11}. After elimination of x_1, the new constant terms are

$$\begin{aligned} b_1' &= a_{11}^{-1} b_1 = l_{11}^{-1} b_1 \\ b_2' &= b_2 - a_{21} b_1' \\ b_3' &= b_3 - a_{31} b_1' \\ &\vdots \\ b_n' &= b_n - a_{n1} b_1'. \end{aligned} \tag{5.19}$$

Assume $a_{22}' \neq 0$ and proceed with the second elimination. We obtain

$$\begin{aligned}
b_1'' &= b_1' \\
b_2'' &= a_{22}'^{-1} b_2' = l_{22}^{-1}(b_2 - l_{21} b_1') \\
b_3'' &= b_3' - a_{32}' b_2'' = b_3 - l_{31} b_1' - l_{32} b_2'' \\
&\quad \cdot \\
&\quad \cdot \\
&\quad \cdot \\
b_n'' &= b_n' - a_{n2}' b_2'' = b_n - l_{n1} b_1' - l_{n2} b_2''.
\end{aligned} \qquad (5.20)$$

Comparison between (5.20) and (5.14) should make the emerging pattern clear, but let us make one more elimination. Assume $a_{33}'' \neq 0$ and pivot on a_{33}''. We obtain

$$\begin{aligned}
b_1''' &= b_1'' = b_1' \\
b_2''' &= b_2'' \\
b_3''' &= a_{33}''^{-1} b_3'' = l_{33}^{-1}(b_3 - l_{31} b_1' - l_{32} b_2'') \\
&\quad \cdot \\
&\quad \cdot \\
&\quad \cdot \\
b_n''' &= b_n'' - a_{n3}'' b_3''' = b_n - l_{n1} b_1' - l_{n2} b_2'' - l_{n3} b_3'''.
\end{aligned} \qquad (5.21)$$

It is not difficult to see that this will generate the solution given in (5.14). From this, the substitution in the backward course gives the solution in (5.15).

For solving a single system of linear equations there is no difference in the work required whether we use Gaussian elimination or the lower-upper triangular factorization. But if we wish to solve several systems, all with the same coefficient matrix, then each additional solution can be obtained with the n^2 steps required in (5.14) and (5.15).

One might ask, "If we intend to solve many systems of equations, all with the same coefficient matrix, wouldn't it be better to obtain A^{-1}? Then for each system $AX = B$, the solution is merely $X = A^{-1}B$." This is a reasonable question. But it takes n^2 multiplicative steps to compute $A^{-1}B$, just the same number as it takes to solve both (5.14) and (5.15). Finding the inverse of A requires n^3 multiplicative steps, and computing the lower-upper triangular factorization requires less than $\frac{n^3}{3}$ steps.

Other factorizations are possible. For example, a square matrix A (subject to assumptions about being able to carry out all the steps indicated) can be written in the form $A = UL$, where U is upper triangular and L is lower triangular with 1's in the main diagonal. This would amount to using Gaussian elimination starting with the lower right corner.

We can also write A in the form $A = LU$, where L is lower triangular with 1's along the main diagonal and U is upper triangular. This could be obtained from the lower-upper triangular factorization already given with n^2 additional multiplicative steps. However, it is more instructive to see how to obtain it independently.

Assume $a_{11} \neq 0$ so that we can pivot on a_{11}. However, we do not divide the first row by a_{11}. For the first type II operation, we compute

$$m_{i1} = a_{i1}/a_{11}. \tag{5.22}$$

We then add $-m_{i1}$ times row 1 to row i. When these operations are completed we obtain

$$A_1 = \begin{bmatrix} a_{11} & a_{12} & a_{13} & \cdots & a_{1n} \\ 0 & a_{22}' & a_{23}' & \cdots & a_{2n}' \\ \cdot & \cdot & \cdot & \cdots & \cdot \\ \cdot & \cdot & \cdot & \cdots & \cdot \\ \cdot & \cdot & \cdot & \cdots & \cdot \\ 0 & a_{m2}' & a_{m2}' & \cdots & a_{mn}' \end{bmatrix} \tag{5.23}$$

The matrix Q_1, a product of elementary matrices described in Section 3-4, such that $A_1 = Q_1 A$ is

$$Q_1 = \begin{bmatrix} 1 & 0 & 0 & \cdots & 0 \\ -m_{21} & 1 & 0 & \cdots & 0 \\ -m_{31} & 0 & 1 & \cdots & 0 \\ \cdot & \cdot & \cdot & \cdots & \cdot \\ \cdot & \cdot & \cdot & \cdots & \cdot \\ \cdot & \cdot & \cdot & \cdots & \cdot \\ -m_{m1} & 0 & 0 & \cdots & 1 \end{bmatrix} \tag{5.24}$$

The inverse of Q_1 can be obtained without further calculations.

$$Q_1^{-1} = \begin{bmatrix} 1 & 0 & 0 & \cdots & 0 \\ m_{21} & 1 & 0 & \cdots & 0 \\ m_{31} & 0 & 1 & \cdots & 0 \\ \cdot & \cdot & \cdot & \cdots & \cdot \\ \cdot & \cdot & \cdot & \cdots & \cdot \\ \cdot & \cdot & \cdot & \cdots & \cdot \\ m_{m1} & 0 & 0 & \cdots & 1 \end{bmatrix}. \tag{5.25}$$

A comparison of (5.23) with (5.7), and of (5.25) with (5.2), should reveal what is going to develop. The first column of (5.7) contains the elements of the first column of A, and the first row of (5.2) contains the ratios a_{1j}/a_{11}. The first row of (5.23) contains the elements of the first row of A, and the first column of (5.25) contains the ratios a_{i1}/a_{11}. The subsequent steps of the process are similar to those already described. The end result will be a factorization of A into a product $A = LU$. But in this factorization the diagonal elements of L will be 1's. Conceptually, the difference is that this time we postpone the elementary operations of type I to the backward course. The number of arithmetic operations is the same for either factorization.

In the discussion of the lower-upper triangular factorization, we assumed that $a_{11} \neq 0$ to obtain (5.2). Then we assumed $a_{22}' \neq 0$ to obtain (5.4), and so on. We want to consider what can be done without these assumptions.

Actually, no matter how the pivot elements are chosen a factorization can be obtained. If A' is the matrix obtained after the elementary row operations are performed, and Q is the product of the corresponding elementary matrices, then $A = PA'$. If the elementary row operations operate only on rows not containing a previously selected pivot element, the matrix P can be written down without further calculations. Such a factorization is just as easy to obtain as the lower-upper triangular factorization, and it can be used to solve systems of linear equations in the same way.

To illustrate the procedures described here we will give a few numerical examples. First let us factor a square matrix of order 4 into a product of a lower triangular matrix with 1's along the main diagonal and an upper triangular matrix. We start by writing A in the form $A = IA$ and develop L as we develop U. Follow the steps.

3-5 A Useful Factorization of a Matrix

$$A = \begin{bmatrix} 1 & 0 & 0 & 0 \\ 0 & 1 & 0 & 0 \\ 0 & 0 & 1 & 0 \\ 0 & 0 & 0 & 1 \end{bmatrix} \begin{bmatrix} 7 & 3 & -3 & 5 \\ 3 & 3 & 1 & -3 \\ -5 & 3 & 5 & 1 \\ 3 & -1 & 5 & 5 \end{bmatrix}$$

$$= \begin{bmatrix} 1 & 0 & 0 & 0 \\ \frac{3}{7} & 1 & 0 & 0 \\ -\frac{5}{7} & 0 & 1 & 0 \\ \frac{3}{7} & 0 & 0 & 1 \end{bmatrix} \begin{bmatrix} 7 & 3 & -3 & 5 \\ 0 & \frac{12}{7} & \frac{16}{7} & -\frac{36}{7} \\ 0 & \frac{36}{7} & \frac{20}{7} & \frac{32}{7} \\ 0 & -\frac{16}{7} & \frac{44}{7} & \frac{20}{7} \end{bmatrix}$$

$$= \begin{bmatrix} 1 & 0 & 0 & 0 \\ \frac{3}{7} & 1 & 0 & 0 \\ -\frac{5}{7} & 3 & 1 & 0 \\ \frac{3}{7} & -\frac{4}{3} & 0 & 1 \end{bmatrix} \begin{bmatrix} 7 & 3 & -3 & 3 \\ 0 & \frac{12}{7} & \frac{16}{7} & -\frac{36}{7} \\ 0 & 0 & -4 & 20 \\ 0 & 0 & \frac{28}{3} & -4 \end{bmatrix}$$

$$= \begin{bmatrix} 1 & 0 & 0 & 0 \\ \frac{3}{7} & 1 & 0 & 0 \\ -\frac{5}{7} & 3 & 1 & 0 \\ \frac{3}{7} & -\frac{4}{3} & -\frac{7}{3} & 1 \end{bmatrix} \begin{bmatrix} 7 & 3 & -3 & 5 \\ 0 & \frac{12}{7} & \frac{16}{7} & -\frac{36}{7} \\ 0 & 0 & -4 & 20 \\ 0 & 0 & 0 & \frac{128}{3} \end{bmatrix} \quad (5.26)$$

Let us give an example where we run out of non-zero pivot elements.

$$A = \begin{bmatrix} 1 & 0 & 0 & 0 \\ 0 & 1 & 0 & 0 \\ 0 & 0 & 1 & 0 \\ 0 & 0 & 0 & 1 \end{bmatrix} \begin{bmatrix} 1 & 2 & 3 & 4 \\ 5 & 6 & 7 & 8 \\ 7 & 6 & 5 & 4 \\ 3 & 2 & 1 & 0 \end{bmatrix}$$

$$= \begin{bmatrix} 1 & 0 & 0 & 0 \\ 5 & 1 & 0 & 0 \\ 7 & 0 & 1 & 0 \\ 3 & 0 & 0 & 1 \end{bmatrix} \begin{bmatrix} 1 & 2 & 3 & 4 \\ 0 & -4 & -8 & -12 \\ 0 & -8 & -16 & -24 \\ 0 & -4 & -8 & -12 \end{bmatrix} \quad (5.27)$$

$$= \begin{bmatrix} 1 & 0 & 0 & 0 \\ 5 & 1 & 0 & 0 \\ 7 & 2 & 1 & 0 \\ 3 & 1 & 0 & 1 \end{bmatrix} \begin{bmatrix} 1 & 2 & 3 & 4 \\ 0 & -4 & -8 & -12 \\ 0 & 0 & 0 & 0 \\ 0 & 0 & 0 & 0 \end{bmatrix}$$

142 • Bases and Coordinate Systems

Finally, let us give an example where neither factor is in triangular form. We will obtain 1's in the pivot positions of the second factor.

$$A = \begin{bmatrix} 1 & 0 & 0 & 0 \\ 0 & 1 & 0 & 0 \\ 0 & 0 & 1 & 0 \\ 0 & 0 & 0 & 1 \end{bmatrix} \begin{bmatrix} 3 & -4 & 7 & -3 \\ 2 & 2 & 8 & 2 \\ -1 & 2 & -3 & 1 \\ -2 & 12 & -3 & 6 \end{bmatrix}$$

$$= \begin{bmatrix} 1 & 0 & 3 & 0 \\ 0 & 1 & 2 & 0 \\ 0 & 0 & -1 & 0 \\ 0 & 0 & -2 & 1 \end{bmatrix} \begin{bmatrix} 0 & 2 & -2 & 0 \\ 0 & 6 & 2 & 4 \\ 1 & -2 & 3 & -1 \\ 0 & 8 & 3 & 4 \end{bmatrix}$$

$$= \begin{bmatrix} 2 & 0 & 3 & 0 \\ 6 & 1 & 2 & 0 \\ 0 & 0 & -1 & 0 \\ 8 & 0 & -2 & 1 \end{bmatrix} \begin{bmatrix} 0 & 1 & -1 & 0 \\ 0 & 0 & 8 & 4 \\ 1 & -2 & 3 & -1 \\ 0 & 0 & 11 & 4 \end{bmatrix} \qquad (5.28)$$

$$= \begin{bmatrix} 2 & 0 & 3 & 0 \\ 6 & 4 & 2 & 0 \\ 0 & 0 & -1 & 0 \\ 8 & 4 & -2 & 1 \end{bmatrix} \begin{bmatrix} 0 & 1 & -1 & 0 \\ 0 & 0 & 2 & 1 \\ 1 & -2 & 3 & -1 \\ 0 & 0 & 3 & 0 \end{bmatrix}$$

$$= \begin{bmatrix} 2 & 0 & 3 & 0 \\ 6 & 4 & 2 & 0 \\ 0 & 0 & -1 & 0 \\ 8 & 4 & -2 & 3 \end{bmatrix} \begin{bmatrix} 0 & 1 & -1 & 0 \\ 0 & 0 & 2 & 1 \\ 1 & -2 & 3 & -1 \\ 0 & 0 & 1 & 0 \end{bmatrix}$$

These factorizations should be checked by carrying out the indicated multiplications.

EXERCISES 3–5

1. How many multiplicative steps are required to find the inverse of a lower triangular matrix with 1's in the main diagonal?

2. How many multiplicative steps are required to find the inverse of an upper triangular matrix with non-zero elements in the main diagonal?

3. Show that the inverse of a lower triangular matrix is also lower triangular. Show that the inverse of an upper triangular matrix is also upper triangular. Show that for an

invertible triangular matrix the inverse has in its main diagonal the inverses of the elements in the main diagonal of the given matrix.

4. Determine the number of multiplicative steps required to compute the product of a lower triangular matrix with 1's in the main diagonal and an upper triangular matrix.

5. Formula (5.10) gives the number of multiplicative steps required to obtain the factorization $A = LU$. Use this information and the answers to the previous exercises to find out how many steps are required to find the inverse of a matrix by computing $A^{-1} = U^{-1}L^{-1}$.

Factor the following matrices.

6. $\begin{bmatrix} 2 & 4 & -6 \\ -3 & -2 & 1 \\ 4 & 9 & -16 \end{bmatrix}$

7. $\begin{bmatrix} 0 & 2 & 3 \\ 2 & -1 & -8 \\ -4 & 0 & 17 \end{bmatrix}$

8. $\begin{bmatrix} 2 & -4 & 6 & -8 \\ 3 & -2 & 1 & -4 \\ -2 & 2 & 1 & -2 \\ -3 & 3 & -1 & 5 \end{bmatrix}$

9. Use the factorization obtained in (5.26) to solve the system of linear equations,

$$7x_1 + 3x_2 - 3x_3 + 5x_4 = -3$$
$$3x_1 + 3x_2 + x_3 - 3x_4 = -7$$
$$-5x_1 + 3x_2 + 5x_3 + x_4 = -3$$
$$3x_1 - x_2 + 5x_3 + 5x_4 = 21.$$

10. Use the factorization obtained in (5.27) to solve the system

$$x_1 + 2x_3 + 3x_3 + 4x_4 = -2$$
$$5x_1 + 6x_2 + 7x_3 + 8x_4 = -2$$
$$7x_1 + 6x_2 + 5x_3 + 4x_4 = 2$$
$$3x_1 + 2x_2 + x_3 = 2.$$

11. Use the factorization obtained in (5.28) to solve the system

$$3x_1 - 4x_2 + 7x_3 - 3x_4 = 1$$
$$2x_1 + 2x_2 + 8x_3 + 2x_4 = -6$$
$$-x_1 + 2x_2 - 3x_3 + x_4 = 1$$
$$-2x_1 + 12x_2 - 3x_3 + 6x_4 = -1.$$

12. Find the inverses of the following lower-triangular matrices.

a) $\begin{bmatrix} 1 & 0 & 0 \\ a & 1 & 0 \\ b & 0 & 1 \end{bmatrix}$

b) $\begin{bmatrix} 1 & 0 & 0 & 0 \\ a & 1 & 0 & 0 \\ b & 0 & 1 & 0 \\ c & 0 & 0 & 1 \end{bmatrix}$

c) $\begin{bmatrix} 1 & 0 & 0 \\ a & 1 & 0 \\ b & c & 1 \end{bmatrix}$

d) $\begin{bmatrix} 1 & 0 & 0 & 0 \\ a & 1 & 0 & 0 \\ b & d & 1 & 0 \\ c & e & f & 1 \end{bmatrix}$

e) $\begin{bmatrix} a & 0 & 0 \\ b & d & 0 \\ c & e & f \end{bmatrix}$ where $adf \neq 0$.

13. Find the inverse of $A = \begin{bmatrix} 1 & 0 & 0 \\ 2 & 1 & 0 \\ 3 & 4 & 1 \end{bmatrix}$.

14. Find the inverse of $B = \begin{bmatrix} 1 & 6 & -7 \\ 0 & 1 & 5 \\ 0 & 0 & 1 \end{bmatrix}$.

15. Let A and B be the matrices given in Exercises 13 and 14. Find the inverse of $AB = \begin{bmatrix} 1 & 6 & -7 \\ 2 & 13 & -9 \\ 3 & 22 & 0 \end{bmatrix}$.

16. Find the inverse of $A = \begin{bmatrix} 1 & 0 & 0 & 0 \\ -2 & 1 & 0 & 0 \\ 1 & 2 & 1 & 0 \\ -3 & 1 & -1 & 1 \end{bmatrix}$.

17. Find the inverse of $B = \begin{bmatrix} 1 & -1 & 1 & -3 \\ 0 & 1 & 2 & 1 \\ 0 & 0 & 1 & -2 \\ 0 & 0 & 0 & 1 \end{bmatrix}$.

18. Let A and B be the matrices given in Exercises 16 and 17.

$$\text{Find the inverse of } AB = \begin{bmatrix} 1 & -1 & 1 & -3 \\ -2 & 3 & 0 & 7 \\ 1 & 1 & 6 & -3 \\ -3 & 4 & -2 & 13 \end{bmatrix}.$$

Summary

For a finite dimensional vector space, a basis permits establishing a bijection between the vector space and a coordinate space of the same dimension. If $\mathscr{A} = \{\alpha_1, \ldots, \alpha_n\}$ is a basis of $\mathscr{V}$ and $\alpha \in \mathscr{V}$ can be written in the form $\alpha = x_1\alpha_1 + \cdots + x_n\alpha_n$, then $X = (x_1, \ldots, x_n)$ *represents* α. Vector addition and scalar multiplication in $\mathscr{V}$ are also represented by these same operations with the representatives in $\mathscr{F}^n$.

If $\mathscr{B} = \{\beta_1, \ldots, \beta_n\}$ is a new basis of $\mathscr{V}$, and $\beta_j = \sum_{i=1}^{n} p_{ij}\alpha_i$, then $P = [p_{ij}]$ is the *matrix of transition* from the basis $\mathscr{A}$ to the basis $\mathscr{B}$. If α is represented by $Y = (y_1, \ldots, y_n)$ with respect to the basis $\mathscr{B}$, then

$$X = PY. \tag{S.1}$$

This is the fundamental formula for a change of coordinates. However, this formula is actually seldom used to change coordinates. We usually write the representations of several vectors in the columns of a matrix A. Then elementary row operations are used to obtain a matrix A' with certain desired properties.

$$A' = P^{-1}A, \tag{S.2}$$

where P is the matrix of transition given above. Use of elementary row operations is a method of changing coordinates. The matrix P^{-1} or P need not be obtained explicitly, but P^{-1} can be obtained by performing the same elementary row operations on the identity matrix.

A useful by-product of the change-of-coordinates methods occurs when A' is obtained in the form of an upper triangular matrix and P is in the form of a lower triangular matrix. Then

$$A = PA' = LU \tag{S.3}$$

is a factorization of A into the product of a lower triangular matrix and an upper triangular matrix.

4
LINEAR TRANSFORMATIONS

Generally, in mathematics one is concerned with functions, or mappings, that are related to the special properties of the sets under consideration. For example, in calculus the central idea is the limit concept. In this case functions with the property

$$\lim_{x \to a} f(x) = f(\lim_{x \to a} x) = f(a) \qquad (0.1)$$

are especially important. A function with the property stated in (0.1) is said to be *continuous* at a. A continuous function "preserves" the limit. That is, the order of taking the limit and evaluating the function can be interchanged. This is the reason for our being interested in continuous functions in calculus.

In algebra we deal with sets in which some kind of algebraic operation is defined. Groups, rings, and fields are names of some types of algebraic structures that are particularly important. In this text we are concerned with vector spaces. In all these subjects, functions that preserve the algebraic structure are of central concern. Generally, such functions are called *homomorphisms*. Homomorphisms of vector spaces are called *linear transformations*, or *linear mappings*. This chapter is concerned with linear transformations.

4-1 LINEAR TRANSFORMATIONS

Definition *Let $\mathcal{U}$ and $\mathcal{V}$ be any two vector spaces over the same field $\mathcal{F}$. Let σ be a function defined on $\mathcal{U}$ with values in $\mathcal{V}$. We say that σ is a **linear transformation**, or a **linear mapping**, if it has the following properties:*

$$\sigma(\alpha + \beta) = \sigma(\alpha) + \sigma(\beta), \qquad (1.1)$$

$$\sigma(a\alpha) = a\sigma(\alpha), \qquad (1.2)$$

where a is an arbitrary scalar and α and β are vectors in $\mathcal{U}$. In equation (1.1) the addition on the left is in $\mathcal{U}$ and that on the right is in $\mathcal{V}$. In equation (1.2) the scalar multiplication on the left is in $\mathcal{U}$ and that on the right is in $\mathcal{V}$.

At first it may seem that a linear transformation is a rather esoteric concept, but in truth it is surprisingly useful and it is almost omnipresent in applied mathematics. We shall mention a few examples of these applications to emphasize the importance of this concept. Some of the most striking examples of linear transformations occur in connection with infinite dimensional vector spaces. Although infinite dimensional vector spaces involve problems we do not wish to consider in this text, we shall give some examples that use infinite dimensional spaces.

It is not entirely inappropriate to give such examples since much of our discussion will apply to vector spaces in general, including infinite dimensional spaces. It is the techniques that are specific to the finite dimensional case that do not generalize.

(1) In calculus one proves that if f and g are differentiable functions and a is a real number, then $f + g$ and af are also differentiable and

$$\frac{d}{dx}(f+g) = \frac{df}{dx} + \frac{dg}{dx}, \quad \frac{d(af)}{dx} = a\frac{df}{dx}. \qquad (1.3)$$

Proving that these relations are valid amounts to showing that the set of all differentiable functions is a vector space, and that the derivative is a linear mapping. In this case $\mathcal{U}$ is the set of differentiable functions and $\mathcal{V}$ is the set of functions.

(2) If f and g are integrable functions and a is a real number, then $f + g$ and af are also integrable and

$$\int (f+g)\,dx = \int f\,dx + \int g\,dx, \quad \int af\,dx = a\int f\,dx. \tag{1.4}$$

This says that the set of all integrable functions is a vector space, and that the integral is a linear mapping. Here, $\mathcal{U}$ is the set of integrable functions. What $\mathcal{V}$ is depends on what kind of integral we have under consideration. In elementary calculus, for the indefinite integral $\mathcal{V}$ is the set of continuous functions. For the definite integral $\mathcal{V}$ is the set of real numbers.

(3) Let $\mathcal{U}$ be the set of real-valued functions of a real variable, and let $\mathcal{V} = \mathcal{R}$, the real numbers. Select and fix a real number p. Define σ by the rule $\sigma(f) = f(p)$. That is, $\sigma(f)$ is just the value of f at p. It is easily seen that

$$\begin{aligned}\sigma(f+g) &= (f+g)(p) = f(p) + g(p) = \sigma(f) + \sigma(g),\\ \sigma(af) &= (af)(p) = a \cdot f(p) = a\sigma(f).\end{aligned} \tag{1.5}$$

This shows that σ is a linear mapping. Although it may appear to be simple, this kind of linear mapping is surprisingly important and useful.

(4) The following example is a little more complex and the reader may have to accept some statements at face value. Let $\mathcal{S} = \{1, \ldots, n\}$ be a finite set. Under certain circumstances, let $p(k)$ be the probability that k is selected from $\mathcal{S}$, and let $X(k)$ be the value of the event if k is selected. This means $p(k) \geq 0$ and $\sum_{k=1}^{n} p(k) = 1$. The *mean value* of X is defined to be

$$M(X) = \sum_{k=1}^{n} X(k)p(k). \tag{1.6}$$

In probability theory the function X is called a *random variable*, $\mathcal{S}$ is called the *sample space* of X, and p is called the *probability function* for X. X is a function on $\mathcal{S}$. We wish to make a vector space out of the set of random variables. The complication is that it is desirable to permit each random variable to have its own sample space and probability function. Thus, let $\mathcal{T} = \{1, \ldots, m\}$ be another finite set, let $q(k)$ be the probability that k is selected from $\mathcal{T}$, and let $Y(k)$ be the value of the event if k is selected. Then $M(Y) = \sum_{k=1}^{m} Y(k)q(k)$.

To evaluate $X + Y$ we must select $s \in \mathcal{S}$ and $t \in \mathcal{T}$, and compute $X(s) + Y(t)$. This means we must determine the probability of selecting both s and t. If s and t can be selected independently, this probability is $p(s)q(t)$. But, generally, these choices might not be independent. In this case let $p(s, t)$ denote the probability of selecting $s \in \mathscr{S}$ and $t \in \mathscr{T}$. Then,

$$M(X+Y) = \sum_{s=1}^{n} \sum_{t=1}^{m} [X(s) + Y(t)] p(s, t)$$

$$= \sum_{s=1}^{n} \sum_{t=1}^{m} X(s) p(s, t) + \sum_{s=1}^{n} \sum_{t=1}^{m} Y(t) p(s, t) \qquad (1.7)$$

$$= \sum_{s=1}^{n} X(s) \sum_{t=1}^{m} p(s, t) + \sum_{t=1}^{m} Y(t) \sum_{s=1}^{n} p(s, t).$$

Now $\sum_{t=1}^{m} p(s, t)$ is the probability of selecting a certain value of s and any value of t. Thus, this sum is just $p(s)$. Similarly, $\sum_{s=1}^{n} p(s, t) = q(t)$. Thus, we have

$$M(X + Y) = \sum_{s=1}^{n} X(s) p(s) + \sum_{t=1}^{m} Y(t) q(t) \qquad (1.8)$$
$$= M(X) + M(Y).$$

For any scalar a we have

$$M(aX) = \sum_{s=1}^{n} (aX)(s) p(s) = \sum_{s=1}^{n} aX(s) p(s) = aM(X). \qquad (1.9)$$

The sum of two random variables is a random variable whose sample space is the set of all pairs (s, t) and the set of all random variables with finite sample spaces is a vector space. M is a linear mapping of the set of random variables into the set of real numbers.

(5) In communication theory, information theory, and control theory, "linear devices" play an important part. Here, $\mathscr{U}$ and $\mathscr{V}$ are sets of functions of time, called signals. A linear device, sometimes called a "black box," is an electrical network, or mechanical system, etc., which operates on the signals in $\mathscr{U}$ called inputs, and produces signals in $\mathscr{V}$, called outputs, and satisfies the linearity conditions of (1.1) and (1.2). Physical devices are, in fact, usually not linear. For example, they will usually not handle arbitrarily large inputs so the

input signals cannot be multiplied by arbitrarily large scalars. The analysis of non-linear devices is extremely difficult. By contrast, a very extensive and satisfactory theory exists for linear devices.

The success of the theory of linear devices stems from the following considerations. One can establish the existence of a restricted collection of signals with the property that every other signal under consideration is a linear combination of the signals in this collection. (This collection corresponds to the concept of a basis that has been under consideration in this text. In an infinite dimensional vector space the situation is more complicated, and it is more difficult to establish the existence of this collection. This is what the theory of Fourier series, for example, is all about.) Then if one can see how the device responds to a signal in the restricted collection, the linearity property allows one to see how the device responds to any other signal.

(6) Many of the differential equations of mathematical physics are linear in that the operators that appear in them are linear in the sense of the definition given here. The linearity of the equation allows one to solve complicated problems by the "principle of superposition." This means that one solves relatively simple problems in detail. Then a more complicated situation is solved by superimposing the solutions of the several parts. For example, in the theory of gravitational attraction one determines the gravitational field of a single isolated particle. Then the field of a collection of particles is the vector sum of the fields of each particle separately. The validity of this principle and linearity of the differential equation are intimately tied together.

(7) Suppose in the analysis of an economic model we are considering the effects of the levels of activities of several manufacturers on the resources they use. For example, the manufacture of automobiles requires the use of steel, copper, plastics, and other materials. The amounts of these materials required depends on the number of automobiles manufactured. Similarly, the manufacture of television sets, or ships, requires amounts of these and other materials. Suppose x_i is a number that indicates the level of the i-th industry. Then $X = (x_1, \ldots, x_n)$ can be used to represent the collective activities of n industries. Similarly, let y_j indicate the demand for the j-th resource. Then $Y = (y_1, \ldots, y_m)$ can be used to indicate the collective demand for m resources. In the situation we are considering, we have $Y \in \mathscr{R}^m$ as a function of

$X \in \mathcal{R}^n$. It is natural to assume this functional relationship is linear. That would simply mean that if the activities are added the demands are added, and if the activities are scaled up, the demands will be scaled up by the same ratio.

Virtually every branch of applied mathematics has a linear theory and a non-linear theory. All the linear theories are closely related, and the ideas that we will study in linear algebra apply to all of them. The central concept is that of a linear transformation.

EXERCISES 4-1

1. Which of the following define linear transformations from $\mathcal{R}^3$ into $\mathcal{R}^2$ or $\mathcal{R}^3$?
   a) $\sigma(x_1, x_2, x_3) = (x_1 - x_2, x_3 + x_2)$
   b) $\sigma(x_1, x_2, x_3) = (3x_2, 5x_1 - x_3)$
   c) $\sigma(x_1, x_2, x_3) = (x_1, x_2)$
   d) $\sigma(x_1, x_2, x_3) = (x_2, x_1, 0)$
   e) $\sigma(x_1, x_2, x_3) = (x_2, 1, 0)$
   f) $\sigma(x_1, x_2, x_3) = (x_2, \log x_1, x_2 - x_3)$
   g) $\sigma(x_1, x_2, x_3) = (x_1 + x_2 + x_3, 2x_1 - x_2 + x_3, -x_1 + 3x_2 + 5x_3)$
   h) $\sigma(x_1, x_2, x_3) = (x_1 x_2, x_2 x_3, x_3 x_1)$

2. Many students in elementary mathematics courses make mistakes by treating all functions as if they were linear. Prove that the following functions are *not* linear.
   a) $f(x) = x^2$. Give the correct expression for $f(x + y)$.
   b) $f(x) = \sin x$. Give the correct expressions for $f(x + y)$ and $f(2x)$.
   c) $f(x) = \log x$. Give the correct expression for $f(3x)$.
   d) $f(x) = \tan x$. Give the correct expression for $f(x + y)$.

3. Let D be a differential operator defined as $D(f) = -\dfrac{d^2 f}{dx^2}$, where f is a twice differentiable real valued function. Show that D is a linear mapping of $\mathcal{U}$, the set of twice differentiable real valued functions, into $\mathcal{V}$, the set of real valued functions.

4. Let D be defined by $D(f) = \dfrac{d^2 f}{dx^2} + \dfrac{1}{x} \dfrac{df}{dx} + f$. Show that D is a linear mapping of $\mathcal{U}$ into $\mathcal{V}$, where $\mathcal{U}$ and $\mathcal{V}$ are as specified in Exercise 3.

5. Let $\mathscr{U}$ be a vector space over a field $\mathscr{F}$. For each scalar $a \in \mathscr{F}$, define $\sigma_a(\alpha) = a\alpha$ for all $\alpha \in \mathscr{U}$. Show that σ_a is a linear transformation of $\mathscr{U}$ into itself. σ_a defined in this way is called a *scalar transformation*.

6. Let $\mathscr{U}$ be the set of integrable real valued functions defined on the interval $[0, 1]$. Show that $M(f) = \int_0^1 f(x)\, dx$ defines a linear transformation of $\mathscr{U}$ into $\mathscr{R}$, the field of real numbers considered as a vector space over itself.

7. Let $\mathscr{U}$ be a 3-dimensional vector space over any field $\mathscr{F}$. Let $\mathscr{A} = \{\alpha_1, \alpha_2, \alpha_3\}$ be any basis of $\mathscr{U}$ over $\mathscr{F}$ which will be kept fixed. For any $\alpha \in \mathscr{U}$, represent α as a linear combination of the basis vectors, $\alpha = a_1\alpha_1 + a_2\alpha_2 + a_3\alpha_3$. Define the function σ by the rule, $\sigma(\alpha) = a_1\alpha_1$. Show that σ is a linear transformation of $\mathscr{U}$ into itself.

8. With $\mathscr{U}$ and $\mathscr{A}$ given as in Exercise 7, let τ be defined by the rule; for $\alpha = a_1\alpha_1 + a_2\alpha_2 + a_3\alpha_3$, let $\tau(\alpha) = (a_1 + a_2 + a_3)\alpha_1$. Show that τ is a linear transformation of $\mathscr{U}$ into itself.

9. With $\mathscr{U}$ and $\mathscr{A}$ given as in Exercise 7, let ϕ be defined by the rule; for $\alpha = a_1\alpha_1 + a_2\alpha_2 + a_3\alpha_3$, let $\phi(\alpha) = a_1$. Show that ϕ is a linear transformation of $\mathscr{U}$ into $\mathscr{F}$.

10. Let $\mathscr{U}$ be the set of 1-column 3-row matrices with real elements. Let S be a given 3×3 real matrix. Define the function F by the rule; for $A \in \mathscr{U}$, $F(A) = SA$. Show that F is a linear transformation of $\mathscr{U}$ into itself.

11. Let $\mathscr{P}$ be the set of all polynomials in an indeterminate x with real coefficients. Define the function D by the rule; for any polynomial $p(x) = \sum_{i=0}^{n} a_i x^i$, let $D(p) = \sum_{i=0}^{n} i a_i x^{i-1}$. Show that D is a linear transformation of $\mathscr{P}$ into itself.

12. With $\mathscr{P}$ defined as in Exercise 11, define the function G by the rule; for $p(x) = \sum_{i=0}^{n} a_i x^i$, let $G(p) = \sum_{i=0}^{n} \frac{a_i}{i+1} x^{i+1}$. Show that G is a linear transformation of $\mathscr{P}$ into itself.

13. With $\mathscr{P}$ and D as defined in Exercise 11, define the function H by the rule: $H(p) = pD(p)$. Show that H is linear.

14. With the situation as described in Exercise 10, let
$$S = \begin{bmatrix} 1 & 2 & -1 \\ -1 & 3 & 4 \\ 2 & -1 & 2 \end{bmatrix}.$$
Evaluate $F(A) = SA$ for $A = (2, 1, 2)$.

15. With the situation as described in Exercise 10, let
$$S = \begin{bmatrix} 1 & -6 & 2 \\ -1 & -4 & 3 \\ 2 & -2 & -1 \end{bmatrix}.$$
Evaluate $F(A) = SA$ for $A = (2, 1, 2)$.

16. Let $\mathscr{U}$ and $\mathscr{V}$ be vector spaces over the same field $\mathscr{F}$. Let $\mathscr{A} = \{\alpha_1, \alpha_2, \ldots, \alpha_n\}$ be any basis of $\mathscr{U}$. Let $\{\beta_1, \ldots, \beta_n\}$ be any set of n vectors in $\mathscr{V}$. Show there is a linear transformation σ of $\mathscr{U}$ into $\mathscr{V}$ such that $\sigma(\alpha_j) = \beta_j$ for each j.

4-2 MATRIX REPRESENTATION OF LINEAR TRANSFORMATIONS

For the purpose of making actual computations in problems involving linear transformations it is necessary to represent the linear transformations by something appropriate for these calculations. Linear transformations between finite dimensional vector spaces can be conveniently and effectively represented by matrices. In infinite dimensional vector spaces other types of representations are used. Generally, that part of our discussion dealing with abstract linear transformations will generalize to infinite dimensional vector spaces while that part of our discussion concerning matrices will not.

Let $\mathscr{U}$ be a vector space of dimension n, and let $\mathscr{V}$ be a vector space of dimension m over the same field. Let σ be a linear transformation of $\mathscr{U}$ into $\mathscr{V}$. Let $\mathscr{A} = \{\alpha_1, \ldots, \alpha_n\}$ be any basis of $\mathscr{U}$, and let $\mathscr{B} = \{\beta_1, \ldots, \beta_m\}$ be any basis of $\mathscr{V}$. Each $\sigma(\alpha_j)$ is an element of $\mathscr{V}$ and has a representation in the form

$$\sigma(\alpha_j) = \sum_{i=1}^{m} a_{ij}\beta_i; \; j = 1, \ldots, n. \qquad (2.1)$$

These coefficients can be written in the form of a matrix

$$A = \begin{bmatrix} a_{11} & a_{12} & \cdots & a_{1n} \\ a_{21} & a_{22} & \cdots & a_{2n} \\ \cdot & \cdot & & \cdot \\ \cdot & \cdot & & \cdot \\ \cdot & \cdot & & \cdot \\ a_{m1} & a_{m2} & \cdots & a_{mn} \end{bmatrix} \qquad (2.2)$$

We say that A *represents* σ with respect to the bases $\mathscr{A}$ in $\mathscr{U}$ and $\mathscr{B}$ in $\mathscr{V}$.

The matrix representing σ is uniquely determined by σ since the coefficients in (2.1) are uniquely determined by $\sigma(\alpha_j)$. Conversely, every $m \times n$ matrix represents a linear transformation of $\mathscr{U}$ into $\mathscr{V}$. To see this, suppose the matrix in (2.2) is given. Then we can define $\sigma(\alpha_j)$ by means of the sum (2.1) for each α_j. Since $\mathscr{A}$ is a basis of $\mathscr{U}$, every $\xi \in \mathscr{U}$ can be represented uniquely in the form

$$\xi = \sum_{j=1}^{n} x_j \alpha_j. \qquad (2.3)$$

Since we hope to define a linear transformation, it will be necessary to have

$$\begin{aligned} \sigma(\xi) &= \sigma\left(\sum_{j=1}^{n} x_j \alpha_j\right) \\ &= \sum_{j=1}^{n} x_j \sigma(\alpha_j) \\ &= \sum_{j=1}^{n} x_j \left(\sum_{i=1}^{m} a_{ij} \beta_i\right) \\ &= \sum_{i=1}^{m} \left(\sum_{j=1}^{n} a_{ij} x_j\right) \beta_i. \end{aligned} \qquad (2.4)$$

The mapping defined by the sum (2.4) is actually linear. It is necessary only to carry through the computations in detail to show that it is.

We have now shown that when bases $\mathscr{A}$ in $\mathscr{U}$ and $\mathscr{B}$ in $\mathscr{V}$ are given there is a one-to-one correspondence between linear transformations and $m \times n$ matrices. Notice how the

4-2 Matrix Representation of Linear Transformations

properties of the bases enter into this correspondence. Since $\mathscr{B}$ spans $\mathscr{V}$ the representation (2.1), and hence (2.2), is possible. Since $\mathscr{B}$ is linearly independent the representation is unique. Since $\mathscr{A}$ spans $\mathscr{U}$ the representation (2.3) is possible and (2.4) can be defined for all ξ. Since $\mathscr{A}$ is linearly independent the coefficients (2.3) and (2.4) are unique, and hence σ is well defined.

There is much more to the idea of representing a linear transformation by a matrix than this one-to-one correspondence. The representing matrix can be used computationally. Consider the functional notation for the linear transformation in the form

$$\sigma(\xi) = \eta. \tag{2.5}$$

Because of formula (2.3), ξ is represented by the n-tuple $(x_1, \ldots, x_n)$. Let η be represented as a linear combination of the vectors in $\mathscr{B}$ in the form

$$\eta = \sum_{i=1}^{m} y_i \beta_i. \tag{2.6}$$

Then η is represented by the m-tuple $(y_1, \ldots, y_m)$. But (2.4) also gives a representation of η as a linear combination of the vectors in $\mathscr{B}$. Since $\mathscr{B}$ is a basis the representation is unique and

$$y_i = \sum_{j=1}^{n} a_{ij} x_j;\ i = 1, \ldots, m. \tag{2.7}$$

If we now let

$$X = \begin{bmatrix} x_1 \\ x_2 \\ \vdots \\ x_n \end{bmatrix}, \quad Y = \begin{bmatrix} y_1 \\ y_2 \\ \vdots \\ y_m \end{bmatrix}, \tag{2.8}$$

then formula (2.7) is equivalent to

$$AX = Y. \tag{2.9}$$

This is the matrix representation of the abstract formula (2.5).

156 • Linear Transformations

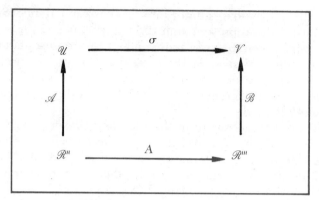

Figure 4–2.1 The matrix A representing σ is "filled in" in this mapping diagram so that the diagram commutes. That is, $\sigma \mathscr{A} = \mathscr{B} A$.

Mapping diagrams for the coordinatizations help here. In Figure 4–2.1, the arrow $\mathscr{A}$ represents the coordinatization of $\mathscr{U}$ with respect to the basis $\mathscr{A}$, the arrow σ represents the linear transformation of $\mathscr{U}$ into $\mathscr{V}$, and the arrow $\mathscr{B}$ represents the coordinatization of $\mathscr{V}$ with respect to the basis $\mathscr{B}$. These three mappings are represented by the following formulas,

$$\mathscr{A} X = \xi, \qquad (2.10)$$

$$\sigma(\xi) = \eta, \qquad (2.11)$$

$$\mathscr{B} Y = \eta. \qquad (2.12)$$

Combining these formulas, we get

$$\mathscr{B} Y = \eta = \sigma(\xi) = \sigma(\mathscr{A} X). \qquad (2.13)$$

The arrow A represents the mapping of $X \in \mathscr{R}^n$ onto $Y = AX \in \mathscr{R}^m$. It is to be "filled in" so that the diagram commutes. By this we mean that $\sigma \mathscr{A} = \mathscr{B} A$, the mapping from the lower-left corner to the upper-right corner is the same whether through $\mathscr{U}$ or through $\mathscr{R}^m$. With this condition we can extend formula (2.13) as follows,

$$\mathscr{B} Y = \sigma(\mathscr{A} X) = (\sigma \mathscr{A}) X = (\mathscr{B} A) X = \mathscr{B}(AX). \qquad (2.14)$$

The left side of (2.14) is the expression in formula (2.6) and the right side is the expression in formula (2.4). Just as there, since $\mathscr{B}$ is a basis, we can conclude that

4-2 Matrix Representation of Linear Transformations

$$Y = AX. \tag{2.9}$$

It is interesting and instructive to compare formula (2.6) in Chapter 3 with formula (2.9) of this section. In formula (2.6) (Chapter 3), X represents the given vector, P is the matrix of transition representing the change of basis, and Y represents the same vector with respect to the new coordinate system. The formula is

$$X = PY. \qquad 3\text{-}(2.6)$$

In formula (2.9), X represents the given vector, A represents a linear transformation, and Y represents the image vector. For the purpose of the comparison we wish to make, let us consider the case where the domain and co-domain are the same. That is, the linear transformation σ is a mapping of a vector space into itself. Then X and Y represent vectors with respect to the same basis.

The contrast we wish to point out is that X is "old" and Y is "new" in both formulas. But X appears on the left side of formula (2.6) in Chapter 3, and on the right side of formula (2.9). A good way to see the reason for this difference is to picture the vector space as a plane plastic sheet lying on a piece of paper on which is drawn a set of coordinate axes. In (2.6) (Chapter 3), the formula for a coordinate change, we are leaving the plane fixed and moving the coordinate system under the plane. In (2.9), the formula for a linear transformation, we are leaving the coordinate system fixed and moving the plane over the axis system. If, for example, the plane is rotated counterclockwise over the coordinate system, the same relative displacement would be achieved by rotating the coordinate system clockwise under the plane. In fact, we can see that the two formulas amount to the same thing if we take $A = P^{-1}$.

Usually, a linear transformation is not given abstractly, but is specified by giving the representing matrix A without reference to any basis in $\mathscr{U}$ or in $\mathscr{V}$. For example, suppose we are asked to consider the linear transformation represented by

$$\begin{bmatrix} 2 & 3 \\ 1 & -2 \end{bmatrix}. \tag{2.15}$$

We are expected to realize that this implies there is a basis $\mathscr{A} = \{\alpha_1, \alpha_2\}$ and a basis $\mathscr{B} = \{\beta_1, \beta_2\}$, and that $\sigma(\alpha_1) = 2\beta_1 + \beta_2$ and $\sigma(\alpha_2) = 3\beta_1 - 2\beta_2$. If we are told that A represents a linear transformation of a vector space $\mathscr{U}$ into itself, we are expected to take $\mathscr{A} = \mathscr{B}$.

Sometimes a linear transformation is specified in terms of the equations of the form (2.7). For example, through (2.9) the matrix in (2.15) is equivalent to

$$y_1 = 2x_1 + 3x_2$$
$$y_2 = x_1 - 2x_2. \qquad (2.16)$$

It is worth the effort to learn to interpret a matrix representation of a linear transformation in geometric terms. We shall illustrate what we have in mind by giving several examples of linear transformations of the real plane into itself. We shall take the basis vectors α_1, α_2 to be oriented as in the following figure.

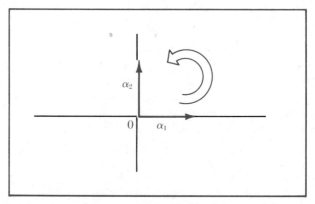

Figure 4–2.2   A coordinate system in $\mathscr{R}^2$. The location of the origin 0 is arbitrary. Two linearly independent vectors $\{\alpha_1, \alpha_2\}$ are selected for a basis. In this drawing they are shown mutually perpendicular and of length 1 to illustrate the geometric significance of some special linear transformations. The curved arrow indicates a sense of orientation or rotation determined by α_1 and α_2: start in the direction of α_1 and turn toward the side containing α_2.

α_1 and α_2 are drawn to be of length 1 and mutually perpendicular for the sake of illustration in familiar terms. Neither length nor angle has any meaning in the present context since all the definitions given so far do not involve these concepts.

(1) Consider the linear transformation σ represented by the matrix

$$\begin{bmatrix} -1 & 0 \\ 0 & 1 \end{bmatrix}. \qquad (2.17)$$

This means $\sigma(\alpha_1) = -\alpha_1$ and $\sigma(\alpha_2) = \alpha_2$. Equivalently, ex-

4–2 Matrix Representation of Linear Transformations • 159

pressed in terms of coordinates, this means σ sends $(1, 0)$ onto $(-1, 0)$ and it sends $(0, 1)$ onto $(0, 1)$. If there is any question where these interpretations come from, compare carefully (2.2) and (2.1). The first column of the representing matrix contains the coordinates of the image of the first basis vector under σ. The second column contains the coordinates of the image of the second basis vector, and so on. If we let $\alpha_1' = \sigma(\alpha_1)$ and $\alpha_2' = \sigma(\alpha_2)$, this means α_1' and α_2' are located as indicated in Figure 4–2.3.

The matrix (2.17) represents a *reflection*. Every point on one side of the vertical axis is mapped onto a point on the other side of the vertical axis, such that the line joining the point with its image is bisected by and perpendicular to the vertical axis. An examination of formula (2.7) or (2.9) shows that (x_1, x_2) is mapped onto $(-x_1, x_2)$. This is another way of seeing that (2.17) represents a reflection with respect to the vertical axis.

All the following examples start with the same figure, Figure 4.2.2. The result of the mapping represented by the given matrix will be shown in the same context as Figure 4.2.3.

(2) Consider the linear transformation σ represented by the matrix

$$\begin{bmatrix} 0 & 1 \\ 1 & 0 \end{bmatrix}. \tag{2.18}$$

This means $\sigma(\alpha_1) = \alpha_2$ and $\sigma(\alpha_2) = \alpha_1$. If we let $\alpha_1' = \sigma(\alpha_1)$

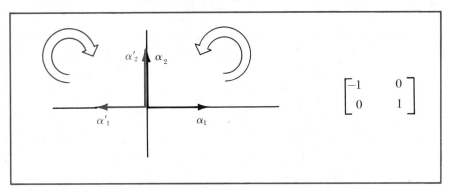

Figure 4–2.3 A reflection. This is a reflection in the vertical axis. Every point on the line of reflection is left fixed. Every point on one side of the line of reflection is mapped onto a point on the other side in such a way that the line joining the point with its image is bisected by and perpendicular to the line of reflection. The sense of orientation is reversed. The point $(1, 0)$ is mapped onto the point $(-1, 0)$, the first column of the representing matrix. The point $(0, 1)$ is left fixed, the second column of the representing matrix.

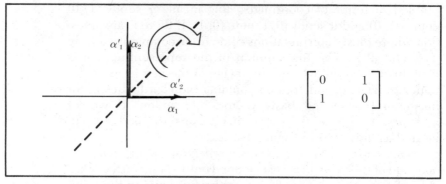

Figure 4-2.4 A reflection. The line of reflection is the 45° line, which is shown as a dotted line. The point (1, 0) is mapped onto the point (0, 1), the first column of the representing matrix. The point (0, 1) is mapped onto the point (1, 0), the second column of the representing matrix. The sense of orientation is reversed.

and $\alpha_2' = \sigma(\alpha_2)$, this means α_1' and α_2' are located as indicated in the above figure.

The matrix (2.18) represents a reflection in the 45° line, which is shown as a dotted line in Figure 4-2.4.

(3) Consider the linear transformation σ represented by the matrix

$$\begin{bmatrix} 1 & c \\ 0 & 1 \end{bmatrix}. \tag{2.19}$$

This means $\sigma(\alpha_1) = \alpha_1$ and $\sigma(\alpha_2) = c\alpha_1 + \alpha_2$. If we let $\alpha_1' = \sigma(\alpha_1)$ and $\alpha_2' = \sigma(\alpha_2)$, this means α_1' and α_1' are located as indicated in the following figure.

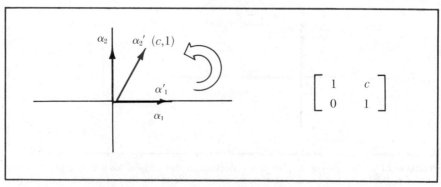

Figure 4-2.5 A shear. The point (1, 0) is fixed, and this gives the first column of the representing matrix. The point (0, 1) is mapped onto the point (c, 1), the second column of the representing matrix. The figure shows the case where $c > 0$. The sense of orientation is preserved.

In general, (x_1, x_2) is mapped onto $(x_1 + cx_2, x_2)$. Since the second coordinate of the image is the same as the second coordinate of the preimage, σ moves the point to the right (if cx_2 is positive) or to the left (if cx_2 is negative). cx_2 is the distance the point is moved, and this is proportional to x_2. To illustrate this mapping, imagine that we are looking at one of the edges of a deck of playing cards and that we slide the cards relative to each other so that the edge (as a whole) takes on the appearance of a parallelogram. Each card is moved a distance that is proportional to its height from the bottom of the deck. This kind of mapping is called a *shear*.

(4) Consider the linear transformation σ represented by the matrix.

$$\begin{bmatrix} \cos\theta & -\sin\theta \\ \sin\theta & \cos\theta \end{bmatrix}. \tag{2.20}$$

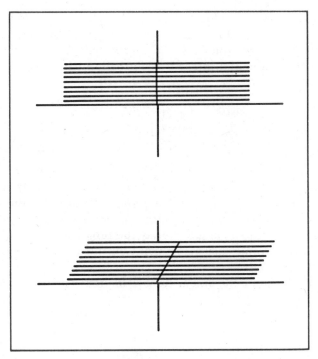

Figure 4–2.6 The shear given in Figure 4–2.5. The horizontal axis is left fixed. Each horizontal line is moved within itself right (or left) a distance proportional to its distance above (or below) the x_1-axis. Imagine a deck of playing cards racked so that the edge forms a parallelogram.

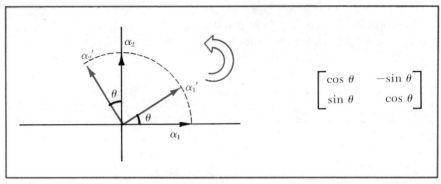

Figure 4-2.7 A rotation. The center of the rotation is the origin. The point (1, 0) is rotated to the point (cos θ, sin θ), the first column of the representing matrix. The point (0, 1) is rotated to the point ($-\sin \theta$, $\cos \theta$), the second column of the representing matrix. The sense of orientation is preserved.

This means the end point of $\sigma(\alpha_1) = \alpha_1'$ is located at (cos θ, sin θ) and the end point of $\sigma(\alpha_2) = \alpha_2'$ is located at ($-\sin \theta$, $\cos \theta$). α_1' and α_2' are located as shown in the following figure. In this case σ performs a *rotation* of the coordinate plane counterclockwise around the origin through an angle of θ.

(5) Consider the linear transformation σ represented by the matrix

$$\begin{bmatrix} 1 & 0 \\ 0 & 0 \end{bmatrix}. \tag{2.21}$$

Here, $\alpha_1' = \sigma(\alpha_1) = \alpha_1$, and $\alpha_2' = \sigma(\alpha_2) = 0$. The vector with representation (x_1, x_2) is mapped onto the vector with representation $(x_1, 0)$. Every point is *projected* vertically onto the point on the x_1-axis with the same x_1 coordinate. However, to avoid referring to an angle, we say the projection is parallel to the x_2-axis. Notice that every point on the x_2-axis is mapped onto the origin.

(6) Consider the linear transformation σ represented by the matrix

$$\begin{bmatrix} -5 & 15 \\ -2 & 6 \end{bmatrix}. \tag{2.22}$$

Although it is not immediately evident, σ is also a projection

4-2 Matrix Representation of Linear Transformations • 163

just like that in example (5). To see this, choose ξ_1 and ξ_2 as basis vectors where ξ_1 is represented by (5, 2) and ξ_2 is represented by (3, 1). Calculating with the representing matrix and the pairs representing ξ_1 and ξ_2, we see that $\sigma(\xi_1) = \xi_1$ and $\sigma(\xi_2) = 0$. Thus, using $\{\xi_1, \xi_2\}$ as a basis, σ is represented by the matrix

$$\begin{bmatrix} 1 & 0 \\ 0 & 0 \end{bmatrix}. \qquad (2.23)$$

When considered with respect to the same basis, the linear transformation represented by the matrix (2.21) and the linear transformation represented by the matrix (2.22) are not the same. However, when a different basis is permitted for each one, it is possible to choose the bases so that the matrices representing them are the same. This means that as far as any intrinsic properties are concerned the two linear transformations are indistinguishable. Each is a projection. Each is a projection onto a different line. Each is a projection in a different direction.

While the linear transformations represented by the matrices (2.21) and (2.22) are distinct, the fact that they are both projections is a meaningful geometric fact. One of the things that can be revealed by a change of coordinates is a similarity of this type. It can reveal that two linear transformations are both rotations, or both shears, or share some other common property.

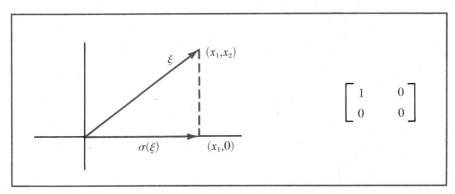

Figure 4-2.8 A projection. The direction of the projection is vertical, onto the horizontal axis. The point (1, 0) is left fixed, and gives the first column of the representing matrix. The point (0, 1) is mapped onto (0, 0), the second column of the representing matrix. (The basis vectors are not shown in this figure.)

EXERCISES 4-2

The matrices in the following exercises represent linear transformations of a 2-dimensional vector space into itself. A basis is implied but not given.

1. Let $\begin{bmatrix} 2 & -3 \\ 5 & -7 \end{bmatrix}$ represent a linear transformation σ. If a vector ξ is represented by $(-4, 3)$, find the representation of $\sigma(\xi)$.

2. Let $\begin{bmatrix} 0 & 1 \\ 1 & 0 \end{bmatrix}$ represent a linear transformation σ of $\mathscr{U}$ into itself. For ξ represented by $(1, 1)$ and ξ_2 represented by $(1, -1)$, find the representations of $\sigma(\xi_1)$ and $\sigma(\xi_2)$. Find the representation of σ with respect to the basis $\{\xi_1, \xi_2\}$.

3. Let $\begin{bmatrix} -4 & 6 \\ -3 & 5 \end{bmatrix}$ represent a linear transformation σ of $\mathscr{U}$ into itself. For ξ represented by $(1, 2)$ find the representation of $\sigma(\xi)$. For ξ_1 represented by $(1, 1)$ and ξ_2 represented by $(-2, -1)$, find the representation of $\sigma(\xi_1)$ and $\sigma(\xi_2)$. Find the representation of σ with respect to the basis $\{\xi_1, \xi_2\}$.

4. Let $\begin{bmatrix} 6 & -2 \\ 15 & -5 \end{bmatrix}$ represent a linear transformation σ. Find a representation of σ with respect to the basis $\{(2, 5), (1, 3)\}$ and describe the linear transformation in geometric terms.

5. Let $\begin{bmatrix} -7 & 3 \\ -16 & 7 \end{bmatrix}$ represent a linear transformation σ. Find a representative of σ with respect to the basis $\{(2, 5), (1, 3)\}$ and describe the linear transformation in geometric terms.

6. Let $\begin{bmatrix} -9 & 4 \\ -25 & 11 \end{bmatrix}$ represent a linear transformation σ. Find a representation of σ with respect to the basis $\{(2, 5), (1, 3)\}$ and describe the linear transformation in geometric terms.

7. Let $\begin{bmatrix} 1 & 0 & 0 \\ 0 & 1 & 0 \\ 0 & 0 & 0 \end{bmatrix}$ represent a linear transformation of a 3-dimensional vector space into itself. This transformation is a generalization to three dimensions of one of the types described in detail in this section. Describe it.

8. Let $\begin{bmatrix} -2 & -4 & -6 \\ 6 & 9 & 12 \\ -3 & -4 & -5 \end{bmatrix}$ represent a linear transformation σ. Let α_1 be a vector represented by $(-2, 0, 1)$, and find a representation of $\sigma(\alpha_1)$. Let α_2 be a vector represented by $(0, 3, -2)$, and find a representation of $\sigma(\alpha_2)$. Let α_3 be represented by $(1, -2, 1)$, find a representation of $\sigma(\alpha_3)$. Finally, find a representation of σ with respect to the basis $\{(-2, 0, 1), (0, 3, -2), (1, -2, 1)\}$, and describe this linear transformation in geometric terms.

9. Let $\begin{bmatrix} 1 & 0 & 0 \\ 0 & 1 & 0 \\ 0 & 0 & -1 \end{bmatrix}$ represent a linear transformation σ of a 3-dimensional vector space into itself. This transformation is a generalization to three dimensions of one of the types described in detail in this section. Describe it.

10. Use the commutativity of the diagram in Figure 4.2.1 to establish the representation of a linear transformation σ by the matrix A in (2.2). Start by writing
$\mathscr{B}A = \sigma\mathscr{A} = \sigma[\alpha_1 \alpha_2 \ldots \alpha_n] = [\sigma\alpha_1 \sigma\alpha_2 \ldots \sigma\alpha_n]$.

11. Show that if two linear transformations of a vector space $\mathscr{U}$ into $\mathscr{V}$ assign the same values to the vectors in a basis of $\mathscr{U}$, then the two linear transformations are identical. That is, let σ and τ be two linear transformations of $\mathscr{U}$ into $\mathscr{V}$. Let $\mathscr{A} = \{\alpha_1, \ldots, \alpha_n\}$ be a basis of $\mathscr{U}$. Show that if $\sigma(\alpha_j) = \tau(\alpha_j)$ for $j = 1, \ldots, n$, then $\sigma = \tau$.

4–3 GENERAL PROPERTIES OF LINEAR TRANSFORMATIONS

Let σ be a linear transformation of $\mathscr{U}$ into $\mathscr{V}$. $\mathscr{U}$ is called the *domain* of σ and $\mathscr{V}$ is called the *codomain* of σ. For a vector $\xi \in \mathscr{U}$, the vector $\sigma(\xi) \in \mathscr{V}$ that the mapping associates with ξ is called the *image* of ξ. For a subset $\mathscr{S} \subset \mathscr{U}$, the set in $\mathscr{V}$ of images of all elements in $\mathscr{S}$ is called the image of $\mathscr{S}$, and this image is denoted by $\sigma(\mathscr{S})$. For any $\eta \in \mathscr{V}$, the set of all $\xi \in \mathscr{U}$ such that $\sigma(\xi) = \eta$ is called the *preimage* of η, and we denote this set by $\sigma^{-1}(\eta)$. Note that $\sigma^{-1}(\eta)$ is a set, not an element, and it might be empty.

Notice that the definition in Section 4–1 of a linear trans-

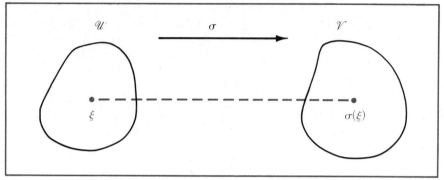

Figure 4–3.1   The linear transformation σ has domain $\mathscr{U}$ and codomain $\mathscr{V}$. It associates with each vector $\xi \in \mathscr{U}$ a vector $\sigma(\xi) \in \mathscr{V}$. $\sigma(\xi)$ is the image of ξ.

formation requires specification of the domain and codomain as well as the correspondence defined by the mapping. For example, consider the zero mapping of $\mathscr{U}$ into $\mathscr{V}$, that is, the mapping that maps every element of $\mathscr{U}$ onto the zero element of $\mathscr{V}$. Suppose $\mathscr{V}'$ is a proper subspace of $\mathscr{V}$. Since 0 is an element of every subspace, there is also a zero mapping of $\mathscr{U}$ into $\mathscr{V}'$. As far as the correspondence defined by these zero mappings is concerned, they determine the same correspondence. But as linear transformations we must consider them to be distinct because they have different codomains. This may seem like a pretty minor distinction. Both mappings would be represented by matrices with zeros for all elements. But the number of rows in the representing matrix is equal to the dimension of the codomain. Thus, these two zero linear transformations would be represented by different matrices.

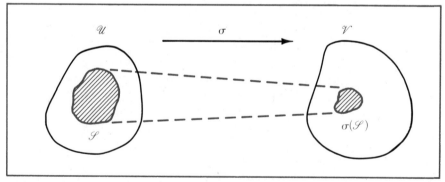

Figure 4–3.2   For each subset $\mathscr{S}$ of $\mathscr{U}$, $\sigma(\mathscr{S})$ is the set of all images of vectors in $\mathscr{S}$. That is, $\eta \in \sigma(\mathscr{S})$ if and only if there is a vector $\xi \in \mathscr{S}$ such that $\sigma(\xi) = \eta$. $\sigma(\mathscr{S})$ is the image of $\mathscr{S}$.

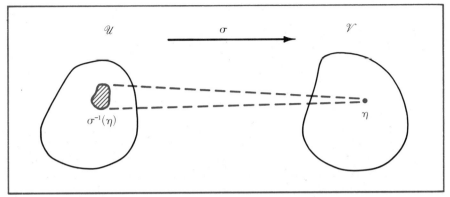

Figure 4-3.3 For each $\eta \in \mathscr{V}$, $\sigma^{-1}(\eta)$ is the set of all $\xi \in \mathscr{U}$ such that $\sigma(\xi) = \eta$. $\sigma^{-1}(\eta)$ is the pre-image of η. $\sigma^{-1}(\eta)$ might be the empty set.

Notice, also, that the mapping of $\mathscr{S} \subset \mathscr{U}$ onto $\sigma(\mathscr{S})$ described above is not the same mapping as σ. The domain of this mapping is the set of subsets of $\mathscr{U}$, and the codomain is the set of subsets of $\mathscr{V}$.

Theorem 3.1 *If $\mathscr{S}$ is a subspace of $\mathscr{U}$, $\sigma(\mathscr{S})$ is a subspace of $\mathscr{V}$.*

Proof. If $\alpha, \beta \in \sigma(\mathscr{S})$, there exists a $\xi \in \mathscr{U}$ such that $\sigma(\xi) = \alpha$, and there exists an $\eta \in \mathscr{U}$ such that $\sigma(\eta) = \beta$. Since $\mathscr{S}$ is a subspace $\xi + \eta \in \mathscr{S}$ and $a\xi \in \mathscr{S}$ for any scalar a. Thus $\alpha + \beta = \sigma(\xi) + \sigma(\eta) = \sigma(\xi + \eta) \in \sigma(\mathscr{S})$ and $a\alpha = a\sigma(\xi) = \sigma(a\xi) \in \sigma(\mathscr{S})$. This shows that $\sigma(\mathscr{S})$ is a subspace. □

Corollary 3.2 *$\sigma(\mathscr{U})$ is a subspace of $\mathscr{V}$.* □

$\sigma(\mathscr{U})$ is also called the *image* of σ, and is denoted by $\text{Im}(\sigma)$. Since $\mathscr{V}$ is finite dimensional, $\text{Im}(\sigma)$ has a dimension. The dimension of $\text{Im}(\sigma)$ is called the *rank* of σ. The set of all $\xi \in \mathscr{U}$ such that $\sigma(\xi) = 0$ is called the *kernel* of σ, and it is denoted by $K(\sigma)$.

Theorem 3.3 *$K(\sigma)$ is a subspace of $\mathscr{U}$.*

Proof. Let $\xi, \eta \in K(\sigma)$. Then $\sigma(\xi + \eta) = \sigma(\xi) + \sigma(\eta) = 0 + 0 = 0$. Also, for any scalar a, $\sigma(a\xi) = a\sigma(\xi) = 0$. Thus, $\xi + \eta \in K(\sigma)$ and $a\xi \in K(\sigma)$, and $K(\sigma)$ is a subspace. □

168 • Linear Transformations

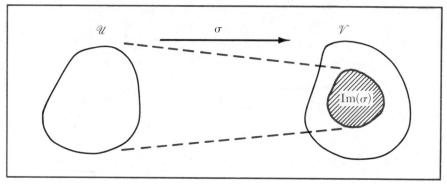

Figure 4–3.4   The image of a linear transformation is the image of the domain $\mathscr{U}$. Thus, $\eta \in \text{Im}(\sigma)$ if and only if there is a $\xi \in \mathscr{U}$ such that $\sigma(\xi) = \eta$. $\text{Im}(\sigma)$ is a subspace of the codomain $\mathscr{V}$.

The dimension of $K(\sigma)$ is called the *nullity* of σ. $K(\sigma)$ is called the *null space* of σ.

Theorem 3.4 $\text{Rank}(\sigma) \leq \min\{\dim \mathscr{U}, \dim \mathscr{V}\}$.

Proof.   Let $\mathscr{A} = \{\alpha_1, \ldots, \alpha_n\}$ be a basis of $\mathscr{U}$. Since every $\xi \in \mathscr{U}$ can be expressed as a linear combination of the form $\xi = \sum_{j=1}^{n} x_j \alpha_j$, every element $\sigma(\xi) \in \text{Im}(\xi)$ can be expressed in the form

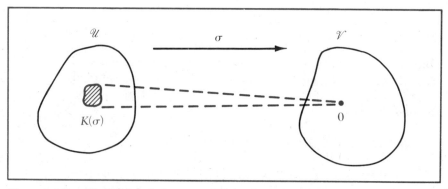

Figure 4–3.5   The kernel of a linear transformation is the preimage of 0. Thus, $\xi \in K(\sigma)$ if and only if $\sigma(\xi) = 0$. $K(\sigma)$ is a subspace of the domain $\mathscr{U}$.

$$\sigma(\xi) = \sum_{j=1}^{n} x_j \sigma(\alpha_j). \tag{3.1}$$

This means that $\{\sigma(\alpha_1), \ldots, \sigma(\alpha_n)\}$ spans $\text{Im}(\sigma)$. Thus, $\text{rank}(\sigma) = \dim \text{Im}(\sigma) \leq n = \dim \mathscr{U}$. Also, since $\text{Im}(\sigma)$ is a subspace of $\mathscr{V}$, $\dim \text{Im}(\sigma) \leq \dim \mathscr{V}$. □

Corollary 3.5 If $\mathscr{S}$ is a subspace of $\mathscr{U}$, $\dim \sigma(\mathscr{S}) \leq \min\{\dim \mathscr{S}, \dim \mathscr{V}\}$. □

Corollary 3.6 $\sigma(0) = 0$.

Proof. $\{0\}$ is a subspace of $\mathscr{U}$ of dimension 0. By Corollary 3.5, $\sigma(0)$ is a subspace of $\mathscr{V}$ of dimension 0. Thus, $\sigma(0) = 0$. □

Actually, a somewhat more general observation than Corollary 3.5 is possible. If $\{\gamma_1, \ldots, \gamma_k\}$ is a linearly dependent set in $\mathscr{U}$, then $\{\sigma(\gamma_1), \ldots, \sigma(\gamma_k)\}$ is also linearly dependent. This follows easily from the fact that if $0 = \sum_{i=1}^{k} c_i \gamma_i$, then $0 = \sigma(0) = \sum_{i=1}^{k} c_i \sigma(\gamma_i)$. Thus, linear transformations preserve linear relations. This implies that for any concept depending on linear independence or linear dependence, such as rank, dimension, and so forth, changes can go only in one direction when a linear transformation is applied. For example, we see in Corollary 3.5 that $\dim \sigma(\mathscr{S}) \leq \dim \mathscr{S}$.

In a numerical problem a linear transformation is specified by giving a matrix representing it, usually without mention of the bases in $\mathscr{U}$ and $\mathscr{V}$ with respect to which the representing matrix is determined.

These bases are implied and all other objects occurring in the problem, including any solutions, are to be represented with respect to the same basis in $\mathscr{U}$ and $\mathscr{V}$.

Suppose, for example, the matrix

$$A = \begin{bmatrix} a_{11} & a_{12} & \cdots & a_{1n} \\ \cdot & \cdot & & \cdot \\ \cdot & \cdot & & \cdot \\ \cdot & \cdot & & \cdot \\ a_{m1} & a_{m2} & \cdots & a_{mn} \end{bmatrix} \tag{3.2}$$

is given as representing a linear transformation σ. Since A is an $m \times n$ matrix, σ is a linear transformation of an n-dimensional vector space into an m-dimensional space. There are bases $\mathscr{A} = \{\alpha_1, \ldots, \alpha_n\}$ in $\mathscr{U}$ and $\mathscr{B} = \{\beta_1, \ldots, \beta_m\}$ in $\mathscr{V}$ such that $\sigma(\alpha_j) = \sum_{i=1}^{m} a_{ij}\beta_i$. $\sigma(\alpha_j)$ is represented by the m-tuple $(a_{1j}, a_{2j}, \ldots, a_{mj})$ with respect to the basis $\mathscr{B}$. These m-tuples are the columns of A. In the proof of Theorem 3.4 we observed that $\{\sigma(\alpha_1), \ldots, \sigma(\alpha_n)\}$ spans $\text{Im}(\sigma)$. Thus, the columns of A span a subspace of the coordinate space $\mathscr{R}^m$ isomorphic to $\text{Im}(\sigma)$.

The dimension of $\text{Im}(\sigma)$ is therefore equal to the maximal number of linearly independent columns of A. Accordingly, we call the maximal number of linearly independent columns of A the *rank* of A, denoted rank(A). A linear transformation and a matrix representing the linear transformation have the same rank.

We wish to find a basis for $\text{Im}(\sigma)$. What we will actually do is find a basis for the corresponding subspace of the coordinate space. Finding such a basis is a problem we have considered in detail in Section 2-6. Performing elementary row operations amounts to changing the coordinate system for $\mathscr{V}$. Throughout these changes column j remains a representation of $\sigma(\alpha_j)$ with respect to each successive coordinate system. Using pivot operations we reduce the matrix A to basic form. In the basic form the columns corresponding to the basic variables (the basic columns) represent vectors in a basis for $\text{Im}(\sigma)$. These columns are linearly independent and the other columns are linearly dependent on them. In the basic form the number of non-zero rows is equal to the number of basic columns, which is the dimension of $\text{Im}(\sigma)$ and the rank of σ.

Let us examine a specific numerical example. Let

$$A = \begin{bmatrix} 2 & 1 & 3 & 2 & 2 \\ 1 & 1 & 1 & -1 & 4 \\ -1 & 2 & -4 & 0 & 3 \\ 2 & 3 & 1 & -2 & 10 \end{bmatrix}. \tag{3.3}$$

By a sequence of pivot operations this matrix can be reduced to the basic form

$$\begin{bmatrix} 1 & 0 & 2 & 0 & 1 \\ 0 & 1 & -1 & 0 & 2 \\ 0 & 0 & 0 & 1 & -1 \\ 0 & 0 & 0 & 0 & 0 \end{bmatrix}. \tag{3.4}$$

The columns of (3.4) represent the vectors $\sigma(\alpha_j)$, but not in the coordinate system with which we started. From the matrix (3.4) we see that $\{\sigma(\alpha_1), \sigma(\alpha_2), \sigma(\alpha_4)\}$ is a basis for Im(σ). We refer to the matrix A in (3.3) to find the coordinates of these vectors in the original coordinate system. Thus, $\{(2, 1, -1, 2), (1, 1, 2, 3), (2, -1, 0, -2)\}$ are the representatives of a basis of Im(σ). We consider this to be a satisfactory answer to the question of finding a basis for Im(σ) since we have expressed these vectors in terms of the same basis used in specifying A. We observe that the rank of A is 3.

What about the problem of finding the kernel of σ? We are seeking all ξ such that $\sigma(\xi) = 0$. In matrix form, this amounts to finding all X such that $AX = 0$. Translated further, this is the familiar problem of solving a system of homogeneous linear equations. The information needed to write down a basis for this solution set is available from the reduced basic form (3.4). From it we see that 5-tuples representing a basis for the kernel are $\{(-2, 1, 1, 0, 0), (-1, -2, 0, 1, 1)\}$. The nullity of σ, then, is 2.

Generally, we find the kernel of a linear transformation σ by solving the homogeneous equation $AX = 0$, where A is a matrix representing σ. The dimension of $K(\sigma)$ is the same as the number of linearly independent vectors in the solution subspace of $AX = 0$. It is appropriate, therefore, to define the *nullity* of a matrix A to be the dimension of the solution subspace of $AX = 0$. The nullity of a linear transformation is equal to the nullity of any matrix that represents it. We denote the nullity of a matrix A by nullity(A).

Theorem 3.7 *If σ is a linear transformation of $\mathscr{U}$ into $\mathscr{V}$ and $\mathscr{U}$ is of dimension n, then* rank(σ) + nullity(σ) = n.

Proof. Let $\mathscr{K}$ be the kernel of σ, and $\{\alpha_1, \ldots, \alpha_k\}$ any basis of $\mathscr{K}$. Extend this linearly independent set to a basis $\{\alpha_1, \ldots, \alpha_k, \alpha_{k+1}, \ldots, \alpha_n\}$ of $\mathscr{U}$. Since $\sigma(\alpha_1) = \ldots = \sigma(\alpha_k) = 0$, Im($\sigma$) is spanned by $\{\sigma(\alpha_{k+1}), \ldots, \sigma(\alpha_n)\}$. We wish to show that this set is also linearly independent. Thus, suppose we have a linear relation of the form

$$\sum_{j=k+1}^{n} c_j \sigma(\alpha_j) = 0. \tag{3.5}$$

This means that $\sum_{j=k+1}^{n} c_j \alpha_j \in \mathscr{K}$, and can be expressed as a linear combination of the basis for $\mathscr{K}$. That is,

$$\sum_{j=k+1}^{n} c_j \alpha_j = \sum_{j=1}^{k} d_j \alpha_j. \qquad (3.6)$$

By transposing the terms on the right to the left side we get a linear relation among the elements of $\{\alpha_1, \ldots, \alpha_n\}$. But this set is linearly independent and hence all $c_j = 0$. This shows $\{\sigma(\alpha_{k+1}), \ldots, \sigma(\alpha_n)\}$ is a basis for $\text{Im}(\sigma)$. Thus $\text{rank}(\sigma) = n - k = n - \text{nullity}(\sigma)$. □

Theorem 3.7 is quite useful, for it has a number of interesting and important implications. For example, σ is one-to-one if and only if $\text{nullity}(\sigma) = 0$. For if $\sigma(\xi) = \sigma(\eta)$ with $\xi \neq \eta$, then $\sigma(\xi - \eta) = \sigma(\xi) - \sigma(\eta) = 0$ with $\xi - \eta \neq 0$. This would imply $\mathcal{K} \neq \{0\}$, and hence $\text{nullity}(\sigma) \neq 0$. On the other hand, if $\text{nullity}(\sigma) \neq 0$, there is a $\xi \in \mathcal{K}$ such that $\xi \neq 0$. Then $\sigma(\xi) = 0 = \sigma(0)$ and σ is not one-to-one. Theorem 3.7 can be used to conclude that $\text{nullity}(\sigma) = 0$ or $\text{nullity}(\sigma) \neq 0$ depending on whether $\text{rank}(\sigma) = n = \dim \mathcal{U}$ or $\text{rank}(\sigma) \neq n$.

EXERCISES 4-3

1. Let $\begin{bmatrix} 1 & 0 & 2 & -1 \\ 2 & 1 & 1 & 3 \\ 3 & -1 & 1 & 2 \end{bmatrix}$ represent a linear transformation σ of a 4-dimensional space $\mathcal{U}$ into a 3-dimensional space $\mathcal{V}$. Show that we can conclude without making any calculations, that $K(\sigma)$ is of positive dimension. Find a basis for $K(\sigma)$. Without finding a basis for $\text{Im}(\sigma)$ determine the rank of σ.

2. Let $\begin{bmatrix} 1 & 2 & 3 \\ 2 & 1 & 0 \\ 3 & 4 & 5 \end{bmatrix}$ represent a linear transformation σ of a 3-dimensional space $\mathcal{U}$ into a 3-dimensional space $\mathcal{V}$. Find a basis for $\text{Im}(\sigma)$, and thus determine the rank of σ.

3. Let $\begin{bmatrix} 1 & 2 & 0 & 3 \\ 2 & 1 & 3 & 2 \\ -1 & 2 & -4 & 1 \\ 3 & 1 & 5 & 2 \end{bmatrix}$ represent a linear transformation σ of a 4-dimensional space $\mathcal{U}$ into a 4-dimensional space $\mathcal{V}$. Find a basis for $K(\sigma)$ and a basis for $\text{Im}(\sigma)$.

4. Let σ be a linear transformation of $\mathcal{U}$ into $\mathcal{V}$. If $\dim \mathcal{U} = 6$

and dim $\mathscr{V} = 3$, $K(\sigma)$ contains at least one non-zero vector. In fact, find a lower bound for the nullity of σ. Show that σ cannot be a one-to-one mapping (that is, at least two distinct vectors in $\mathscr{U}$ have the same image in $\mathscr{V}$).

5. Let σ be a linear transformation of $\mathscr{U}$ into $\mathscr{V}$. If dim $\mathscr{U} = 3$ and dim $\mathscr{V} = 6$, show that $\text{Im}(\sigma)$ is a proper subspace of $\mathscr{V}$. Find an upper bound for the rank of σ. Show that σ cannot be an onto mapping (that is, not every vector in $\mathscr{V}$ is the image of a vector in $\mathscr{U}$).

6. In Exercise 3 of Section 4–1, the vector spaces are not finite dimensional and D cannot be represented by a matrix. But the concepts kernel and image are still meaningful. Find the kernel of D.

7. In Exercise 7 of Section 4–1, find the kernel and image of σ (that is, find bases for the kernel and image).

8. In Exercise 8 of Section 4–1, find the kernel and image of τ.

9. Find the kernel of the linear transformation D in Exercise 11 of Section 4–1.

10. Find the image of the linear transformation G in Exercise 12 of Section 4–1.

11. Let σ be a linear transformation of $\mathscr{U}$ into $\mathscr{V}$. For any subset $\mathscr{T} \in \mathscr{V}$, let $\sigma^{-1}(\mathscr{T})$ denote the set of all vectors in $\mathscr{U}$ that are mapped into $\mathscr{T}$ by σ. Show that if $\mathscr{T}$ is a subspace of $\mathscr{V}$, then $\sigma^{-1}(\mathscr{T})$ is a subspace of $\mathscr{U}$.

4–4 COMPOSED TRANSFORMATIONS AND INVERSES

Let σ be a linear transformation of $\mathscr{U}$ into $\mathscr{V}$, and let τ be a linear transformation of $\mathscr{V}$ into $\mathscr{W}$. We can define a mapping of $\mathscr{U}$ into $\mathscr{W}$ in the following way. For $\xi \in \mathscr{U}$, $\sigma(\xi) \in \mathscr{V}$. Since $\mathscr{V}$ is the domain of τ, τ maps $\sigma(\xi)$ onto $\tau(\sigma(\xi))$. We denote this mapping of $\xi \in \mathscr{U}$ onto $\tau(\sigma(\xi))$ by $\tau\sigma$. That is,

$$(\tau\sigma)(\xi) = \tau(\sigma(\xi)) \qquad (4.1)$$

for all $\xi \in \mathscr{U}$. $\tau\sigma$ is a linear transformation since $(\tau\sigma)(a\xi + b\eta) = \tau(\sigma(a\xi + b\eta)) = \tau(a\sigma(\xi) + b\sigma(\eta)) = a\tau(\sigma(\xi)) + b\tau(\sigma(\eta)) = a(\tau\sigma)(\xi) + b(\tau\sigma)(\eta)$. To simplify notation, we will usually write $\tau\sigma(\xi) = (\tau\sigma)(\xi)$. But it should be understood that $\tau\sigma$ is then the notation for a single linear transformation with domain $\mathcal{U}$ and codomain $\mathcal{W}$. The process of forming $\tau\sigma$ is called the operation of *composition* of linear transformations. Sometimes $\tau \circ \sigma$ is used to denote the composition of σ and τ.

The computation of $\tau(\sigma(\xi))$ is possible if the domain of τ includes the image of σ. However, we are going to show later in this section that the matrix representing the composition of two linear transformations is the product of the representing matrices. To form the product of two matrices, BA, it is necessary that the number of rows in A be the same as the number of columns of B. This corresponds to the condition that the dimension of the codomain of σ be the same as the dimension of the domain of τ. Thus, we shall insist that the domain of τ and the codomain of σ must be the same before we can discuss the "composition" of σ and τ.

If σ and τ are linear transformations of $\mathcal{U}$ into $\mathcal{V}$, we define $\sigma + \tau$ by the rule

$$(\sigma + \tau)(\xi) = \sigma(\xi) + \tau(\xi). \tag{4.2}$$

We define $a\sigma$ by the rule

$$(a\sigma)(\xi) = a\sigma(\xi). \tag{4.3}$$

We have defined the addition of linear transformations and multiplication by scalars. With these two operations, the set of all linear transformations of $\mathcal{U}$ into $\mathcal{V}$ forms a vector space.

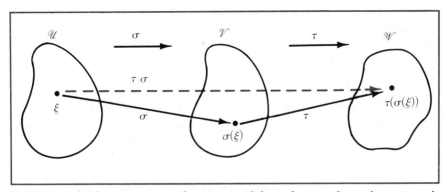

Figure 4–4.1 The composition of mappings. If the codomain of σ is the same as the domain of τ, these mappings can be composed. For each $\xi \in \mathcal{U}$, the domain of σ, $\tau\sigma(\xi)$ is the τ-image of $\sigma(\xi)$. If σ and τ are linear, $\tau\sigma$ is linear.

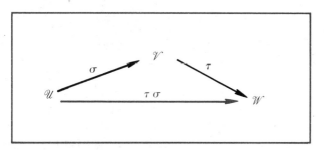

Figure 4–4.2 The composition of mappings. The mapping $\tau\sigma$ is defined so that the diagram commutes. That is, for each $\xi \in \mathcal{U}$, $(\tau\sigma)(\xi) = \tau(\sigma(\xi))$.

This vector space is denoted by $\mathrm{Hom}(\mathcal{U}, \mathcal{V})$. The structure of $\mathrm{Hom}(\mathcal{U}, \mathcal{V})$ as a vector space is quite useful in the study of linear transformations, but it is of minor importance for the particular topics to which this text is limited. Accordingly, we leave to the reader the task of verifying that $\mathrm{Hom}(\mathcal{U}, \mathcal{V})$ is, in fact, a vector space.

Let σ_1 and σ_2 be linear transformations of $\mathcal{U}$ into $\mathcal{V}$, and let τ be a linear transformation of $\mathcal{V}$ into $\mathcal{W}$. Then

$$\tau(\sigma_1 + \sigma_2) = \tau\sigma_1 + \tau\sigma_2. \qquad (4.4)$$

To prove this distributive law, three things must be established. We must show $\tau(\sigma_1 + \sigma_2)$ and $\tau\sigma_1 + \tau\sigma_2$ have the same domains, the same codomains, and define the same mappings. The first two requirements are easily seen to be satisfied. The third requirement is established as follows.

$$\begin{aligned}(\tau(\sigma_1 + \sigma_2))(\xi) &= \tau((\sigma_1 + \sigma_2)(\xi))\\ &= \tau(\sigma_1(\xi) + \sigma_2(\xi))\\ &= \tau(\sigma_1(\xi)) + \tau(\sigma_2(\xi))\\ &= (\tau\sigma_1)(\xi) + (\tau\sigma_2)(\xi)\\ &= (\tau\sigma_1 + \tau\sigma_2)(\xi).\end{aligned} \qquad (4.5)$$

Similarly, one can show that if σ is a linear transformation of $\mathcal{U}$ into $\mathcal{V}$ and τ_1 and τ_2 are linear transformations of $\mathcal{V}$ into $\mathcal{W}$, then

$$(\tau_1 + \tau_2)\sigma = \tau_1\sigma + \tau_2\sigma. \qquad (4.6)$$

All the linear transformations of $\mathcal{U}$ into itself have the same domain and codomain. Thus, any pair of linear transformations in $\text{Hom}(\mathcal{U}, \mathcal{U})$ can be composed in either order, and the result is an element of $\text{Hom}(\mathcal{U}, \mathcal{U})$. Relative to the addition operation in $\text{Hom}(\mathcal{U}, \mathcal{U})$, composition plays the role of a multiplication as far as the various distributive laws are concerned, as shown in (4.4) and (4.6). This multiplication is associative but not commutative. A vector space with a multiplication is called an *algebra*. $\text{Hom}(\mathcal{U}, \mathcal{U})$ is a *linear algebra*. We will not pursue this potential line of discussion beyond giving this definition. We are concerned more with the properties of particular linear transformations and with special kinds of linear transformations and only incidentally with the properties of the set of all linear transformations.

Theorem 4.1 $\text{Rank}(\tau\sigma) \le \min\{\text{rank}(\tau), \text{rank}(\sigma)\}$.

Proof. $\text{Im}(\tau\sigma) = \tau(\text{Im}(\sigma))$. By Corollary 3.5, $\dim \text{Im}(\tau\sigma) \le \dim \text{Im}(\sigma) = \text{rank}(\sigma)$. Since $\text{Im}(\tau\sigma) \subset \text{Im}(\tau)$, $\dim \text{Im}(\tau\sigma) \le \dim \text{Im}(\tau) = \text{rank}(\tau)$. □

The implication of this theorem is that in the composition of linear transformations, rank can change in only one direction—down.

For each scalar a there is a mapping of the vector space $\mathcal{U}$ into itself that maps each $\xi \in \mathcal{U}$ onto $a\xi$. This mapping is linear since $a(\xi + \eta) = a\xi + a\eta$ and $a(b\xi) = b(a\xi)$ for each b. A linear transformation of this kind is called a *scalar* transformation, and we shall denote it by the symbol a_u, or just by a if there is no possibility of confusion.

Of particular importance is the scalar transformation corresponding to the unit 1. This linear transformation is called the *identity* transformation. Thus, $1_\mathcal{U}(\xi) = \xi$ for all $\xi \in \mathcal{U}$. If σ is a linear transformation of $\mathcal{U}$ into $\mathcal{V}$, then $\sigma 1_\mathcal{U} = \sigma$ and $1_\mathcal{V} \sigma = \sigma$. The identity transformation acts like a unit in the composition of linear transformations.

Let σ be a given linear transformation of $\mathcal{U}$ into $\mathcal{V}$. An *inverse* of σ is a linear transformation τ of $\mathcal{V}$ into $\mathcal{U}$ such that $\tau\sigma = 1_\mathcal{U}$ and $\sigma\tau = 1_\mathcal{V}$. Not all linear transformations have inverses, and we are interested in establishing conditions on σ that are sufficient to guarantee that it has an inverse.

Clearly, if $\sigma(\xi) = \eta$, τ should be defined so that $\tau(\eta) = \xi$. First of all, this requires that σ should be a *one-to-one* mapping. By this term we mean that if $\xi_1 \ne \xi_2$, then $\sigma(\xi_1) \ne \sigma(\xi_2)$. A more modern term is to call σ *injective*. If for any $\xi_1 \ne \xi_2$ we have $\sigma(\xi_1) = \sigma(\xi_2) = \eta$, then $\tau(\eta)$ cannot be well defined. Second, the domain of τ is $\mathscr{V}$, which means that every $\eta \in \mathscr{V}$ must also be in $\text{Im}(\sigma)$. Thus, $\text{Im}(\sigma) = \mathscr{V}$. If $\text{Im}(\sigma) = \mathscr{V}$, we say that σ is *onto* or *surjective*.

In algebra we typically study sets in which one or more binary relations are defined. A *binary relation* $*$ defined in a set $\mathscr{S}$ is a function of pairs of elements in $\mathscr{S}$ which associates with the pair some element in $\mathscr{S}$. That is, given a and b in $\mathscr{S}$, $a * b$ is an element of $\mathscr{S}$. This binary relation may have several properties, depending on the particular kind of relation. Examples of binary relations are addition and multiplication of integers, and the several associative, commutative, and distributive laws are properties that these relations have. In this book we are studying vector spaces, and vector addition is a binary relation in a vector space. We also have scalar multiplication, which is a binary relation defined on a pair where one element of the pair is in the field and the other is in the vector space.

Generally, we are interested in mappings that preserve the relations under study. In other words, if $\mathscr{S}$ were a set with a binary relation $*$, and $\mathscr{T}$ were a set with a binary relation $\#$, we would be interested in studying functions from $\mathscr{S}$ into $\mathscr{T}$ for which

$$f(a * b) = f(a) \# f(b). \tag{4.7}$$

There might also be other binary relations which are carried over by the mapping. In this book, we have two binary relations and we are studying linear transformation, mappings that preserve both vector addition and scalar multiplication.

Mappings that preserve the relations under consideration are called *homomorphisms*. This is a generic term. A homomorphism from one vector space to another is called a linear transformation. A surjective homomorphism is called an *epimorphism*. An injective homomorphism is called a *monomorphism*. A mapping that is both surjective and injective is called *bijective*. A bijective homomorphism is called an *isomorphism*.

An isomorphism has an inverse which is also an isomorphism. To see this, consider the mapping f from $\mathscr{S}$ to $\mathscr{T}$ satisfying relationship (4.7), and suppose f is bijective. The inverse g from $\mathscr{T}$ to $\mathscr{S}$ can be defined: for each $y \in \mathscr{T}$ there is an $x \in \mathscr{S}$ such that $f(x) = y$, and we define $g(y) = x$. It is easily seen that g is also bijective. We must show that g carries over the relations under consideration. For y_1 and $y_2 \in \mathscr{T}$, let $x_1 = g(y_1)$ and $x_2 = g(y_2)$. Then $g(y_1 \# y_2) = g(f(x_1) \# f(x_2)) = g(f(x_1 * x_2)) = x_1 * x_2 = g(y_1) * g(y_2)$. Thus, g is an isomorphism. In particular, if σ is a linear transformation from $\mathscr{U}$ into $\mathscr{V}$ which is injective and surjective (bijective) an inverse mapping τ can be defined, and τ is a linear transformation which is also bijective.

Theorem 4.2 σ *is one-to-one if and only if* $K(\sigma) = \{0\}$.

Proof. If $K(\sigma) \neq \{0\}$, there is a $\xi \in K(\sigma)$ such that $\xi \neq 0$. Then $\sigma(\xi) = 0 = \sigma(0)$ and σ is not one-to-one. On the other hand, if σ is not one-to-one, there exist $\xi_1 \neq \xi_2$ such that $\sigma(\xi_1) = \sigma(\xi_2)$. Then $\sigma(\xi_1 - \xi_2) = \sigma(\xi_1) - \sigma(\xi_2) = 0$. Thus $\xi_1 - \xi_2 \neq 0$ and $\xi_1 - \xi_2 \in K(\sigma)$. □

Thus, according to the discussion preceding Theorem 4.2 σ has an inverse if and only if nullity$(\sigma) = 0$ and rank$(\sigma) = $ dim $\mathscr{V}$ ($K(\sigma) = \{0\}$ and Im$(\sigma) = \mathscr{V}$). Let dim $\mathscr{U} = n$ and dim $\mathscr{V} = m$. There are three possibilities $m < n$, $m > n$, and $m = n$. If $m < n$, nullity$(\sigma) = n - $ rank$(\sigma) \geq n - m > 0$. In this case σ cannot have an inverse. If $m > n$, rank$(\sigma) \leq $ dim $\mathscr{U} < $ dim $\mathscr{V}$. In this case, too, σ cannot have an inverse. If $m = n$, rank$(\sigma) = $ dim $\mathscr{V}$ if and only if nullity$(\sigma) = 0$. Summary: If dim $\mathscr{U} \neq $ dim $\mathscr{V}$, σ cannot have an inverse. If dim $\mathscr{U} = $ dim $\mathscr{V}$, σ can have an inverse and either the condition nullity$(\sigma) = 0$ or the condition rank$(\sigma) = $ dim $\mathscr{V}$ is sufficient.

Suppose σ is a linear transformation of $\mathscr{U}$ into $\mathscr{V}$ and τ is a linear transformation of $\mathscr{V}$ into $\mathscr{W}$. Since $\sigma(\mathscr{U}) \subset \mathscr{V}$, Im$(\tau\sigma) = \tau(\sigma(\mathscr{U})) \subset \tau(\mathscr{V}) = $ Im(τ). If $\tau\sigma$ is onto, $\mathscr{W} = $ Im$(\tau\sigma) \subset $ Im$(\tau) \subset \mathscr{W}$. Hence τ is onto. Similarly, $(\tau\sigma)(K(\sigma)) = \tau(\sigma K(\sigma)) = \tau(0) = 0$. Thus, $K(\sigma) \subset K(\tau\sigma)$. If $\tau\sigma$ is one-to-one, $K(\sigma) = \{0\}$ and σ is one-to-one.

As a particular application of the discussion of the previous paragraph, consider the situation in which τ is an inverse of σ. Since $1_{\mathscr{U}} = \tau\sigma$ is one-to-one and onto, τ is onto and σ is one-to-one. Since $1_{\mathscr{V}} = \sigma\tau$ is one-to-one and onto, σ is onto and τ is one-to-one.

Theorem 4.3 *If* $\tau_1\sigma = \tau_2\sigma$ *and σ is onto, then* $\tau_1 = \tau_2$. *If* $\tau\sigma_1 = \tau\sigma_2$ *and τ is one-to-one, then* $\sigma_1 = \sigma_2$.

Proof. Suppose $\tau_1\sigma = \tau_2\sigma$ and σ is onto. If $\tau_1 \neq \tau_2$, there is an $\eta \in \mathscr{V}$ such that $\tau_1(\eta) \neq \tau_2(\eta)$. Since σ is onto, there is a $\xi \in \mathcal{U}$ such that $\eta = \sigma(\xi)$. Then $\tau_1(\eta) = \tau_1\sigma(\xi) = \tau_2\sigma(\xi) = \tau_2(\eta)$ is a contradiction.

Suppose $\tau\sigma_1 = \tau\sigma_2$ and τ is one-to-one. If $\sigma_1 \neq \sigma_2$, there is a $\xi \in \mathcal{U}$ such that $\sigma_1(\xi) \neq \sigma_2(\xi)$. Since τ is one-to-one $\tau\sigma_1(\xi) \neq \tau\sigma_2(\xi)$, which is a contradiction. □

Corollary 4.4 *If τ_1 and τ_2 are inverses of σ, $\tau_1 = \tau_2$.*

Proof. $1_{\mathcal{U}} = \tau_1\sigma = \tau_2\sigma$. Since σ is onto, $\tau_1 = \tau_2$. □

Let σ be a linear transformation of $\mathcal{U}$ into $\mathcal{V}$ and let τ be a linear transformation of $\mathcal{V}$ into $\mathcal{W}$. Let $\mathcal{A} = \{\alpha_1, \ldots, \alpha_n\}$ be a basis of $\mathcal{U}$, $\mathcal{B} = \{\beta_1, \ldots, \beta_m\}$ be a basis of $\mathcal{V}$, and $\mathcal{C} = \{\gamma_1, \ldots, \gamma_r\}$ be a basis of $\mathcal{W}$. Let σ be represented by the matrix A with respect to $\mathcal{A}$ and $\mathcal{B}$, and let τ be represented by B with respect to $\mathcal{B}$ and $\mathcal{C}$. Computing,

$$\sigma(\alpha_j) = \sum_{i=1}^{m} a_{ij}\beta_i,$$

$$\tau(\beta_j) = \sum_{i=1}^{r} b_{ij}\gamma_i. \tag{4.8}$$

Then

$$(\tau\sigma)(\alpha_j) = \tau(\sigma(\alpha_j))$$

$$= \tau\left(\sum_{k=1}^{m} a_{kj}\beta_k\right)$$

$$= \sum_{k=1}^{m} a_{kj}\tau(\beta_k) \tag{4.9}$$

$$= \sum_{k=1}^{m} a_{kj}\left(\sum_{i=1}^{r} b_{ik}\gamma_i\right)$$

$$= \sum_{i=1}^{r}\left(\sum_{k=1}^{m} b_{ik}a_{kj}\right)\gamma_i.$$

This means $\tau\sigma$ is represented by the matrix $C = [c_{ij}]$ where

$$c_{ij} = \sum_{k=1}^{m} b_{ik}a_{kj}. \tag{4.10}$$

This is the formula for matrix multiplication. That is, $\tau\sigma$ is represented by BA.

Let σ and τ be linear transformations of $\mathcal{U}$ into $\mathcal{V}$, and let σ be represented by $A = [a_{ij}]$ and let τ be represented by $B = [b_{ij}]$. Then

$$(\sigma + \tau)(\alpha_j) = \sigma(\alpha_j) + \tau(\alpha_j)$$
$$= \sum_{i=1}^{m} a_{ij}\beta_i + \sum_{i=1}^{m} b_{ij}\beta_i \qquad (4.11)$$
$$= \sum_{i=1}^{m} (a_{ij} + b_{ij})\beta_i.$$

This means $\sigma + \tau$ is represented by $A + B$. Also, for any scalar a

$$(a\sigma)(\alpha_j) = a\sigma(\alpha_j)$$
$$= a \sum_{i=1}^{m} a_{ij}\beta_i \qquad (4.12)$$
$$= \sum_{i=1}^{m} (aa_{ij})\beta_i.$$

Thus, $a\sigma$ is represented by aA.

The discussion of this section has immediate application to matrices. In particular, if A and B are any two matrices for which AB can be formed, Theorem 4.1 says that $\text{rank}(AB) \leq \min\{\text{rank}(A), \text{rank}(B)\}$. First, suppose A is a square matrix of order n and B is a matrix such that $AB = I$. This implies that B is a square matrix of order n. Since $\text{rank}(I) = n$, we also see that $\text{rank}(A) = \text{rank}(B) = n$. This means a matrix of rank less than n cannot have an inverse. Second, suppose $\text{rank}(A) = n$. Then any linear transformation that A represents also has $\text{rank} = n$ and it must have an inverse. This, in turn, implies that A has an inverse. We wish to point up two conclusions. A square matrix A of order n has an inverse if and only if its rank is n. If a square matrix has an inverse on one side, it has an inverse on both sides and its inverse is unique.

It should be pointed out, however, that it is possible for a non-square matrix to have an inverse on one side without its having an inverse. For example, consider

$$\begin{bmatrix} 1 & 0 & 1 \\ 0 & 1 & 1 \end{bmatrix} \begin{bmatrix} 1 & 0 \\ 0 & 1 \\ 0 & 0 \end{bmatrix} = \begin{bmatrix} 1 & 0 \\ 0 & 1 \end{bmatrix}. \tag{4.13}$$

The first factor is of rank 2 and the product in this order is the identity of order 2. The product in the other order is a 3×3 matrix which cannot be of rank 3, thus it cannot be an identity.

We have defined three operations on linear transformations: addition, multiplication by scalars, and composition. When linear transformations are represented by matrices these operations correspond to addition, multiplication by scalars, and multiplication of matrices. Thus, the algebra of linear transformations is represented by the algebra of matrices. With these three operations the set of all square matrices of order n constitutes what is called a *complete matrix algebra* of order n.

The correspondence between linear transformations and matrices can be used to resolve several questions which were postponed in Chapter 1. In Section 4-2 we showed that the correspondence between linear transformations of $\mathscr{U}$ into $\mathscr{V}$ and the set of $m \times n$ matrices (where $n = \dim \mathscr{U}$ and $m = \dim \mathscr{V}$) is a bijection. In this section we have shown that the correspondence is a homomorphism: If ρ is a linear transformation represented by A and σ is a linear transformation represented by B, then $\rho + \sigma$ is represented by $A + B$, $\rho\sigma$ is represented by AB, and if a is a scalar $a\rho$ is represented by aA. We presumed in the previous sentence that each of the binary relations mentioned is defined. Specifically, the sum $\rho + \sigma$ is defined if and only if both have the same domain and codomain, which implies that A and B must be matrices of the same size. The composition $\rho\sigma$ is defined if and only if the domain of ρ is the codomain of σ, which implies A must have the same number of columns as B has rows.

Suppose τ is another linear transformation such that the composition $\sigma\tau$ is defined. It is easily shown that $(\rho\sigma)\tau = \rho(\sigma\tau)$. Simply apply the definition of the composition of mappings and observe that the conclusion does not depend on the linearity at all. If C is a matrix representing τ, it then follows that $(AB)C = A(BC)$. The associative law for matrix multiplication can be shown directly, but it is more tedious than showing the associative law for the composition of mappings. The other laws for operating with matrices which were stated in Section 1-1 without proof can now be established quite easily. In particular, formulas (4.4) and (4.6) imply corresponding distributive laws for matrix multiplication.

EXERCISES 4-4

1. Let $A = \begin{bmatrix} 6 & -2 \\ 15 & -5 \end{bmatrix}$ represent a linear transformation σ of a 2-dimensional vector space $\mathcal{U}$ into itself. Compute the matrix representing σ^2. A is the matrix appearing in Exercise 4 of Section 4-2. Find the kernel and image of σ^2.

2. Let $A = \begin{bmatrix} -7 & 3 \\ -16 & 7 \end{bmatrix}$ represent a linear transformation σ of a 2-dimensional vector space $\mathcal{U}$ into itself, and let $B = \begin{bmatrix} 7 & -3 \\ -16 & 7 \end{bmatrix}$ represent a linear transformation τ. Find the matrix representing $\tau\sigma$ and describe $\tau\sigma$ in geometric terms.

3. With σ and τ defined in Exercise 2, you can show that $(\tau\sigma)^2 = 1$. Use this observation to find the inverse of σ (or, equivalently, of the matrix A.)

4. Let $A = \begin{bmatrix} -2 & -4 & -6 \\ 6 & 9 & 12 \\ -3 & -4 & -5 \end{bmatrix}$ represent a linear transformation of a 3-dimensional vector space $\mathcal{U}$ into itself. Find the matrix representing σ^2. A is the matrix appearing in Exercise 8 of Section 4-2. Find the kernel and image of σ^2.

5. Let $A = \begin{bmatrix} -10 & 4 \\ -25 & 10 \end{bmatrix}$ represent a linear transformation τ of a 2-dimensional vector space $\mathcal{U}$ into itself. Find the matrix representing τ^2. Find the rank of τ^2. Compare the rank of τ^2 in this exercise with the rank of σ^2 in Exercise 1.

6. Show that if σ has an inverse, then σ^{-1} and σ have the same rank.

7. Show that if σ has an inverse and $\sigma\tau$ is defined, then rank τ = rank $\sigma\tau$.

8. Show that if σ has an inverse and $\tau\sigma$ is defined, then rank τ = rank $\tau\sigma$.

9. Show that if σ is a surjective linear transformation (an epimorphism), then rank $(\tau\sigma)$ = rank (τ).

10. Show that if τ is an injective linear transformation (a monomorphism), then rank $(\tau\sigma) =$ rank (σ).

11. Show that if σ is an epimorphism and $\tau_1\sigma = \tau_2\sigma$, then $\tau_1 = \tau_2$.

12. Show that if τ is a monomorphism and $\tau\sigma_1 = \tau\sigma_2$, then $\sigma_1 = \sigma_2$.

13. A *projection* π of a vector space into itself is a linear transformation with the property that $\pi^2 = \pi$. Show that if π is a projection, then $1 - \pi$ is also a projection.

14. Show that if π is a projection, $K(\pi) = \text{Im}(1-\pi)$, and $\text{Im}(\pi) = K(1-\pi)$.

15. Show that if π is a projection, then $K(\pi) \cap \text{Im}(\pi) = \{0\}$.

16. Show that the linear transformation σ discussed in Exercise 1 is a projection. Show for this example that $K(\sigma) \cap \text{Im}(\sigma) = \{0\}$.

17. Let π be a projection of a vector space $\mathcal{U}$ into itself. Show that every vector $\alpha \in \mathcal{U}$ can be written in the form $\alpha = \alpha_1 + \alpha_2$ where $\alpha_1 \in \text{Im}(\pi)$ and $\alpha_2 \in K(\pi)$. Furthermore, show that this representation is unique.

4-5 GEOMETRIC EXAMPLES

In Section 4-2 several examples of matrices representing linear transformations of $\mathcal{R}^2$ into itself were given. Example (1) is a reflection of $\mathcal{R}^2$ about the x_2-axis. Example (2) is a reflection of $\mathcal{R}^2$ about the 45° line. What is the result of performing one of these operations after the other? There are several ways a question of this type can be answered.

Let σ denote the reflection about the x_2-axis and let τ denote the reflection about the line $x_1 = x_2$. Let us compute both $\sigma\tau$ and $\tau\sigma$. Since τ maps α_1 onto α_2 and σ leaves α_2 fixed, $\sigma\tau$ maps α_1 onto α_2. τ maps α_2 onto α_1 and σ maps α_1 onto $-\alpha_1$. Thus $\sigma\tau$ maps α_2 onto $-\alpha_1$. This shows that $\sigma\tau$ is represented by the matrix

$$\begin{bmatrix} 0 & -1 \\ 1 & 0 \end{bmatrix}. \tag{5.1}$$

The same result can be obtained by multiplying the matrix representations. Thus, $\sigma\tau$ is represented by

$$\begin{bmatrix} -1 & 0 \\ 0 & 1 \end{bmatrix} \begin{bmatrix} 0 & 1 \\ 1 & 0 \end{bmatrix} = \begin{bmatrix} 0 & -1 \\ 1 & 0 \end{bmatrix}. \tag{5.2}$$

Comparison of (5.1) with (2.15) shows that $\sigma\tau$ is a rotation counterclockwise through 90°.

Clearly, for a reflection σ, $\sigma\sigma = 1$ so that a reflection is its own inverse. Then $(\tau\sigma)(\sigma\tau) = \tau(\sigma\sigma)\tau = \tau 1 \tau = \tau\tau = 1$, so that $\tau\sigma$ is the inverse of $\sigma\tau$. Since $\sigma\tau$ is a rotation 90° counterclockwise, $\tau\sigma$ is a 90° rotation clockwise. This can be verified by direct calculation analogous to (5.2).

Now let σ be a rotation counterclockwise through an angle of θ, and let τ be a rotation counterclockwise through an angle of ϕ. Then $\tau\sigma$ is a rotation counterclockwise through an angle of $(\theta + \phi)$. σ and τ can be represented by matrices in the form of (2.15) and we can calculate the representation of $\tau\sigma$. Thus, $\tau\sigma$ is represented by

$$\begin{bmatrix} \cos\phi & -\sin\phi \\ \sin\phi & \cos\phi \end{bmatrix} \begin{bmatrix} \cos\theta & -\sin\theta \\ \sin\theta & \cos\theta \end{bmatrix}$$
$$= \begin{bmatrix} \cos\phi\cos\theta - \sin\phi\sin\theta & -\cos\phi\sin\theta - \sin\phi\cos\theta \\ \sin\phi\cos\theta + \cos\phi\sin\theta & -\sin\phi\sin\theta + \cos\phi\cos\theta \end{bmatrix}. \tag{5.3}$$

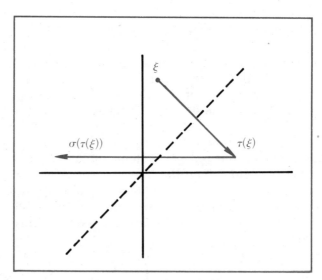

Figure 4–5.1 A composition of mappings. The first mapping τ is a reflection about the 45° line, and σ is a reflection about the vertical axis. (See Figures 4–2.4 and 4–2.3.) The point ξ is shown in a general (non-special) position. $\tau(\xi)$ is its mirror image with respect to the 45° line. $\sigma\tau(\xi)$ is the image of $\tau(\xi)$ reflected in the vertical axis.

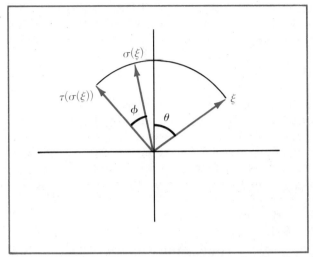

Figure 4–5.2 A composition of two rotations. A general point ξ is rotated by σ through an angle θ about the origin to $\sigma(\xi)$, and then through an angle ϕ to $\tau(\sigma(\xi))$.

But, $\tau\sigma$ must be represented by the matrix

$$\begin{bmatrix} \cos(\theta+\phi) & -\sin(\theta+\phi) \\ \sin(\theta+\phi) & \cos(\theta+\phi) \end{bmatrix}. \quad (5.4)$$

Thus,

$$\begin{aligned} \cos\theta\cos\phi - \sin\theta\sin\phi &= \cos(\theta+\phi), \\ \sin\theta\cos\phi + \cos\theta\sin\phi &= \sin(\theta+\phi). \end{aligned} \quad (5.5)$$

If σ is a rotation counterclockwise through an angle of θ, the inverse of σ is a rotation clockwise through an angle of θ (or counterclockwise through an angle of $-\theta$). Verify by computing the matrix product that

$$\begin{bmatrix} \cos\theta & \sin\theta \\ -\sin\theta & \cos\theta \end{bmatrix} \quad (5.6)$$

is the inverse of

$$\begin{bmatrix} \cos\theta & -\sin\theta \\ \sin\theta & \cos\theta \end{bmatrix}. \quad (5.7)$$

What is a rotation followed by a reflection? Let σ be a rotation represented by (5.7) and let τ be the reflection with respect to the x_1-axis. Then $\tau\sigma$ is represented by

$$\begin{bmatrix} 1 & 0 \\ 0 & -1 \end{bmatrix} \begin{bmatrix} \cos\theta & -\sin\theta \\ \sin\theta & \cos\theta \end{bmatrix} = \begin{bmatrix} \cos\theta & -\sin\theta \\ -\sin\theta & -\cos\theta \end{bmatrix}. \tag{5.8}$$

Although one might guess, it is not very evident what kind of linear transformation is represented by the matrix (5.8). With a little thought one can see that some points are left fixed by $\tau\sigma$. (See Figure 4–5.4.) The points on the line $\theta/2$ clockwise from the x_1-axis are first rotated by σ to the line $\theta/2$ counter-clockwise from the x_1-axis, and then reflected by τ back to their former positions.

To verify this observation, compute the effect of $\tau\sigma$ on $(\cos\theta/2, -\sin\theta/2)$.

$$\begin{bmatrix} \cos\theta & -\sin\theta \\ -\sin\theta & -\cos\theta \end{bmatrix} \begin{bmatrix} \cos\theta/2 \\ -\sin\theta/2 \end{bmatrix}$$

$$= \begin{bmatrix} \cos\theta\cos\theta/2 + \sin\theta\sin\theta/2 \\ -\sin\theta\cos\theta/2 + \cos\theta\sin\theta/2 \end{bmatrix} \tag{5.9}$$

$$= \begin{bmatrix} \cos(\theta - \theta/2) \\ -\sin(\theta - \theta/2) \end{bmatrix} = \begin{bmatrix} \cos\theta/2 \\ -\sin\theta/2 \end{bmatrix}.$$

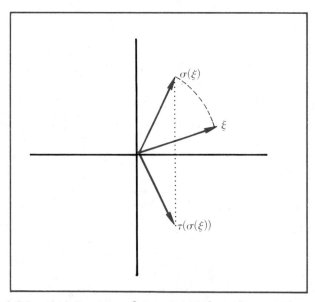

Figure 4–5.3 A composition of a rotation and a reflection. The general point ξ is rotated through an angle σ about the origin to $\sigma(\xi)$, and then reflected about the horizontal axis to $\tau(\sigma(\xi))$.

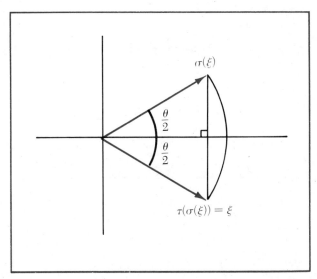

Figure 4–5.4 A fixed point for $\tau\sigma$. A point ξ on the line $\theta/2$ counterclockwise from the horizontal axis is rotated through an angle θ to a position on the line $\theta/2$ clockwise from the horizontal axis. Then the reflection about the horizontal axis returns it to its original position.

Thus, the vector ξ_1 represented by $(\cos\theta/2, -\sin\theta/2)$ is left fixed by $\tau\sigma$. In fact, the entire line through the origin and the end point of ξ_1 is left fixed. We suspect that $\tau\sigma$ is a reflection with respect to this line, so let us test its effect on the point $(\sin\theta/2, \cos\theta/2)$ which is on a line perpendicular to the fixed line.

$$\begin{bmatrix} \cos\theta & -\sin\theta \\ -\sin\theta & -\cos\theta \end{bmatrix} \begin{bmatrix} \sin\theta/2 \\ \cos\theta/2 \end{bmatrix}$$

$$= \begin{bmatrix} \cos\theta\sin\theta/2 - \sin\theta\cos\theta/2 \\ -\sin\theta\sin\theta/2 - \cos\theta\cos\theta/2 \end{bmatrix} \qquad (5.10)$$

$$= \begin{bmatrix} -\sin\theta/2 \\ -\cos\theta/2 \end{bmatrix} = - \begin{bmatrix} \sin\theta/2 \\ \cos\theta/2 \end{bmatrix}.$$

Now let ξ_2 be the vector represented by $(\sin\theta/2, \cos\theta/2)$. Then $(\tau\sigma)(\xi_1) = \xi_1$ and $(\tau\sigma)(\xi_2) = -\xi_2$. The matrix representing $\tau\sigma$ with respect to the basis $\{\xi_1, \xi_2\}$ is

$$\begin{bmatrix} 1 & 0 \\ 0 & -1 \end{bmatrix}. \qquad (5.11)$$

This computation shows that $\tau\sigma$ is a reflection with respect to the line through the origin and the point $(\cos\theta/2, -\sin\theta/2)$.

188 • Linear Transformations

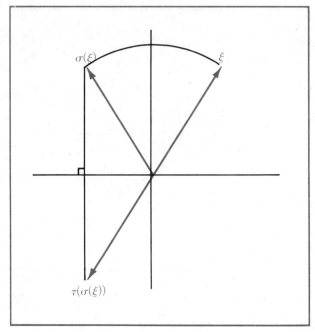

Figure 4–5.5 A point reflected through the origin by $\tau\sigma$. A point ξ on the line $\theta/2$ counterclockwise from the vertical axis is rotated through an angle θ to a position on the line $\theta/2$ clockwise from the vertical axis. Then the reflection about the horizontal axis puts it on the original line, but at an equal distance on the opposite side of the origin. This line and the fixed line in Figure 4–5.4 are orthogonal. The composite $\tau\sigma$ is a reflection about the fixed line.

A few simple facts about these linear transformations may help to unify the ideas contained in these examples. *Orientation* in $\mathscr{R}^2$ is a sense of rotation, clockwise or counterclockwise. Rotations preserve orientation and reflections reverse orientation. A mapping is *distance preserving* if for any two points ξ, η, the distance between $\sigma(\xi)$ and $\sigma(\eta)$ is the same as the distance between ξ and η. All orientation preserving, distance preserving mappings of the plane into itself that leave the origin fixed are rotations. All orientation reversing, distance preserving mappings of the plane into itself that leave the origin fixed are reflections. (As we already noted, the concept of distance is used here only for description, so we will not attempt to prove these assertions. The interested reader can find satisfactory proofs for these assertions for himself.) By these principles, the composition of two reflections is orientation preserving and, hence, a rotation.

The composition of a rotation and a reflection is orientation reversing and, hence, a reflection.

The situation in three dimensions is somewhat more complicated. But at least every distance preserving mapping that preserves orientation and leaves the origin fixed is a rotation. Distance preserving, orientation reversing mappings that leave the origin fixed are either reflections or a combination of a reflection and rotation. A typical rotation-reflection is a rotation about a vertical axis followed by a reflection with respect to the horizontal plane. This opens an interesting line of discussion but we shall resist pursuing it.

Consider $\mathcal{R}^3$, which we draw with three mutually perpendicular axes. Let α_1, α_2, α_3 be vectors of length one in the direction of each axis. A rotation about the x_3-axis leaves α_3 fixed. It rotates the x_1, x_2-plane as in the previous examples. If σ is a rotation through θ about the x_3-axis, σ is represented by

$$\begin{bmatrix} \cos\theta & -\sin\theta & 0 \\ \sin\theta & \cos\theta & 0 \\ 0 & 0 & 1 \end{bmatrix}. \qquad (5.12)$$

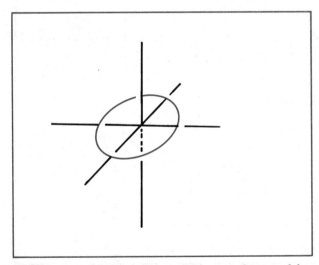

Figure 4–5.6 A rotation in $\mathcal{R}^3$. The vertical axis is the axis of the rotation. The horizontal plane is rotated within itself through an angle θ about this axis.

To simplify matters take $\theta = \pi/2$. Then (5.12) takes the form

$$\begin{bmatrix} 0 & -1 & 0 \\ 1 & 0 & 0 \\ 0 & 0 & 1 \end{bmatrix}. \tag{5.13}$$

Similarly, a rotation through $\pi/2$ about the x_2-axis would be represented by

$$\begin{bmatrix} 0 & 0 & -1 \\ 0 & 1 & 0 \\ 1 & 0 & 0 \end{bmatrix}. \tag{5.14}$$

Thus, the rotation represented by (5.13) followed by the rotation represented by (5.14) would be represented by

$$\begin{bmatrix} 0 & 0 & -1 \\ 1 & 0 & 0 \\ 0 & -1 & 0 \end{bmatrix}. \tag{5.15}$$

(5.15) also represents a rotation. If ρ is the linear transformation represented by (5.15), then $\rho(\alpha_1) = \alpha_2$, $\rho(\alpha_2) = -\alpha_3$, and $\rho(-\alpha_3) = \alpha_1$. Thus, the rotation ρ is the one represented in Figure 4–5.7. The axis of this rotation goes through the

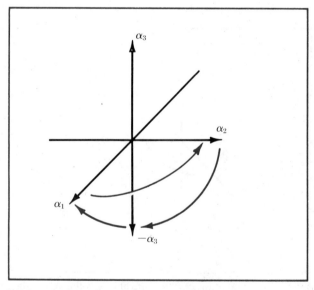

Figure 4–5.7 A rotation in $\mathscr{R}^3$. This rotation sends α_1 to the position of α_2, α_2 to the position of $-\alpha_3$, and $-\alpha_3$ to the position of α_1. It permutes $\{\alpha_1, \alpha_2, -\alpha_3\}$ cyclically. The axis of this rotation extends into the lower-right-front orthant at an equal angle from all coordinate axes.

origin and the point $(1, 1, -1)$, and the angle of this rotation is through $\theta = 2\pi/3$.

EXERCISES 4-5

1. Analogous to (5.12), write down the matrix representing a rotation through θ about the x_2-axis.

2. If one knows that a linear transformation is a rotation, the axis of the rotation can be determined by finding points left fixed by the transformation. Suppose that we know the composition of two rotations is a rotation. Show how we can find the axis of the composition of the rotations represented by (5.13) and (5.14).

3. Suppose we know that every distance preserving, orientation preserving linear transformation of a 3-dimensional space is a rotation. Show that the composition of two rotations is a rotation.

4. Let us accept as a fact that an orientation reversing distance preserving linear transformation of a 3-dimensional space into itself is of the form of a reflection in a plane followed by a rotation about an axis perpendicular to the plane of rotation. For example, if the plane of reflection is the x_1, x_2-plane, the matrix representing this transformation takes the form

$$\begin{bmatrix} \cos\phi & -\sin\phi & 0 \\ \sin\phi & \cos\phi & 0 \\ 0 & 0 & -1 \end{bmatrix} \quad (5.16)$$

In particular, for $\phi = \pi$, (5.16) takes the form

$$\begin{bmatrix} -1 & 0 & 0 \\ 0 & -1 & 0 \\ 0 & 0 & -1 \end{bmatrix} \quad (5.17)$$

Compose (5.17) with (5.12) and show that the result is of the form of (5.16). Compose (5.16) with (5.12) and show the result is in the form of (5.16).

4-6 CHANGE OF BASIS

Whenever anything has a representation in a mathematical model we must always answer the question as to how this representation is changed if any arbitrary choices are changed. We have faced this problem in Chapter 3, in which we used coordinates to represent vectors. Now, in this chapter we have represented linear transformations by matrices. This representation depends on the choice of bases, and the representation will be changed if different bases are chosen. What is the relation between the various representations?

Let σ be a linear transformation of $\mathscr{U}$ into $\mathscr{V}$. Let $\mathscr{A} = \{\alpha_1, \ldots, \alpha_n\}$ be a given basis in $\mathscr{U}$ and $\mathscr{B} = \{\beta_1, \ldots, \beta_m\}$ be a given basis in $\mathscr{V}$. As in Section 4-2, σ is represented by $A = [a_{ij}]$ where

$$\sigma(\alpha_j) = \sum_{i=1}^{m} a_{ij}\beta_i. \tag{6.1}$$

Let $\mathscr{A}' = \{\alpha_1', \ldots, \alpha_n'\}$ be a new basis in $\mathscr{U}$ and let $P = [p_{ij}]$ be the matrix of transition from $\mathscr{A}$ to $\mathscr{A}'$. As defined in Section 3-2 this means

$$\alpha_j' = \sum_{i=1}^{n} p_{ij}\alpha_i. \tag{6.2}$$

Now, if $A' = [a_{ij}']$ is the matrix representing σ with respect to $\mathscr{A}'$ and $\mathscr{B}$, this means

$$\sigma(\alpha_j') = \sum_{i=1}^{m} a_{ij}'\beta_i$$

$$= \sigma\left(\sum_{k=1}^{n} p_{kj}\alpha_k\right)$$

$$= \sum_{k=1}^{n} p_{kj}\sigma(\alpha_k) \tag{6.3}$$

$$= \sum_{k=1}^{n} p_{kj}\left(\sum_{i=1}^{m} a_{ik}\beta_i\right)$$

$$= \sum_{i=1}^{m} \left(\sum_{k=1}^{n} a_{ik}p_{kj}\right)\beta_i.$$

Since the representation of $\sigma(\alpha_j')$ is unique we have

$$a_{ij}' = \sum_{k=1}^{n} a_{ik}p_{kj}. \tag{6.4}$$

(6.4) is equivalent to the matrix equation

$$A' = AP. \tag{6.5}$$

(6.5) tells us what we want to know. If A is given and P is the matrix of transition, the matrix representing σ in the new coordinate system is readily computed from (6.5).

In Figure 4–6.1, the left triangle is from Figure 3–2.2. We showed in Section 3–2 that this triangle is commutative, which gives the formula $\mathcal{A}' = \mathcal{A}P$. The rectangle is from Figure 4–2.1. It is also commutative and gives the formula $\sigma\mathcal{A} = \mathcal{B}A$. Then

$$\begin{aligned}\eta = \sigma(\xi) &= \sigma(\mathcal{A}'X') = \sigma(\mathcal{A}PX') = (\sigma\mathcal{A})(PX') \\ &= (\mathcal{B}A)(PX') = \mathcal{B}(APX').\end{aligned} \tag{6.6}$$

Since the coordinatization of $\mathcal{V}$ means $\eta = \mathcal{B}Y$ and $\mathcal{B}$ is an isomorphism, we have

$$Y = APX'. \tag{6.7}$$

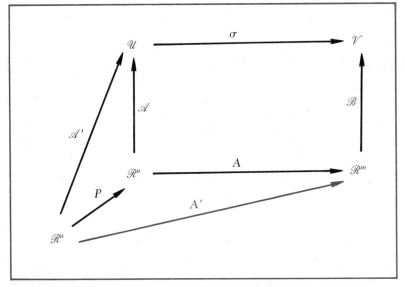

Figure 4–6.1 The triangle to the left is the triangle of Figure 3–2.2, with a change of notation. The rectangle is the rectangle of Figure 4–2.1. Both the triangle and the rectangle are commutative. The matrix A' fills in the diagram so that the lower triangle commutes. This gives $A' = AP$, which is formula (6.5).

Now the commutativity of the lower triangle, which is Formula (6.5), is equivalent to $Y = A'X'$. Thus, A' represents σ with respect to $\mathscr{A}'$ and $\mathscr{B}$.

Now suppose we change the basis in $\mathscr{V}$. Let $\mathscr{B}' = \{\beta_1', \ldots, \beta_m'\}$ be a new basis in $\mathscr{V}$ with matrix of transition $Q = [q_{ij}]$. That is,

$$\beta_j' = \sum_{i=1}^{m} q_{ij}\beta_i. \tag{6.8}$$

If $A'' = [a_{ij}'']$ is the matrix representing σ with respect to $\mathscr{A}$ and $\mathscr{B}'$, we have

$$\sigma(\alpha_j) = \sum_{i=1}^{m} a_{ij}''\beta_i'. \tag{6.9}$$

Substituting (6.8) in (6.9) we have

$$\sigma(\alpha_j) = \sum_{i=1}^{m} a_{ij}'' \left(\sum_{k=1}^{m} q_{ki}\beta_k \right)$$
$$= \sum_{k=1}^{m} \left(\sum_{i=1}^{m} q_{ki}a_{ij}'' \right) \beta_k. \tag{6.10}$$

Again, since the representation of $\sigma(\alpha_j)$ in terms of the basis $\mathscr{B}$ is unique we have

$$a_{kj} = \sum_{i=1}^{m} q_{ki}a_{ij}''. \tag{6.11}$$

Equation (6.11) is equivalent to the matrix relation

$$A = QA''. \tag{6.12}$$

Again, (6.12) tells us what we wish to know. But here an additional calculation is required. If A and Q are given,

$$A'' = Q^{-1}A. \tag{6.13}$$

In Figure 4–6.2, the rectangle is from Figure 3–2.1, and the triangle at the right is from Figure 3–2.2. Both are commutative and give the formulas $\sigma\mathscr{A} = \mathscr{B}A$ and $\mathscr{B}Q = \mathscr{B}'$. Then

$$\eta = \sigma(\xi) = \sigma(\mathscr{A}X) = (\sigma\mathscr{A})X = (\mathscr{B}A)X = \mathscr{B}(AX)$$
$$= (\mathscr{B}'Q^{-1})(AX) = \mathscr{B}'(Q^{-1}AX). \tag{6.14}$$

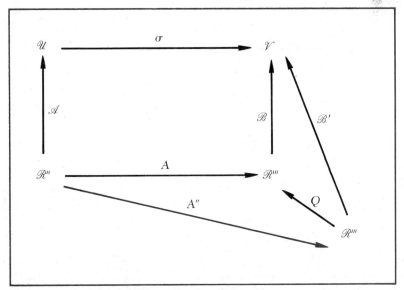

Figure 4-6.2 The rectangle is the rectangle of Figure 4-2.1. The triangle at the right is the triangle of Figure 3-2.2, with a change of notation. Both the rectangle and the triangle are commutative. The matrix A'' fills in the diagram so the lower triangle commutes. This gives $A = QA''$, or $A'' = Q^{-1}A$.

The coordination of $\mathscr{V}$ with respect to $\mathscr{B}'$ means $\eta = \mathscr{B}'Y'$. Since $\mathscr{B}'$ is an isomorphism, we have

$$Y' = Q^{-1}AX. \qquad (6.15)$$

Now, the commutativity of the lower triangle, which is formula (6.13), is equivalent to $Y' = A''X$. Thus, A'' represents σ with respect to $\mathscr{A}$ and $\mathscr{B}'$.

It is possible to change bases in both $\mathscr{U}$ and $\mathscr{V}$ at the same time. If both changes are made the matrix representing σ is

$$Q^{-1}AP. \qquad (6.16)$$

The situation in the full generality discussed here is rather simple. Let a basis for $\mathscr{U}$ be chosen as in the proof of Theorem 3.7, except that we will list the elements of this basis in a different order. Let $\{\alpha_1, \ldots, \alpha_r, \alpha_{r+1}, \ldots, \alpha_n\}$ be a basis for $\mathscr{U}$ where $\{\alpha_{r+1}, \ldots, \alpha_n\}$ is a basis for $K(\sigma)$. As in the proof of Theorem 3.7, $\{\sigma(\alpha_1), \ldots, \sigma(\alpha_r)\}$ is linearly independent. Let $\sigma(\alpha_j) = \beta_j$ for $j = 1, \ldots, r$, and extend this set to a basis $\{\beta_1, \ldots, \beta_r, \beta_{r+1}, \ldots, \beta_m\}$ of $\mathscr{V}$. Since $\sigma(\alpha_j) =$

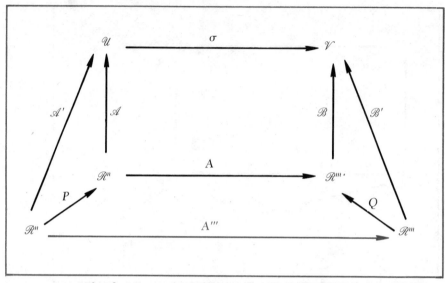

Figure 4-6.3 This diagram combines Figures 4-6.1 and 4-6.2. The matrix A''' fills in the diagram so the lower rectangle commutes. The commutativity of the lower rectangle gives $AP = QA'''$, or $A''' = Q^{-1}AP$, which is formula (6.16).

β_j for $j = 1, \ldots, r$ and $\sigma(\alpha_j) = 0$ for $j > r$, σ is represented by the matrix

$$\begin{bmatrix} 1 & 0 \cdots 0 & 0 \cdots 0 \\ 0 & 1 \cdots 0 & 0 \cdots 0 \\ \cdot & \cdot & \cdot \\ \cdot & \cdot & \cdot \\ \cdot & \cdot & \cdot \\ 0 & 0 \cdots 1 & 0 \cdots 0 \\ 0 & 0 \cdots 0 & 0 \cdots 0 \\ \cdot & \cdot & \cdot \\ \cdot & \cdot & \cdot \\ \cdot & \cdot & \cdot \\ 0 & 0 \quad 0 & 0 \quad 0 \end{bmatrix} \quad (6.17)$$

The matrix in (6.17) has 1's in the first r positions of the main diagonal and zeros elsewhere.

This means that, given any linear transformation σ of $\mathscr{U}$ into $\mathscr{V}$ of rank r, there exists a choice of bases for which σ is represented by (6.17). Thus, any two linear transformations of $\mathscr{U}$ into $\mathscr{V}$ of the same rank are "essentially" alike. They may be different when compared to each other, and they may have different representations with respect to any given choice of bases. But they cannot be distinguished from each other by reference to any internal or intrinsic properties.

The situation is far more complicated and interesting when we take $\mathscr{U} = \mathscr{V}$. In this case it is implicit that the bases $\mathscr{A}$ and $\mathscr{B}$ are the same. This also means that if $\mathscr{A}$ is replaced by the basis $\mathscr{A}'$, where P is the matrix of transition from $\mathscr{A}$ to $\mathscr{A}'$, formula (6.16) for the new matrix representing a linear transformation becomes

$$P^{-1}AP. \tag{6.18}$$

We say that A and $P^{-1}AP$ are *similar*. In other words, A and A' are similar if and only if there exists a non-singular P such that $A' = P^{-1}AP$.

The matrix in (6.17) is called a *normal form*, or a *canonical form*. In mathematics these words mean that the form is "standard" in some sense or other. In the first two chapters we studied matrices that represented systems of linear equations. In the context of that discussion, matrices were equivalent if and only if they represented equivalent systems. We found out that this meant that two matrices were equivalent if and only if one could be obtained from the other by a sequence of elementary row operations. The normal form we obtained there was the reduced basic form.

The problem we face now is that of finding a normal form under similarity. That is, given any square matrix A, find a matrix P with an inverse P^{-1} such that $P^{-1}AP$ is in normal form. Or, find the normal form without finding P, if possible. This turns out to be more complicated than the two cases discussed so far in this text (the reduced basic form in Chapter 1 and the form in (6.17)). To find this form is the primary objective of Chapter 6.

What properties should a normal form have? And why should we bother? The normal form should be relatively simple. It should be possible to find it by direct and effective methods. The normal form should reveal information that we wish to know, by mere inspection if possible. By taking another look at the first two chapters we should see that the reduced basic form we obtained there could be found by direct and effective methods, and the general solution of the corresponding system of linear equations could be read off by inspection. In the case of the normal form in (6.17), it is not necessary to obtain the matrices P and Q such that $Q^{-1}AP$ is in normal form. There the normal form is known as soon as the rank of A is known. But that is really all the information the normal form can give anyway. Thus, there is little point in trying to obtain (6.17) as a normal form. It is quite sufficient to determine the rank of A.

EXERCISES 4-6

1. In Exercise 3 of Section 4-2, we gave the matrix $\begin{bmatrix} -4 & 6 \\ -3 & 5 \end{bmatrix}$ as representing a linear transformation σ, and asked for the matrix representing σ with respect to the basis $\{(1, 1), (-2, -1)\}$. Use $P = \begin{bmatrix} 1 & -2 \\ 1 & -1 \end{bmatrix}$ as a matrix of transition to find the required representation.

2. Do Exercise 4 of Section 4-2 in the same manner, using $\begin{bmatrix} 2 & 1 \\ 5 & 3 \end{bmatrix}$ as the matrix of transition.

3. Do Exercise 5 of Section 4-2 in the same manner with the appropriate matrix of transition.

4. Do Exercise 6 of Section 4-2 in the same manner with the appropriate matrix of transition.

5. Do Exercise 8 of Section 4-2 in the same manner, using $\begin{bmatrix} -2 & 0 & 1 \\ 0 & 3 & -2 \\ 1 & -2 & 1 \end{bmatrix}$ as the matrix of transition.

6. Show the mapping of A into $A' = P^{-1}AP$ is an isomorphism of the set of $n \times n$ matrices into themselves. This means the mapping preserves all three operations defined for matrices and is a bijection.

7. Use the isomorphism described in Exercise 6 to compute A^{10} where $A = \begin{bmatrix} -4 & 6 \\ -3 & 5 \end{bmatrix}$ is the matrix given in Exercise 1. (Hint: Use the matrix of transition P given in Exercise 1 to obtain the simpler matrix $P^{-1}AP$, compute $(P^{-1}AP)^{10} = P^{-1}A^{10}P$, and then compute A^{10}.)

8. Show that A and $P^{-1}AP$ (where P^{-1} exists) have the same rank.

9. Show that if $A' = P^{-1}AP$ and $B' = P^{-1}BP$, then $A'B' = P^{-1}(AB)P$. Show that $A' + B' = P^{-1}(A+B)P$. Show that for any scalar a, $aA' = P^{-1}(aA)P$.

10. Let $f(x) = a_n x^n + \cdots + a_0$ be a polynomial with coefficients in F. To substitute a square matrix A for x in $f(x)$ we must replace the constant term a_0 by $a_0 I$, where I is the identity of the same order as A. Show that $f(P^{-1}AP) = P^{-1}f(A)P$.

11. Show that if a matrix A satisfies a polynomial equation with scalar coefficients, then every matrix similar to A satisfies the same equation.

4-7 LINEAR TRANSFORMATIONS AND LINEAR PROBLEMS

In Chapter 1 we considered the problem of solving systems of linear equations, and there we introduced matrix notation so that such a system could be written in the form

$$AX = B, \qquad (7.1)$$

where A is an $m \times n$ matrix, X is a one-column n-tuple, and B is a one-column m-tuple. In view of what we have discussed in this chapter, we can also consider A as representing a linear transformation σ of an n-dimensional space $\mathscr{U}$ into an m-dimensional space $\mathscr{V}$. We consider B as representing a vector $\beta = \sum_{i=1}^{m} b_i \beta_i \in \mathscr{V}$, and X as representing a vector $\xi = \sum_{j=1}^{n} x_j \alpha_j \in \mathscr{U}$. In this form, solving (7.1) is equivalent to solving the equation

$$\sigma(\xi) = \beta, \qquad (7.2)$$

where σ and β are given. We are to find any and all $\xi \in \mathscr{U}$ satisfying (7.2). This is called a *linear problem*.

As in Chapter 1, there are really two questions here. Are there any solutions at all? And, if there are any solutions, what does the solution set look like? For the first question, it is clear that (7.2) has a solution if and only if $\beta \in \text{Im}(\sigma)$. The problem of determining whether $\beta \in \text{Im}(\sigma)$ was discussed in Section 2-6. In terms of representations in the coordinate spaces, the columns of A span $\text{Im}(\sigma)$, and we need merely determine whether B is a linear combination of these columns.

Suppose we determine that the problem has a solution and we proceed to find it. We would find that we would have to duplicate much of the same sequence of steps in determining whether $\beta \in \text{Im}(\sigma)$. There is so little difference in the work of answering these two questions that it is not worth while trying to answer the first question as a separate question. If we direct our attention to a technique for answering the second question, the answer to the first question will be available by observation.

Suppose that ξ_1 and ξ_2 are solutions to equation (7.2). Then

$$\sigma(\xi_1 - \xi_2) = \sigma(\xi_1) - \sigma(\xi_2) = \beta - \beta = 0. \tag{7.3}$$

The equation

$$\sigma(\xi) = 0 \tag{7.4}$$

is called the *associated homogeneous problem*. Its solution is the kernel of σ, $K(\sigma)$. From (7.3) we see that $\xi_1 - \xi_2$ is a solution of the associated homogeneous problem. Thus, if ξ_1 and ξ_2 are solutions of equation (7.2), then $\xi_1 - \xi_2 \in K(\sigma)$.

Any solution of equation (7.2) is called a *particular solution*. Let ξ_0 denote any particular solution, and let $\mathscr{S}$ denote the solution subset for (7.2). Let $\xi_0 + K(\sigma)$ denote the set of all vectors of the form $\xi_0 + \eta$, where $\eta \in K(\sigma)$. If ξ is any solution of equation (7.2), then $\xi - \xi_0 \in K(\sigma)$, or $\xi \in \xi_0 + K(\sigma)$. This shows that $\mathscr{S} \subset \xi_0 + K(\sigma)$.

On the other hand, if $\xi \in \xi_0 + K(\sigma)$, then $\xi = \xi_0 + \eta$ where $\eta \in K(\sigma)$. Then $\sigma(\xi) = \sigma(\xi_0) + \sigma(\eta) = \beta + 0 = \beta$. Thus, $\xi_0 + K(\sigma) \subset \mathscr{S}$. This shows that $\mathscr{S} = \xi_0 + K(\sigma)$. To solve a linear problem *we find the general solution of the associated homogeneous problem and any particular solution. Then, add the particular solution to the general solution of the associated homogeneous problem.*

Consider the linear problem $AX = B$ in equation form

$$\begin{aligned} a_{11}x_1 + a_{12}x_2 + \cdots + a_{1n}x_n &= b_1 \\ a_{21}x_1 + a_{22}x_2 + \cdots + a_{2n}x_n &= b_2 \\ &\vdots \\ a_{m1}x_1 + a_{m2}x_2 + \cdots + a_{mn}x_n &= b_m. \end{aligned} \tag{7.5}$$

The augmented matrix for this system is

$$\begin{bmatrix} a_{11} & a_{12} & \cdots & a_{1n} & b_1 \\ a_{21} & a_{22} & \cdots & a_{2n} & b_2 \\ \cdot & \cdot & & \cdot & \cdot \\ \cdot & \cdot & & \cdot & \cdot \\ \cdot & \cdot & & \cdot & \cdot \\ a_{m1} & a_{m2} & \cdots & a_{mn} & b_m \end{bmatrix} \qquad (7.6)$$

Symbolically, this augmented matrix can be represented in the form $[A\ B]$. Reduce (7.6) to basic form, attempting at each step to choose a pivot element from one of the first n columns (i.e., not from the last column). There are two possibilities: either it is possible to reduce $[A\ B]$ to basic form this way or it is eventually necessary to choose a pivot element from the last column. If it is necessary to choose a pivot element from the last column, the last column is not a linear combination of the preceding columns. This means $\beta \notin \text{Im}(\sigma)$ and the linear problem has no solution. If it is possible to reduce $\lfloor A\ B \rfloor$ to basic form by choosing pivot elements from the first n columns, $\beta \in \text{Im}(\sigma)$ and the linear problem has a solution. Furthermore, A will have been reduced to basic form separately. The solution of the associated homogeneous problem $AX = 0$ can be obtained from the reduced basic form, as discussed in Chapter 2. There remains the problem of finding a particular solution for $AX = B$. But the last column of $[A\ B]$ is a linear combination of the preceding columns, and any such linear combination gives a particular solution. How this linear combination can be obtained from the reduced basic form of $[A\ B]$ is explained in Section 2–6. This is the recommended way to solve a system of linear equations.

We shall give a numerical example to illustrate the method and the form of the solution. In Section 1–5, Example 4, we considered a system of linear equations,

$$\begin{aligned} 2x_1 \phantom{{}+2x_2} + 6x_3 + x_4 + 2x_5 + 4x_6 &= 5 \\ 2x_1 + 2x_2 + 10x_3 \phantom{{}+x_4} + x_5 + 3x_6 &= 8 \\ x_1 + x_2 + 5x_3 + x_4 + 3x_5 + x_6 &= 8 \\ 3x_1 \phantom{{}+2x_2} + 9x_3 + x_4 + x_5 + 7x_6 &= 4 \\ 5x_1 + 3x_2 + 21x_3 + 5x_4 + 2x_5 + 18x_6 &= 10. \end{aligned} \qquad (7.7)$$

The augmented matrix is

$$\begin{bmatrix} 2 & 0 & 6 & 1 & 2 & 4 & 5 \\ 2 & 2 & 10 & 0 & 1 & 3 & 8 \\ 1 & 1 & 5 & 1 & 3 & 1 & 8 \\ 3 & 0 & 9 & 1 & 1 & 7 & 4 \\ 5 & 3 & 21 & 5 & 2 & 18 & 10 \end{bmatrix} \quad (7.8)$$

We can reduce this matrix to a basic form. To be specific, we obtain the row-echelon form. Thus, we have

$$\begin{bmatrix} 1 & 0 & 3 & 0 & 0 & 2 & 1 \\ 0 & 1 & 2 & 0 & 0 & 0 & 2 \\ 0 & 0 & 0 & 1 & 0 & 2 & -1 \\ 0 & 0 & 0 & 0 & 1 & -1 & 2 \\ 0 & 0 & 0 & 0 & 0 & 0 & 0 \end{bmatrix} \quad (7.9)$$

The basic elements are shown in color. Since the last column does not contain a basic element, the system has a solution.

If $\mathscr{A} = \{\alpha_1, \ldots, \alpha_n\}$ is a basis of $\mathscr{U}$, then $\{\sigma(\alpha_1), \ldots, \sigma(\alpha_n)\}$ spans $\text{Im}(\sigma)$. If $\beta \in \text{Im}(\sigma)$, then β can be represented in the form

$$\beta = \sum_{j=1}^{n} x_j \sigma(\alpha_j). \quad (7.10)$$

For any such representation, let $\xi = \sum_{j=1}^{n} x_j \alpha_j$. Because of the linearity of σ, $\sigma(\xi) = \beta$ and ξ is a solution of the linear problem. All solutions can be obtained in this way, but at the moment we are interested in finding a particular solution. Thus, we are interested in finding a representation in the form of (7.10) that is easy to obtain.

In (7.9) the columns containing the basic elements are the representations of a basis for $\text{Im}(\sigma)$. Thus, in this example $\{\sigma(\alpha_1), \sigma(\alpha_2), \sigma(\alpha_4), \sigma(\alpha_5)\}$ is a basis for $\text{Im}(\sigma)$. Using this basis a representation of β in the form of (7.10) can be read off—the numbers in the last column are the required coefficients. Thus, $\beta = \sigma(\alpha_1) + 2\sigma(\alpha_2) - \sigma(\alpha_4) + 2\sigma(\alpha_5)$ is such a representation. This means $\xi_0 = \alpha_1 + 2\alpha_2 - \alpha_4 + 2\alpha_5$ is a particular solution. Its representation as a 6-tuple is $(1, 2, 0, -1, 2, 0)$. Since the original problem is given in equation form, and therefore in terms of 6-tuples, the answer should be given in terms of 6-tuples. Thus, one should write down $(1, 2, 0, -1,$

2, 0) as a particular solution and leave out as much of the intervening steps and reasoning as possible.

Next one should find a basis for the kernel of σ. This is available from examination of the first six columns of (7.9), as discussed in detail in Chapter 2. The third and sixth columns correspond to parameters. Thus, $\{(-3, -2, 1, 0, 0, 0), (-2, 0, 0, -2, 1, 1)\}$ represents a basis of $K(\sigma)$. Since the solution set is $\xi_0 + K)\sigma)$, all elements of the solution set can be represented in the form

$$(1, 2, 0, -1, 2, 0) + t_1(-3, -2, 1, 0, 0, 0) \\ + t_2(-2, 0, 0, -2, 1, 1) \qquad (7.11)$$

where t_1 and t_2 are parameters of the solution set.

In this example, $K(\sigma)$ is a two-dimensional subspace, a plane. For every vector η in $K(\sigma)$ there corresponds one vector $\xi_0 + \eta$ in $\xi_0 + K(\sigma)$. A dependable picture of this situation is to think of plane $K(\sigma)$ "translated" parallel to itself to a new position, as in the following figure.

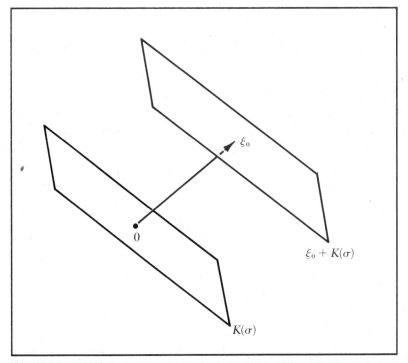

Figure 4-7.1   A coset $K(\sigma)$ is a subspace. For each ξ_0, $\xi_0 + K(\sigma)$ is a coset of $K(\sigma)$. It is the set $K(\sigma)$ translated parallel to itself. Since $0 \in K(\sigma)$, $\xi_0 \in \xi_0 + K(\sigma)$.

A set of the form $\xi + K(\sigma)$ is called a *coset* of $K(\sigma)$. If two cosets have a vector in common they are identical. To see this, suppose ξ is in both $\xi_1 + K(\sigma)$ and $\xi_2 + K(\sigma)$. Then there exist $\eta_1, \eta_2 \in K(\sigma)$ such that $\xi = \xi_1 + \eta_1$ and $\xi = \xi_2 + \eta_2$. Then $\xi_1 = \xi_2 + \eta_2 - \eta_1 \in \xi_2 + K(\sigma)$, and $\xi_1 + \eta = \xi_2 + (\eta_2 - \eta_1 + \eta) \in \xi_2 + K(\sigma)$. This shows that $\xi_1 + K(\sigma) \subset \xi_2 + K(\sigma)$. Symmetrically, it can be shown that $\xi_2 + K(\sigma) \subset \xi_1 + K(\sigma)$. The cosets of $K(\sigma)$ thus form a "stack" of parallel planes. A reasonably good picture is to think of it like a stack of playing cards. If $K(\sigma)$ were 1-dimensional, the cosets of $K(\sigma)$ would look like a bundle of parallel straight lines. The transformation σ maps all points in a coset onto the same vector in $\mathscr{V}$. Vectors in different cosets are mapped onto different vectors in $\mathscr{V}$. There is a one-to-one correspondence between cosets of $K(\sigma)$ and vectors in $\text{Im}(\sigma)$.

EXERCISES 4–7

1. Let $\mathscr{V}$ be the vector space of differentiable real valued functions in the real numbers, and let D be the linear

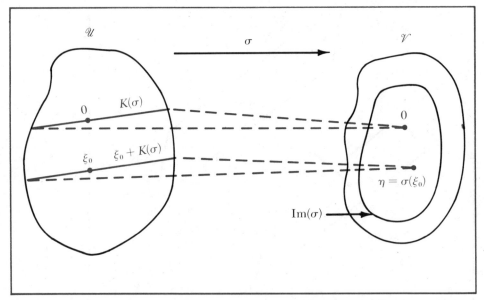

Figure 4–7.2 A linear problem $\sigma(\xi) = \eta$ has a solution if and only if $\eta \in \text{Im}(\sigma)$. If $\eta \in \text{Im}(\sigma)$ the solution set is a coset of the kernel. All elements of a coset of the kernel are mapped onto a single vector in $\text{Im}(\sigma)$.

operator $Df = \dfrac{df}{dx}$. Show that this linear problem. "Find f such that $\dfrac{df}{dx} = \sin x$," has a solution set of the form described in this section (a particular solution + the solution of the homogeneous problem).

2. For each of the exercises in Section 1–4, show that the solution set is of the form described in this section.

3. Let us examine the solvability of this system

$$2x_1 - 3x_2 - 8x_3 = 7$$
$$-x_1 + 2x_2 + 5x_3 = -4$$
$$3x_1 + 4x_2 + 5x_3 = 2$$

more closely. Determine the rank of the linear transformation represented by the matrix $\begin{bmatrix} 2 & -3 & -8 \\ -1 & 2 & 5 \\ 3 & 4 & 5 \end{bmatrix}$. Show that if σ represents a linear transformation of $\mathscr{U}$ into $\mathscr{V}$, where both are 3-dimensional vector spaces, then $\mathrm{Im}(\sigma)$ is a proper subspace of $\mathscr{V}$. That is, there are vectors $\beta \in \mathscr{V}$ for which the problem $\sigma(\xi) = \beta$ has no solution.

4. Consider the system

$$2x_1 - 3x_2 - 8x_3 = y_1$$
$$-x_1 + 2x_2 + 5x_3 = y_2$$
$$3x_1 + 4x_2 + 5x_3 = y_3,$$

where (y_1, y_2, y_3) is arbitrary and we consider solving the system (7.13) for each choice of (y_1, y_2, y_3). Write the augmented matrix for the system (7.13) and reduce it to basic form, except for the last column. The last row will yield an equation for the y_i that must be satisfied in order that the system (7.13) have a solution. Show that this equation has $\mathrm{Im}(\sigma)$ as its solution set.

For any (y_1, y_2, y_3) for which the system (7.13) has a solution, the solution set of (7.13) is a plane. Show that, for each such choice of (y_1, y_2, y_3), we get a distinct solution set (no common elements for different solution sets). Show that there is one distinct solution set of each element of $\mathrm{Im}(\sigma)$. In other words, the correspondence between cosets of $K(\sigma)$ and elements of $\mathrm{Im}(\sigma)$ is a bijection.

5. Go through the analysis outlined in Exercises 3 and 4 for the system

$$\begin{aligned} x_1 - 3x_2 + x_3 + x_4 &= y_1 \\ -x_1 + 3x_2 + 2x_3 - x_4 &= y_2 \\ 2x_1 - 6x_2 - x_3 + 2x_4 &= y_3 \\ -2x_1 + 6x_2 + x_3 - x_4 &= y_4 \end{aligned}$$

6. Go through the analysis outlined in Exercises 3 and 4 for the system

$$\begin{aligned} 2x_1 \phantom{{}+2x_2} + 6x_3 + x_4 + 2x_5 + 4x_6 &= 5 \\ 2x_1 + 2x_2 + 10x_3 \phantom{{}+x_4} + x_5 + 3x_6 &= 8 \\ x_1 + x_2 + 5x_3 + x_4 + 3x_5 + x_6 &= 8 \\ 3x_1 \phantom{{}+2x_2} + 9x_3 + x_4 + x_5 + 7x_6 &= 4 \\ 5x_1 + 3x_2 + 21x_3 + 5x_4 + 2x_5 + 18x_6 &= 10. \end{aligned}$$

Summary

The concept of a linear transformation is perhaps the most important single idea treated in this book. Given a domain $\mathscr{U}$ and a codomain $\mathscr{V}$, a mapping σ of $\mathscr{U}$ into $\mathscr{V}$ is *linear* if for all $\alpha, \beta \in \mathscr{U}, a \in \mathscr{F}$, we have

$$\sigma(\alpha + \beta) = \sigma(\alpha) + \sigma(\beta), \tag{S.1}$$

$$\sigma(a\alpha) = a\sigma(\alpha). \tag{S.2}$$

Let $\mathscr{A} = \{\alpha_1, \ldots, \alpha_n\}$ be a basis of $\mathscr{U}$ and $\mathscr{B} = \{\beta_1, \ldots, \beta_m\}$ be a basis of $\mathscr{V}$. Then

$$\sigma(\alpha_j) = \sum_{i=1}^{m} a_{ij} \beta_i \tag{S.3}$$

defines a matrix $A = [a_{ij}]$ which represents σ. Also, A represents σ in the sense that the equation

$$\sigma(\xi) = \eta \tag{S.4}$$

has the matrix form

$$AX = Y, \tag{S.5}$$

where X represents ξ with respect to the basis $\mathscr{A}$ and Y represents η with respect to the basis $\mathscr{B}$.

Two important subspaces associated with each linear transformation σ are the *kernel* $K(\sigma)$ and the *image* $\text{Im}(\sigma)$. The kernel is the set of all vectors in $\mathscr{U}$ mapped onto 0, and the image is the set of all vectors in $\mathscr{V}$ that are images of vectors from $\mathscr{U}$. The dimension of the image is the *rank* of σ and the dimension of the kernel is the *nullity* of σ. There are several important equalities and inequalities associated with these numbers.

$$\text{rank}(\sigma) \leq \min\{\dim \mathscr{U}, \dim \mathscr{V}\}, \qquad (S.6)$$

$$\text{rank}(\sigma) + \text{nullity}(\sigma) = \dim \mathscr{U}, \qquad (S.7)$$

$$\text{rank}(\tau\sigma) \leq \min\{\text{rank}(\tau), \text{rank}(\sigma)\}. \qquad (S.8)$$

A linear transformation σ is *injective* (one-to-one) if and only if $K(\sigma) = \{0\}$. A linear transformation σ is *surjective* (onto) if and only if $\text{Im}(\sigma) = \mathscr{V}$. A linear transformation is *invertible* if and only if it is both injective and surjective.

Let A represent σ with respect to the basis $\mathscr{A}$ in $\mathscr{U}$ and the basis $\mathscr{B}$ in $\mathscr{V}$. Let P be the matrix of transition from $\mathscr{A}$ to $\mathscr{A}'$ in $\mathscr{U}$, and let Q be the matrix of transition from $\mathscr{B}$ to $\mathscr{B}'$ in $\mathscr{V}$. Then

$$A' = Q^{-1}AP \qquad (S.9)$$

is the new matrix representing σ with respect to $\mathscr{A}'$ and $\mathscr{B}'$. The most interesting case occurs when $\mathscr{U} = \mathscr{V}$ and $Q = P$. Then

$$P^{-1}AP \qquad (S.10)$$

is the new matrix representing σ. In this case we say that A and $P^{-1}AP$ are *similar*.

Let a linear transformation σ of $\mathscr{U}$ into $\mathscr{V}$ and a vector $\beta \in \mathscr{V}$ be given. A linear problem is to find $\xi \in \mathscr{U}$ such that

$$\sigma(\xi) = \beta. \qquad (S.11)$$

There is a close connection between linear problems and systems of linear equations studied earlier in this book. If A represents σ, X represents ξ, and B represents β, then the linear problem given in (S.11) takes the form

$$AX = B \qquad (S.12)$$

in matrix form. The problem has a solution if and only if $\beta \in \text{Im}(\sigma)$. If $\beta \in \text{Im}(\sigma)$, the solution set is a coset of $K(\sigma)$.

5

DETERMINANTS

In the chapter following this one we shall need a function assigning scalar values to a square matrix which will tell us whether the matrix is singular or non-singular. Furthermore, the value of this function must be computable explicitly in terms of the elements of the matrix. In this chapter we define such a function, the determinant.

The determinant has many interesting properties and a number of applications besides the one we will make in the next chapter. We shall develop the most fundamental of these properties and give them a geometric interpretation.

5-1 THE DETERMINANT AS A VOLUME

We need to determine when a matrix A is singular and when it is non-singular. We are looking for a function of the matrix for which the desired information is contained in the value of the function. There is an obvious function. Assign a value of 1 to the matrix A if it is non-singular and assign a value of 0 if it is singular. The trouble with such a function is that we would have to know what we wish to conclude in order to evaluate the function. We need a function that can be evaluated explicitly in terms of the elements of the matrix.

Either the columns or the rows of an $n \times n$ matrix can be thought of as n-tuples representing vectors in an n-dimensional coordinate space. We know, then, that the matrix is non-singular if and only if the n columns or rows are linearly independent. Therefore, it will suffice to find a function that will have a zero value when the vectors are linearly dependent and have a non zero value when the vectors are linearly independent.

For a 2×2 matrix the rows or columns represent two vectors in a plane. For

$$\begin{bmatrix} a_1 & b_1 \\ a_2 & b_2 \end{bmatrix}, \tag{1.1}$$

(a_1, a_2) and (b_1, b_2) represent two vectors. Consider Figure 5-1.1. The vectors are linearly dependent if and only if they lie in a common line. They are linearly independent if and only if the parallelogram with vertices at the origin, the points (a_1, a_2), (b_1, b_2), and $(a_1 + b_1, a_2 + b_2)$, has a non-zero area.

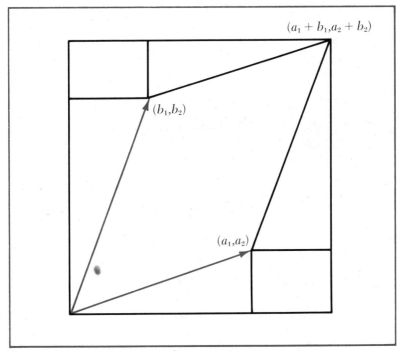

Figure 5-1.1 In $\mathscr{R}^2$, two position vectors determine a parallelogram. The area of the entire rectangle shown is $(a_1 + b_1)(a_2 + b_2) = a_1 a_2 + a_1 b_2 + b_1 a_2 + b_1 b_2$. The area of the four triangles is $a_1 a_2 + b_1 b_2$. The area of the two small rectangles is $2 a_2 b_1$. Thus, the area of the parallelogram is $a_1 b_2 - a_2 b_1$.

For a 3×3 matrix, the three columns represent three vectors in space. For

$$\begin{bmatrix} a_1 & b_1 & c_1 \\ a_2 & b_2 & c_2 \\ a_3 & b_3 & c_3 \end{bmatrix}, \qquad (1.2)$$

consider Figure 5–1.2. These three vectors are linearly dependent if and only if they lie in a common 2-dimensional subspace, a plane through the origin. They are linearly independent if and only if the parallelepiped they form has a non-zero volume.

This suggests that a function of n vectors in an n-dimensional space that gives the n-dimensional analogue of volume would have the properties we need. If the "volume" is zero the vectors are linearly dependent and if the "volume" is non-zero the vectors are linearly independent. The determinant can be interpreted as a volume, and this is how the information can be used to tell us whether a matrix is singular.

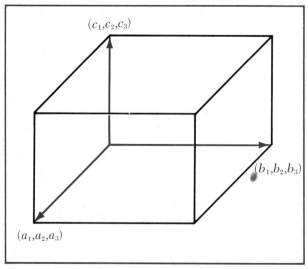

Figure 5–1.2 In $\mathscr{R}^3$, three position vectors determine a parallelopiped. The parallelopiped has a non-zero volume if and only if the three vectors are linearly independent.

5-1 The Determinant as a Volume

We all have a good idea of what area in the plane means and what volume in three dimensions means. Without too much strain we could extend our intuition to give us an idea of what "volume" means in an n-dimensional space. However, in this development of linear algebra we have not defined length, area, or volume yet, and we must be careful not to use such ideas to prove anything. This is not a perverse desire to limit ourselves and make things more difficult. It is no more difficult one way or the other. Since using volume as a basis for the determinant is unnecessary, there is no point in doing it. We use these ideas only for motivation and descriptive purposes.

Let A be an $n \times n$ matrix. We shall think of the n columns of $A = [a_{ij}]$ as representing n vectors in an n-dimensional coordinate space. Let these n vectors be denoted by $\{\alpha_1, \alpha_2, \ldots, \alpha_n\}$. The function we wish to define can be defined equally well for A and for the set $\{\alpha_1, \alpha_2, \ldots, \alpha_n\}$. Thus, $\det A$ and $\det\{\alpha_1, \ldots, \alpha_n\}$ will be thought of as being equivalent.

The n columns of the identity matrix I represent a basis $\mathscr{D} = \{\delta_1, \delta_2, \ldots, \delta_n\}$, which we have called the natural basis. In the plane the parallelogram formed by the basis $\{\delta_1, \delta_2\}$ is the "unit square." In space, the parallelepiped formed by the basis $\{\delta_1, \delta_2, \delta_3\}$ is the "unit cube." Motivated by this consideration we define

$$\det I = \det\{\delta_1, \delta_2, \ldots, \delta_n\} = 1. \tag{1.3}$$

We shall state other properties that the determinant function should have for n vectors in an n-dimensional space. But to motivate these statements we shall illustrate the situation in the 2-dimensional plane.

Consider a parallelepiped determined by $\{\alpha_1, \ldots, \alpha_n\}$ as in Figure 5-1.3. Suppose we multiply one of the vectors in this set by a scalar c. The volume of the new parallelepiped is c times the former volume, as in Figure 5-1.4. For the determinant function of n vectors, this idea would be expressed in the form

$$\det\{\alpha_1, \ldots, c\alpha_1, \ldots, \alpha_n\} = c \cdot \det\{\alpha_1, \ldots, \alpha_n\}, \tag{1.4}$$

where only one vector, the i-th vector, in the set on the left side is multiplied by c.

Suppose α_i is a sum of the form $\alpha_i = \beta_i + \gamma_i$. In a 2-dimensional space suppose $\alpha_2 = \beta_2 + \gamma_2$. We have in mind

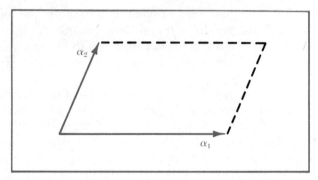

Figure 5-1.3 A parallelogram, which we use to imagine a general parallelopiped in $\mathscr{R}^n$.

the situation illustrated in Figure 5-1.5. The triangles OAB and $CA'B'$ are congruent and have the same area. Thus, the area of the parallelogram $OCA'A$ is the sum of the areas of $OCB'B$ and $BB'A'A$. We express this observation in the form $\det\{\alpha_1, \alpha_2\} = \det\{\alpha_1, \beta_2 + \gamma_2\} = \det\{\alpha_1, \beta_2\} + \det\{\alpha_1, \gamma_2\}$. For n dimensions this idea would be expressed in the form

$$\det\{\alpha_1, \ldots, \alpha_i, \ldots, \alpha_n\} = \det\{\alpha_1, \ldots, \beta_i + \gamma_i, \ldots, \alpha_n\}$$
$$= \det\{\alpha_1, \ldots, \beta_i, \ldots, \alpha_n\}$$
$$+ \det\{\alpha_1, \ldots, \gamma_i, \ldots, \alpha_n\}. \quad (1.5)$$

Conditions (1.4) and (1.5) together say that the determinant function is linear in each vector variable separately.

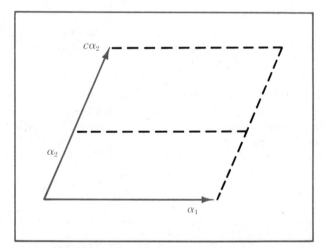

Figure 5-1.4 The base of the parallelogram shown is the same as the base in Figure 5-1.3. The altitude is c times as great. Hence, the area of this parallelogram is c times the area of the parallelogram in Figure 5-1.3.

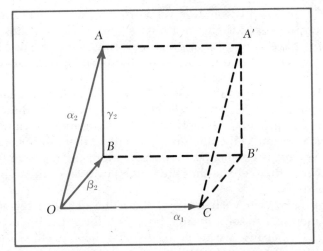

Figure 5-1.5 The figure shown lies in the plane $\mathscr{R}^2$. Triangles OAB and $CA'B'$ are congruent. Consider the odd-shaped area consisting of the two parallelograms $OCB'B$ and $BB'A'A$. If the triangle $CA'B'$ is cut off and replaced by the triangle OAB, we obtain the parallelogram $OCA'A$. Thus, the area of the parallelogram $OCA'A$ is equal to the area of the odd-shaped region.

Notice in formula (1.4) that if c is negative, one of the two volumes must be negative. We are accustomed to thinking of area and volume as being non-negative quantities. But the linearity of the determinant function is so useful that we choose to adjust our thinking about volume to allow the idea of a negative volume. Suppose we start with the situation represented in Figure 5-1.3 and take $c=-1$. Then the parallelogram determined by $\{\alpha_1, -\alpha_2\}$ is as shown in Figure 5-1.6. In the plane, the sense of rotation from α_1 to α_2

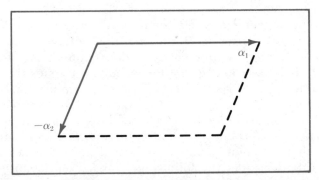

Figure 5-1.6 The parallelogram in this figure has the same area as the parallelogram in Figure 5-1.3. However, the sense of rotation from α_1 to $-\alpha_2$ is opposite to the sense of rotation from α_1 to α_2 in Figure 5-1.3. Thus, we assign one parallelogram an area which is the negative of the area of the other.

in Figure 5-1.3 is counterclockwise, and the sense of rotation from α_1 to $-\alpha_2$ in Figure 5-1.6 is clockwise. We say these two figures have "opposite orientation." In general, we associate the sign of the determinant with the concept of orientation. We say the n vectors $\{\alpha_1, \ldots, \alpha_n\}$ form a positively (negatively) oriented set if $\det\{\alpha_1, \ldots, \alpha_n\}$ is positive (negative). We make no attempt to interpret the sign of the orientation with any particular geometric picture. Assigning the value 1 to $\det I$ in formula (1.3) amounts to deciding that one special set of n vectors is given a positive orientation, and the orientation of every other set is either the same (positive) or the opposite (negative).

Finally, we wish to have the value of the determinant assigned to a set of n vectors be zero if the set is linearly dependent. Because we are going to insist on having the linearity condition it turns out that it is sufficient to state this condition in the following form: "If $\{\alpha_1, \ldots, \alpha_n\}$ contains two identical vectors then $\det\{\alpha_1, \ldots, \alpha_n\} = 0$." Certainly, if $\{\alpha_1, \ldots, \alpha_n\}$ contains two identical vectors it is linearly dependent, and we wish the value to be zero in this case. Conversely, suppose α_n is a linear combination of the other vectors; that is, $\alpha_n = \sum_{i=1}^{n-1} c_i \alpha_i$. Then

$$\det\{\alpha_1, \ldots, \alpha_n\} = \det\left\{\alpha_1, \ldots, \alpha_{n-1}, \sum_{i=1}^{n-1} c_i \alpha_i\right\}$$
$$= \sum_{i=1}^{n-1} c_i \det\{\alpha_1, \ldots, \alpha_{n-1}, \alpha_i\} = 0.$$
(1.6)

The dependency condition is usually given in a different, but equivalent, form: "If the positions of two vectors in $\{\alpha_1, \ldots, \alpha_n\}$ are interchanged, the determinant changes sign." It is easy to see that this condition implies the other condition. If two vectors in $\{\alpha_1, \ldots, \alpha_n\}$ are identical, then interchanging them changes the sign of the determinant but doesn't change the set. If the value of the determinant changes sign and remains the same it must be zero. The implication in the other direction is more complicated since it makes use of linearity. Consider the following sequence of equalities. In each term the vectors $\alpha_3, \ldots, \alpha_n$ are unchanged.

$$\det\{\alpha_1, \alpha_2, \ldots, \alpha_n\}$$
$$= \det\{\alpha_1, \alpha_2, \ldots, \alpha_n\} + \det\{\alpha_1, \alpha_1, \ldots, \alpha_n\}$$
$$= \det\{\alpha_1, \alpha_1 + \alpha_2, \ldots, \alpha_n\}$$
$$= \det\{\alpha_1, \alpha_1 + \alpha_2, \ldots, \alpha_n\}$$
$$\quad - \det\{\alpha_1 + \alpha_2, \alpha_1 + \alpha_2, \ldots, \alpha_n\} \qquad (1.7)$$
$$= \det\{-\alpha_2, \alpha_1 + \alpha_2, \ldots, \alpha_n\}$$
$$= \det\{-\alpha_2, \alpha_1 + \alpha_2, \ldots, \alpha_n\} + \det\{-\alpha_2, -\alpha_2, \ldots, \alpha_n\}$$
$$= \det\{-\alpha_2, \alpha_1, \ldots, \alpha_n\}$$
$$= -\det\{\alpha_2, \alpha_1, \ldots, \alpha_n\}.$$

All this discussion is merely motivation, an attempt to make the following conditions look reasonable. In fact, the conditions do not require any geometric interpretation and are not logically dependent on geometric considerations. We are seeking a scalar-valued function of n vector variables that satisfies the following definition.

Definition *For each n-dimensional vector space $\mathscr{V}$, the **determinant** function* det *satisfies the following conditions.*

    I *(The unit condition).* $\det\{\delta_1, \delta_2, \ldots, \delta_n\} = 1$ *for some specified and fixed basis,* $\{\delta_1, \delta_2, \ldots, \delta_n\}$.

    II *(The linearity condition).* $\det\{\alpha_1, \ldots, \alpha_n\}$ *is linear in each vector variable separately.*

    III *(The dependency condition).* $\det\{\alpha_1, \ldots, \alpha_n\} = 0$ *if two vectors in the set* $\{\alpha_1, \ldots, \alpha_n\}$ *are identical.*

Or III' *(The sign-change condition).* $\det\{\alpha_1, \ldots, \alpha_n\}$ *changes sign if two vectors are interchanged.*

As already pointed out, conditions III and III' are equivalent, if the linearity condition is assumed. Since we do assume linearity, we will treat these conditions as if they were interchangeable.

This leaves us with several unresolved problems. Is there a function that satisfies these conditions? If there is such a function, is it unique? Can we find a formula for calculating $\det\{\alpha_1, \ldots, \alpha_n\}$ explicitly in terms of the coordinates of the α_i? The purpose of this chapter is to provide affirmative answers to these questions.

We have used the notation "det" to emphasize that the determinant is a function, a function of a matrix or a set of vectors. For an explicitly given matrix it is customary to use simpler notation. We use vertical bars in place of the brackets of the matrix notation to indicate that a determinant is to be computed. This notation is used in the following exercises.

EXERCISES 5-1

Use the properties we wish a determinant function to have to evaluate the following determinants.

1. $\begin{vmatrix} 1 & 0 \\ 0 & 1 \end{vmatrix}$

2. $\begin{vmatrix} 1 & 0 \\ 0 & 3 \end{vmatrix}$

3. $\begin{vmatrix} 1 & 0 \\ 0 & d \end{vmatrix}$

4. $\begin{vmatrix} 1 & 3 \\ 0 & 1 \end{vmatrix} = \begin{vmatrix} 1 & 0 \\ 0 & 1 \end{vmatrix} + \begin{vmatrix} 1 & 3 \\ 0 & 0 \end{vmatrix}$

5. $\begin{vmatrix} 1 & b \\ 0 & 1 \end{vmatrix}$

6. $\begin{vmatrix} 1 & 0 \\ c & 1 \end{vmatrix}$

7. $\begin{vmatrix} 1 & 0 \\ c & d \end{vmatrix} = d \begin{vmatrix} 1 & 0 \\ c & 1 \end{vmatrix}$

8. $\begin{vmatrix} a & 0 \\ c & d \end{vmatrix} = ad \begin{vmatrix} 1 & 0 \\ c/a & 1 \end{vmatrix}$, if $a \neq 0$

9. $\begin{vmatrix} 0 & b \\ c & d \end{vmatrix} = - \begin{vmatrix} b & 0 \\ d & c \end{vmatrix}$

10. $\begin{vmatrix} a & b \\ c & d \end{vmatrix} = \begin{vmatrix} a & 0 \\ c & d \end{vmatrix} + \begin{vmatrix} a & b \\ c & 0 \end{vmatrix}$

11. $\begin{vmatrix} a & 0 & 0 \\ 0 & b & 0 \\ 0 & 0 & c \end{vmatrix} = abc \begin{vmatrix} 1 & 0 & 0 \\ 0 & 1 & 0 \\ 0 & 0 & 1 \end{vmatrix}$

12. $\begin{vmatrix} 0 & b & 0 \\ a & 0 & 0 \\ 0 & 0 & c \end{vmatrix} = - \begin{vmatrix} a & 0 & 0 \\ 0 & b & 0 \\ 0 & 0 & c \end{vmatrix}$

13. $\begin{vmatrix} a_1 & 0 & 0 \\ a_2 & b & 0 \\ 0 & 0 & c \end{vmatrix} = \begin{vmatrix} a_1 & 0 & 0 \\ 0 & b & 0 \\ 0 & 0 & c \end{vmatrix} + \begin{vmatrix} 0 & 0 & 0 \\ a_2 & b & 0 \\ 0 & 0 & c \end{vmatrix}$

14. $\begin{vmatrix} a_1 & b_1 & 0 \\ a_2 & b_2 & 0 \\ 0 & 0 & c \end{vmatrix} = \begin{vmatrix} a_1 & 0 & 0 \\ a_2 & b_2 & 0 \\ 0 & 0 & c \end{vmatrix} + \begin{vmatrix} a_1 & b_1 & 0 \\ a_2 & 0 & 0 \\ 0 & 0 & c \end{vmatrix}$

15. $\begin{vmatrix} a_1 & b_1 & 0 \\ a_2 & b_2 & 0 \\ 0 & b_3 & c \end{vmatrix} = \begin{vmatrix} a_1 & b_1 & 0 \\ a_2 & b_2 & 0 \\ 0 & 0 & c \end{vmatrix}$

16. $\begin{vmatrix} a_1 & b_1 & 0 \\ a_2 & b_2 & 0 \\ a_3 & b_3 & c \end{vmatrix} = \begin{vmatrix} a_1 & b_1 & 0 \\ a_2 & b_2 & 0 \\ 0 & b_3 & c \end{vmatrix} = \begin{vmatrix} a_1 & b_1 & 0 \\ a_2 & b_2 & 0 \\ a_3 & 0 & c \end{vmatrix}$

17. $\begin{vmatrix} a_1 & b_1 & c_1 \\ a_2 & b_2 & c_2 \\ a_3 & b_3 & c_3 \end{vmatrix}$

5-2 PERMUTATIONS

Consider the effect of the "sign change" rule when vectors are interchanged. For three vectors in a 3-dimensional space, we have

$$\begin{aligned} \det\{\alpha_1, \alpha_2, \alpha_3\} &= -\det\{\alpha_2, \alpha_1, \alpha_3\} \\ &= \det\{\alpha_2, \alpha_3, \alpha_1\} \\ &= -\det\{\alpha_3, \alpha_2, \alpha_1\} \\ &= \det\{\alpha_3, \alpha_1, \alpha_2\} \\ &= -\det\{\alpha_1, \alpha_3, \alpha_2\}. \end{aligned} \quad (2.1)$$

There is a question whether this makes sense. From $\{\alpha_1, \alpha_2, \alpha_3\}$ to $\{\alpha_1, \alpha_3, \alpha_2\}$ in (2.1) we made 5 interchanges. The same rearrangement could have been achieved in one step, by interchanging α_2 and α_3. Both routes involve an odd number of interchanges, and using either route we would conclude that $\det\{\alpha_1, \alpha_2, \alpha_3\} = -\det\{\alpha_1, \alpha_3, \alpha_2\}$. But how do we know that there isn't some other sequence of interchanges, even in number, that would achieve the same result? If this were to happen the determinant function, satisfying the three conditions of the previous section, could not be well defined. In other words, the conditions would be inconsistent.

The arrangements of the set $\{\alpha_1, \alpha_2, \alpha_3\}$ that appear in (2.1) are permutations. Our immediate goal is to see how to determine the sign associated with the determinant directly in terms of the permutations rather than in terms of the sequence of interchanges. This will make the sign dependent on the end result of the permutation and avoid the ambiguity of the arbitrary choices of interchanges.

218 • Determinants

For the purpose of discussing permutations, it is sufficient to consider rearrangements of the indices instead of the vectors themselves. Thus, consider the finite set of integers $\mathscr{S} = \{1, 2, \ldots, n\}$. Each rearrangement of this set is called a *permutation*. The set $\mathscr{S}$ has $n!$ different orderings, hence $n!$ permutations.

Let π denote a one-to-one function of $\mathscr{S}$ onto itself that performs a reordering of $\mathscr{S}$. For example, if $\mathscr{S} = \{1, 2, 3\}$ and $\{2, 3, 1\}$ is the reordering, then $\pi(1) = 2$, $\pi(2) = 3$, and $\pi(3) = 1$. An advantage of considering the function that performs the permutation is that the special ordering of the elements of $\mathscr{S}$ is no longer important. For example, π maps $\{2, 1, 3\}$ onto $\{3, 2, 1\}$. This allows us to consider composition of permutations, that is, permutations of permutations. For example, if σ maps $\{1, 2, 3\}$ onto $\{2, 1, 3\}$, then $\pi\sigma$ maps $\{1, 2, 3\}$ onto $\{\pi\sigma(1), \pi\sigma(2), \pi\sigma(3)\} = \{\pi(2), \pi(1), \pi(3)\} = \{3, 2, 1\}$.

We say that π performs an *inversion* for each pair of elements of $\mathscr{S}$ such that $i < j$ and $\pi(i) > \pi(j)$. Let $k(\pi)$ be the total number of inversions performed by π, and let sgn $\pi = (-1)^{k(\pi)}$: sgn $\pi = 1$ if $k(\pi)$ is even and sgn $\pi = -1$ if $k(\pi)$ is odd. For $n = 3$, let us compute $k(\pi)$ and sgn π for each π. In Table 5-2.1, $\pi(\mathscr{S})$ denotes the set $\{\pi(1), \pi(2), \pi(3)\}$. Notice how the values of sgn π in Table 5-2.1 correspond to the signs of the determinant function in (2.1). We wish to show that this correspondence is not a mere coincidence.

Theorem 2.1   Sgn $\sigma\pi =$ sgn $\sigma \cdot$ sgn π.

Proof.   We must compare i and j with $\pi(i)$ and $\pi(j)$ to count the inversions in π. Since every element in $\mathscr{S}$ appears as a value of π, we can count the inversions in σ by com-

TABLE 5-2.1

$\pi(\mathscr{S})$	$k(\pi)$	sgn π
$\{1, 2, 3\}$	0	1
$\{2, 1, 3\}$	1	-1
$\{2, 3, 1\}$	2	1
$\{3, 2, 1\}$	3	-1
$\{3, 1, 2\}$	2	1
$\{1, 3, 2\}$	1	-1

paring $\pi(i)$ and $\pi(j)$ with $\sigma\pi(i)$ and $\sigma\pi(j)$. For a given $i<j$ there are four possibilities.
1. $i<j$; $\pi(i)<\pi(j)$; $\sigma\pi(i)<\sigma\pi(j)$: no inversions.
2. $i<j$; $\pi(i)<\pi(j)$; $\sigma\pi(i)>\sigma\pi(j)$: one inversion in σ, one in $\sigma\pi$.
3. $i<j$; $\pi(i)>\pi(j)$; $\sigma\pi(i)>\sigma\pi(j)$: one inversion in π, one in $\sigma\pi$.
4. $i<j$; $\pi(i)>\pi(j)$; $\sigma\pi(i)<\sigma\pi(j)$: one inversion in π, one in σ, and none in $\sigma\pi$.

We can see from comparing these possibilities that $k(\sigma\pi)$ differs from $k(\sigma)+k(\pi)$ by an even number. Thus, sgn $\sigma\pi$ = sgn $\sigma \cdot$ sgn π. □

Theorem 2.2 *If π is a permutation that interchanges exactly two elements of $\mathcal{S}$ and leaves all other elements of $\mathcal{S}$ fixed, then* sgn $\pi = -1$.

Proof. Suppose $i<j$, $\pi(i)=j$, $\pi(j)=i$, and $\pi(k)=k$ for $k \neq i, j$. Clearly, any inversion must involve either i or j, or both. For $k<i$, we have $\pi(k)=k<j=\pi(i)$ and $\pi(k)=k<i=\pi(j)$. Thus, there are no inversions involving such k. Similarly, for $k>j$, there are no inversions involving k. For $i<k<j$ we have $\pi(i)=j>k=\pi(k)>i=\pi(j)$. In this case there are two inversions involving k, an even number. Finally, $i<j$ and $\pi(i)>\pi(j)$. Thus, π has an odd number of inversions and sgn $\pi = -1$. □

Theorems 2.1 and 2.2 tell us that a permutation effected by an even number of interchanges has sgn $\pi = 1$, and a permutation effected by an odd number of interchanges has sgn $\pi = -1$. But since sgn π depends only on the end result, this conclusion is independent of any arbitrary choices in the sequence of interchanges used. Accordingly, a permutation π for which sgn $\pi = 1$ is called an *even* permutation, and a permutation π for which sgn $\pi = -1$ is called an *odd* permutation. There is always at least one even permutation, the identity permutation that leaves every element of $\mathcal{S}$ fixed. For $n \geq 2$ there is at least one odd permutation. Just select any permutation that interchanges exactly two elements.

Table 5-2.1 shows that there are three even permutations for $n=3$, and three odd permutations. The three odd permutations can be classified together by noting that they each involve exactly one interchange of a pair of elements of $\mathcal{S}$. Of

TABLE 5-2.2

$\pi(\mathscr{S})$	
$\{1, 2, 3, 4\}$	the identity.
$\{2, 1, 4, 3\}$	
$\{3, 4, 1, 2\}$	two non-overlapping interchanges.
$\{4, 3, 2, 1\}$	
$\{2, 3, 1, 4\}$	
$\{3, 1, 2, 4\}$	
$\{2, 4, 3, 1\}$	a cyclic permutation of three of the four
$\{4, 1, 3, 2\}$	elements of $\mathscr{S}$.
$\{4, 2, 1, 3\}$	
$\{3, 2, 4, 1\}$	
$\{1, 3, 4, 2\}$	
$\{1, 4, 2, 3\}$	

the three even permutations, one is the identity. The other two are cyclic permutations of the elements of $\mathscr{S}$. (A permutation π of $\mathscr{S}$ is *cyclic if* $\{1, \pi(1), \pi^2(1), \ldots, \pi^{n-1}(1)\}$ runs through all the elements of $\mathscr{S}$. For example, if $\mathscr{S} = \{1, 2, 3, 4, 5\}$ and $\pi(1) = 3$, $\pi(3) = 5$, $\pi(5) = 4$, $\pi(4) = 2$, $\pi(2) = 1$, then π is cyclic.)

For $n = 4$ there are $4! = 24$ permutations. Of these 12 are even and 12 are odd. Table 5-2.2 lists the 12 even permutations and Table 5-2.3 lists the 12 odd permutations.

For the cases $n = 3$ and $n = 4$ that we have examined in some detail, half the permutations were even and half were odd. This is generally true; for all n ($n > 1$) half of the $n!$ permutations are even and half are odd. Let $\{\pi_1, \pi_2, \ldots, \pi_r\}$ be the set of even permutations of the set $\mathscr{S}$. Let σ be any odd permutation. Then by Theorem 2.1, all the permutations in the set

TABLE 5-2.3

$\pi(\mathscr{S})$	
$\{2, 1, 3, 4\}$	
$\{3, 2, 1, 4\}$	
$\{4, 2, 3, 1\}$	a single interchange.
$\{1, 3, 2, 4\}$	
$\{1, 4, 3, 2\}$	
$\{1, 2, 4, 3\}$	
$\{2, 3, 4, 1\}$	
$\{2, 4, 1, 3\}$	
$\{3, 4, 2, 1\}$	a cyclic permutation of all four elements of $\mathscr{S}$.
$\{3, 1, 4, 2\}$	
$\{4, 3, 1, 2\}$	
$\{4, 1, 2, 3\}$	

$\{\sigma\pi_1, \sigma\pi_2, \ldots, \sigma\pi_r\}$ are odd. Since σ is an injective mapping, $\sigma\pi_i = \sigma\pi_j$ implies that $\pi_i = \pi_j$. Thus, there are at least r odd permutations. Conversely, if $\{\sigma_1, \sigma_2, \ldots, \sigma_s\}$ is the set of odd permutations, $\{\sigma\sigma_1, \sigma\sigma_2, \ldots, \sigma\sigma_s\}$ is a set of even permutations. This shows that there are at least as many even permutations as there are odd permutations. Hence, half are even and half are odd.

In the previous section we suggested that the sign of the determinant could be associated with the concept of orientation. We are now in a position to be more explicit. In Figure 5-2.1, six possible configurations of three vectors are divided into two groups of three each. An even permutation will transform one configuration into another in the same group.

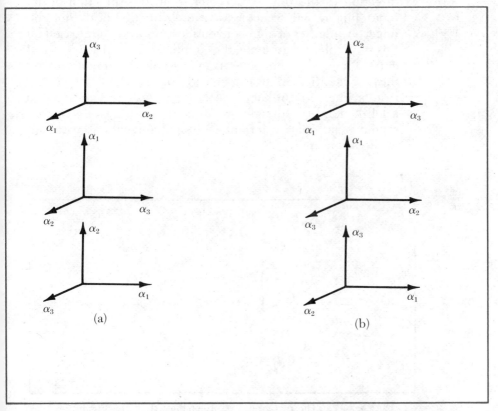

Figure 5-2.1 There are six ways in which three different vectors can be assigned the labels $\{\alpha_1, \alpha_2, \alpha_3\}$. An even permutation of the labels will transform any of the patterns grouped in (a) or (b) into another in the same group. An odd permutation of the labels will transform any of the patterns in one group into a pattern in the other group.

An odd permutation will transform one configuration into one in the other group. Assigning a positive orientation to one particular basis according to the unit condition amounts to designating one of these two groups as being positively oriented and the other as being negatively oriented. Except by being arbitrary it is not possible to decide that one group has the positive orientation and the other group the negative. There is no such ambiguity concerning the sign of the permutations. Regardless of which group is chosen to be positive, an even permutation preserves orientation and an odd permutation changes orientation.

It is customary in vector analysis to assign a positive orientation to the configurations in Figure 5–2.1(a). This is an application of the right-hand rule. Holding the right hand as in Figure 5–2.2, we imagine a vector α_1 pointing in the same direction as the thumb, a vector α_2 in the direction of the index finger, and a vector α_3 in the direction of the middle finger. A number of sign conventions in physics are based on this principle, for example, the coupling between an electrical current and the induced magnetic field. Formulas giving these quantities in terms of coordinates yield the correct results when a right-handed coordinate system is chosen. But a left-handed coordinate system could be used to give consistent results if the formulas or sign conventions were adjusted accordingly.

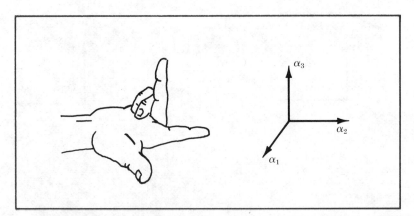

Figure 5–2.2 The two groups of orientations shown in Figure 5–2.1 are described as being right handed or left handed. In a right handed coordinate system the right hand can be placed (approximately) so that the thumb points in the direction of α_1, the index finger points in the direction of α_2, and the middle finger points in the direction of α_3.

EXERCISES 5-2

1. Determine whether $\{2, 3, 5, 4, 1\}$ is an even or an odd permutation of $\{1, 2, 3, 4, 5\}$.

2. Determine whether $\{6, 4, 2, 1, 3, 5\}$ is an even or and odd permutation of $\{2, 3, 6, 5, 1, 4\}$.

3. Let σ be the function that permutes $\{1, 2, 3, 4, 5\}$ to $\{3, 1, 5, 2, 4\}$, and let π be the function that permutes $\{1, 2, 3, 4, 5\}$ to $\{2, 5, 1, 3, 4\}$. Determine the permutations $\pi\sigma$ and $\sigma\pi$.

4. Prove that if a permutation π leaves an element of $\mathscr{S}$ fixed, inversions involving that element need not be considered in determining whether π is even or odd.

5. A permutation of five objects that leaves at least one object fixed can be considered as a permutation of four objects, and the parity of such a permutation is its parity as a permutation of four objects. A permutation that moves every object is a distinctly new type of permutation. Describe all permutations of four objects that leave no object fixed.

6. In Tables 5-2.2 and 5-2.3 we describe all types of permutations of four objects, and we have grouped them into even and odd types. Do the same for all types of permutations of five objects. (There are 120 such permutations and we do not suggest a list of all of them—only a list of the types.)

7. Prove that there are $n!$ permutations of n objects.

8. Show that $\{3, 4, 2, 5, 1\}$ is a cyclic permutation of $\{1, 2, 3, 4, 5\}$.

9. Show that each of the six permutations in Table 5-2.3 that are identified as being cyclic are, in fact, cyclic.

5-3 THE FORMULA FOR THE DETERMINANT

Using the three properties that the determinant function should have, we will derive the formula for calculating the value of the determinant explicitly in terms of the coordinates. This will resolve the question of uniqueness. The formula will satisfy the three conditions, thus resolving the question whether such a function exists. And, of course, we will have the formula

Let $\{\delta_1, \delta_2, \ldots, \delta_n\}$ be the basis distinguished by the unit condition. For the set $\{\alpha_1, \ldots, \alpha_n\}$, let $\alpha_j = \sum_{i=1}^{n} a_{ij}\delta_i$. Then

$$\det\{\alpha_1, \ldots, \alpha_n\} = \det\left\{\sum_{i=1}^{n} a_{i1}\delta_i, \sum_{j=1}^{n} a_{j2}\delta_j, \ldots, \sum_{k=1}^{n} a_{kn}\delta_k\right\}$$

$$= \sum_{i=1}^{n} a_{i1} \sum_{j=1}^{n} a_{j2} \cdots \sum_{k=1}^{n} a_{kn} \det\{\delta_i, \delta_j, \ldots, \delta_k\}, \quad (3.1)$$

by the linearity condition. By the dependency condition, $\det\{\delta_i, \delta_j, \ldots, \delta_k\}$ vanishes whenever any vector in $\mathscr{D}$ appears twice. Thus, the only terms appearing on the right side of (3.1) are those in which all the vectors in the set $\{\delta_i, \delta_j, \ldots, \delta_k\}$ are distinct.

Let π denote a permutation of the indices in the set $\{1, 2, \ldots, n\}$. With this notation, the right side of formula (3.1) takes the form

$$\Sigma_\pi a_{\pi(1),1} a_{\pi(2),2} \cdots a_{\pi(n),n} \det\{\delta_{\pi(1)}, \delta_{\pi(2)}, \ldots, \delta_{\pi(n)}\}, \quad (3.2)$$

where the symbol "Σ_π" means that the sum includes a term for each permutation π. Since there are $n!$ permutations of n objects, there are $n!$ terms in the sum (3.2).

The arrangement of the vectors in $\{\delta_{\pi(1)}, \delta_{\pi(2)}, \ldots, \delta_{\pi(n)}\}$ is obtained from the set $\{\delta_1, \delta_2, \ldots, \delta_n\}$ by applying the permutation π. Thus,

$$\det\{\delta_{\pi(1)}, \delta_{\pi(2)}, \ldots, \delta_{\pi(n)}\}$$
$$= \operatorname{sgn} \pi \det\{\delta_1, \delta_2, \ldots, \delta_n\}. \quad (3.3)$$

By the unit condition, $\det\{\delta_1, \delta_2, \ldots, \delta_n\} = 1$. Thus,

$$\det\{\delta_{\pi(1)}, \delta_{\pi(2)}, \ldots, \delta_{\pi(n)}\} = \operatorname{sgn} \pi. \quad (3.4)$$

Substituting this in (3.2), we have

$$\det\{\alpha_1, \ldots, \alpha_n\} = \Sigma_\pi \operatorname{sgn} \pi \, a_{\pi(1),1} a_{\pi(2),2} \cdots a_{\pi(n),n}. \quad (3.5)$$

Formula (3.5) allows us to define the determinant for any $n \times n$ matrix. For $A = [a_{ij}]$ this amounts to considering the columns of A as representing vectors with respect to the natural basis. Thus, we define

$$\det A = \Sigma_\pi \operatorname{sgn} \pi \, a_{\pi(1),1} a_{\pi(2),2} \cdots a_{\pi(n),n}. \qquad (3.6)$$

This formula may look complicated, but it is mainly a notational problem. Careful examination of a few particular cases such as (3.7) and (3.8) should make it meaningful. Furthermore, we shall show that one seldom computes the value of a determinant by direct use of the formula.

Starting with the three conditions that we would like to have a determinant function satisfy, we have obtained the formula in (3.5) and (3.6). Thus, we have shown that if there is a function satisfying those conditions it must be the formula in (3.5). This resolves the uniqueness question raised at the end of Section 5-1. But does the formula given in (3.5) actually satisfy the conditions we require of a determinant function? This has not been shown. If it does satisfy those conditions, that would resolve the question as to whether such a function exists.

The traditional treatment of determinants is to start with formula (3.6) as the definition of the determinant. When this is taken as a starting point, the three conditions that we have imposed, and other conditions, must be derived. The route we have chosen is considered to be more modern, which merely means that it is the current fashion. At this point, our three conditions on the basis of formula (3.6) must be verified to show that a determinant function exists. This does not mean that the traditional treatment is simpler than our discussion — all the material on permutations must be established in either case. The direct definition of the determinant in terms of formula (3.6) avoids only the motivational discussion.

Let us see what formula (3.6) looks like for a 2×2 matrix and a 3×3 matrix. A customary notation for the determinant is to put vertical bars on each side of the matrix. When a matrix is represented by a single symbol, "$\det A$" is preferable since "$|A|$" looks like an absolute value. When the matrix is displayed, the vertical bar notation is more convenient, as in the following formulas.

$$\begin{vmatrix} a_{11} & a_{12} \\ a_{21} & a_{22} \end{vmatrix} = a_{11}a_{22} - a_{21}a_{12}, \qquad (3.7)$$

$$\begin{vmatrix} a_{11} & a_{12} & a_{13} \\ a_{21} & a_{22} & a_{23} \\ a_{31} & a_{32} & a_{33} \end{vmatrix} = a_{11}a_{22}a_{33} + a_{21}a_{32}a_{13} + a_{31}a_{12}a_{23} \\ - a_{11}a_{32}a_{23} - a_{21}a_{12}a_{33} - a_{31}a_{22}a_{13}. \quad (3.8)$$

Formula (3.7) is simple enough. Formula (3.8) is reasonable. Notice that the signs of the various terms are determined by the parity of the corresponding permutations. (By parity we mean the evenness or oddness of the permutation.) For example, in the third terms the row indices {1, 2, 3} are permuted to the order {3, 1, 2}. This is an even permutation so that sgn $\pi = 1$ for that term.

We could give a similar expansion for a 4×4 matrix. It would involve 24 terms, each a product of four factors. There is little point in writing out such a formula. It is sufficient to remember formula (3.6). However, if a 2×2 or 3×3 matrix is given, there are simple rules for obtaining the values of the determinant that are worth using. Consider the following diagrams.

$$\begin{vmatrix} a_{11} & a_{12} \\ a_{21} & a_{22} \end{vmatrix} \qquad \begin{vmatrix} a_{11} & a_{12} & a_{13} \\ a_{21} & a_{22} & a_{23} \\ a_{31} & a_{32} & a_{33} \end{vmatrix} \begin{matrix} a_{11} & a_{12} \\ a_{21} & a_{22} \\ a_{31} & a_{32} \end{matrix} \quad (3.9)$$

For the 3×3 matrix copy the first two columns to the right of the matrix. For both the 2×2 and 3×3 matrices compute the products along descending diagonals and assign sgn $\pi = 1$, and compute the products along ascending diagonals and assign sgn $\pi = -1$. It is not recommended that you use (3.9) to obtain formulas (3.7) and (3.8). Instead, (3.9) should be used to obtain the expansion of a numerical matrix directly, as in the following examples. This simple scheme does not extend to determinants of higher order.

$$\begin{vmatrix} 1 & 2 & 3 \\ 4 & 5 & 6 \\ 7 & 8 & 10 \end{vmatrix} \begin{matrix} 1 & 2 \\ 4 & 5 \\ 7 & 8 \end{matrix} = 1 \cdot 5 \cdot 10 + 2 \cdot 6 \cdot 7 + 3 \cdot 4 \cdot 8 - 7 \cdot 5 \cdot 3 \\ - 8 \cdot 6 \cdot 1 - 10 \cdot 4 \cdot 2 = -3 \quad (3.10)$$

$$\begin{vmatrix} 1 & -3 & 1 \\ -1 & 2 & -2 \\ 0 & 1 & -1 \end{vmatrix} \begin{matrix} 1 & -3 \\ -1 & 2 \\ 0 & 1 \end{matrix} = -2 + 0 - 1 + 0 + 2 + 3 = 2. \quad (3.11)$$

The work involved in calculating the value of a determinant of large order by the formula is very great. It is even a formidable task for a high-speed electronic computer. There is little value to be gained by developing a skill in computing determinants of large order by hand and we do not encourage the development of such a skill. However, a number of important properties of a determinant can be used to simplify such calculations and these are worth knowing.

The *transpose* of a matrix A, designated by A^T, is the matrix obtained by interchanging rows and columns of A. More formally, if $A = [a_{ij}]$, then $A^T = [a_{ij}']$ where $a_{ij}' = a_{ji}$.

Theorem 3.1 $\det A^T = \det A$.

Proof. Each term of the expansion of $\det A^T$ is of the form

$$\operatorname{sgn} \pi \, a_{\pi(1),1}' a_{\pi(2),2}' \cdots a_{\pi(n),n}'$$
$$= \operatorname{sgn} \pi \, a_{1,\pi(1)} a_{2,\pi(2)} \cdots a_{n,\pi(n)}. \quad (3.12)$$

Let π^{-1} denote the permutation that undoes the permutation effected by π. That is, if $\pi(i) = k$, then $\pi^{-1}(k) = i$. For each factor on the right side of (3.12), wherever the first index is i the second index is $\pi(i)$. Thus, wherever the second index is k the first index is $\pi^{-1}(k)$. Since π^{-1} performs an inversion whenever π does, $\operatorname{sgn} \pi^{-1} = \operatorname{sgn} \pi$. Thus, the right side of (3.12) is

$$\operatorname{sgn} \pi^{-1} \, a_{\pi^{-1}(1),1} a_{\pi^{-1}(2),2} \cdots a_{\pi^{-1}(n),n}. \quad (3.13)$$

The formula for the determinant sums over all permutations. Since every permutation is the inverse of another, a sum over all π^{-1} is the same as a sum over all π. Hence, $\det A^T = \det A$. □

Theorem 3.2 *If A' is the matrix obtained from A by multiplying a row of A by a scalar c, then $\det A' = c \cdot \det A$.*

Proof. Each term of the expansion of $\det A$ contains just one element from each row of A. Thus, multiplying a row of A by c multiplies each term in the expansion of A by c. Thus $\det A' = c \cdot \det A$. □

Theorem 3.3 *If A' is the matrix obtained from A by interchanging two rows of A, then $\det A' = -\det A$.*

Proof. Interchanging two rows of A has the effect of interchanging the corresponding row indices of the elements appearing in A. If σ is the permutation interchanging these two indices, this operation has the effect of replacing each permutation π by the permutation $\sigma\pi$. Since σ is an odd permutation, this has the effect of changing the sign of every term in the expansion of det A. Thus, det $A' = -\det A$. □

Corollary 3.4 *If A has two equal rows, $\det A = 0$.*

Proof. Interchanging the two equal rows of A does not change A. But, by Theorem 3.3, interchanging two rows changes the sign of det A. This implies that det $A = 0$. □

Theorem 3.5 *Let $A = [a_{ij}]$, $B = [b_{ij}]$, $C = [c_{ij}]$, where for some row index k, $c_{kj} = a_{kj} + b_{kj}$ and for $i \neq k$, $a_{ij} = b_{ij} = c_{ij}$. Then $\det C = \det A + \det B$.*

Proof. For each permutation π consider the corresponding term in the expansion of the determinant of C. For that π there is a j such that $\pi(j) = k$. Then $c_{\pi(j),j} = c_{kj} = a_{kj} + b_{kj} = a_{\pi(j),j} + b_{\pi(j),j}$.

$$\begin{aligned}
&\operatorname{sgn} \pi\, c_{\pi(1),1} \cdots c_{\pi(j),j} \cdots c_{\pi(n),n} \\
&= \operatorname{sgn} \pi\, c_{\pi(1),1} \cdots (a_{\pi(j),j} + b_{\pi(j),j}) \cdots c_{\pi(n),n} \\
&= \operatorname{sgn} \pi\, c_{\pi(1),1} \cdots a_{\pi(j),j} \cdots c_{\pi(n),n} \\
&\quad + \operatorname{sgn} \pi\, c_{\pi(1),1} \cdots b_{\pi(j),j} \cdots c_{\pi(n),n} \\
&= \operatorname{sgn} \pi\, a_{\pi(1),1} \cdots a_{\pi(j),j} \cdots a_{\pi(n),n} \\
&\quad + \operatorname{sgn} \pi\, b_{\pi(1),1} \cdots b_{\pi(j),j} \cdots b_{\pi(n),n}
\end{aligned} \quad (3.14)$$

From (3.14) it follows that $\det C = \det A + \det B$. □

Theorems 3.2 and 3.5 say that the determinant function is linear in each row separately. Theorem 3.3 is the sign-change condition and Corollary 3.4 is the dependency condition for rows. By Theorem 3.1 there are corresponding theorems for columns. Thus, the determinant function is linear in each column separately, and the sign-change condition and the dependency condition hold for columns.

Theorem 3.6 $\det I = 1$.

Proof. $I = [\delta_{ij}]$. Hence, each term in the expansion is 0 for any permutation π for which $\pi(j) \neq j$. There is then but one term corresponding to the identity permutation, which is even. Thus det $I = 1$. □

These theorems show that the determinant function (3.5) satisfies the unit condition, the linearity condition, and the dependency condition of Section 5–1. This closes the circle in the sense that we have now shown that there exists a function satisfying those conditions and that this function is uniquely determined by those conditions. The function is given explicitly by formula (3.5) (or (3.6) for matrices).

EXERCISES 5–3

1. Evaluate the following determinates.

   a) $\begin{vmatrix} 2 & 3 \\ 3 & 5 \end{vmatrix}$

   b) $\begin{vmatrix} 2 & -5 \\ 3 & 4 \end{vmatrix}$

   c) $\begin{vmatrix} 1 & 2 & 3 \\ 2 & 3 & 4 \\ 3 & 4 & 5 \end{vmatrix}$

   d) $\begin{vmatrix} 2 & 0 & 1 \\ 0 & 3 & -2 \\ 5 & 4 & 2 \end{vmatrix}$

   e) $\begin{vmatrix} 3 & 2 & 5 \\ -4 & 3 & -2 \\ -6 & 4 & -9 \end{vmatrix}$

   f) $\begin{vmatrix} 1 & -1 & 2 & 1 \\ -1 & 3 & 2 & -2 \\ -2 & 0 & 1 & 0 \\ 1 & 4 & 3 & 5 \end{vmatrix}$

2. Suppose a determinant is evaluated by direct use of formula (3.6). Counting each individual multiplication, addition, subtraction, or division as one operation, how many operations are needed to evaluate a third order determinant? How many are needed to evaluate an nth order determinant?

3. If one element in an $n \times n$ matrix is zero (so that no products involving that term need to be computed), how many operations are needed to evaluate the determinant?

4. If two elements in the same row (or column) of an $n \times n$ matrix are zero (so that the zero factors do not appear in the same product), how many operations are needed to evaluate the determinant?

5-4 ELEMENTARY OPERATIONS AND DETERMINANTS

Theorems 3.2 and 3.3 suggest connections between elementary row operations and the determinant function. Theorem 3.2 says that if a row of a matrix is multiplied by a scalar c the value of the determinant of the matrix is multiplied by c. Theorem 3.3 says that interchanging two rows of a matrix changes the sign of the determinant of the matrix. This immediately establishes a connection between these theorems and the elementary row operations of type I and type III. The effect of an elementary row operation of type II is easy to establish from the linearity condition. (See Section 1-5.)

Let $A = [a_{ij}]$ be an $n \times n$ matrix, and suppose we add c times row r to row k. We then obtain a matrix $A' = [a_{ij}']$ in which

$$a_{ij}' = a_{ij} \quad \text{for} \quad i \neq k,$$
$$a_{kj}' = a_{kj} + c \cdot a_{rj}. \tag{4.1}$$

For notational convenience, let $A'' = [a_{ij}'']$ where

$$a_{ij}'' = a_{ij} \quad \text{for} \quad i \neq k,$$
$$a_{kj}'' = a_{rj}. \tag{4.2}$$

Then by (4.1) and the linearity condition of Theorem 3.5, we have

$$\det A' = \det A + c \cdot \det A''. \tag{4.3}$$

But $\det A'' = 0$ since A'' is a matrix in which rows r and k are the same. Thus, $\det A' = \det A$ and we see that an elementary row operation of type II does not change the value of the determinant.

Let us summarize the various observations we have made about the connection between elementary operations and the value of the determinant.

Theorem 4.1 *Let A be a square matrix.* **I.** *If A' is the matrix obtained from A by multiplying a row of A by a nonzero scalar c, then $\det A' = c \cdot \det A$.* **II.** *If A' is the matrix obtained from A by adding a multiple of row r to row k, then $\det A' = \det A$* **III.** *If A' is the matrix obtained from A by interchanging two rows of A, then $\det A' = -\det A$.* □

Because of Theorem 3.1, the three types of elementary column operations have corresponding effects.

In Section 3–4 the elementary matrices have been described as those obtained by applying elementary operations to the identity matrix. Since $\det I = 1$, Theorem 4.1 permits us to calculate the determinant of the elementary matrices. If E is an elementary matrix of type I, then $\det E = c$; if E is an elementary matrix of type II, then $\det E = 1$; and if E is an elementary matrix of type III, then $\det E = -1$.

Theorem 4.2 *For any elementary matrix E and A any square matrix of the same order, $\det EA = (\det E)(\det A)$.*

Proof. Let E be an elementary matrix. Then $A' = EA$ is the matrix obtained by applying the corresponding elementary row operation to the matrix A. In Theorem 4.1, for each type of elementary operation the factor relating $\det A'$ and $\det A$ is precisely $\det E$. Thus, $\det EA = \det A' = (\det E)(\det A)$. □

Theorem 4.3 *For any square matrices A and B of the same order, $\det AB = (\det A)(\det B)$.*

Proof. Let n be the common order of A and B. If $\operatorname{rank}(A) < n$, then $\det A = 0$. By Theorem 4.1 in Chapter 4, $\operatorname{rank}(AB)$ is also $< n$ and $\det AB = 0$. In this case $\det AB = (\det A)(\det B)$. If $\operatorname{rank}(A) = n$, A is non-singular and by Theorem 4.2 of Chapter 3, A can be written as a product of elementary matrices. By repeated application of Theorem 4.2 we can conclude in this case also that $\det AB = (\det A)(\det B)$. □

This theorem ends the use we will make of elementary matrices in this text. But for computational purposes, Theorem 4.1 is quite useful. In the following numerical example justify each step in terms of the effect of an elementary operation.

$$\begin{vmatrix} 2 & 1 & 7 \\ 1 & 2 & 2 \\ 3 & -1 & 3 \end{vmatrix} = \begin{vmatrix} 0 & -3 & 3 \\ 1 & 2 & 2 \\ 0 & -7 & -3 \end{vmatrix} = 3 \begin{vmatrix} 0 & -1 & 1 \\ 1 & 2 & 2 \\ 0 & -7 & -3 \end{vmatrix}$$

$$= 3 \begin{vmatrix} 0 & -1 & 1 \\ 1 & 2 & 2 \\ 0 & 0 & -10 \end{vmatrix} = -30 \begin{vmatrix} 0 & -1 & 1 \\ 1 & 2 & 2 \\ 0 & 0 & 1 \end{vmatrix} \quad (4.4)$$

$$= 30 \begin{vmatrix} 1 & 2 & 2 \\ 0 & -1 & 1 \\ 0 & 0 & 1 \end{vmatrix} = -30.$$

Many different sequences of choices for the elementary operations are just as convenient. Also, elementary column operations should be considered. For example, once a 1 is obtained in the first column in the second determinant, the remaining elements in the second row can be changed to zeros by column operations. Actually, they will not enter into any further calculations anyway. Each term in the formula for the determinant involves a product of factors, one from each row and one from each column. Since the only non-zero element in the first column is in the second row, the remaining elements in the second row will be multiplied by a zero from the first column in any term in which they appear. Thus, there is little point in reducing the other elements in the second row to zeros.

Notice in the example worked out above that we have systematically attempted to introduce zeros into the array. If a determinant must be evaluated, the most efficient way to proceed (unless the array has some special property) is to use the elementary operations of the forward course of Gaussian elimination. In particular, if the lower-upper triangular factorization of Section 3–5 is obtained we have

$$\det A = (\det L)(\det U). \tag{4.5}$$

Either L or U has 1's along the main diagonal, and the value of that determinant is 1. The value of the other determinant is the product of the diagonal elements. Computing that product requires $n-1$ multiplications. Since the factorization requires $n(n^2-1)/3$ multiplicative steps, the evaluation of det A can be achieved in

$$\frac{n^3 + 2n - 3}{3} \tag{4.6}$$

multiplicative steps. The direct evaluation of a determinant by the formula would require $(n-1)(n!)$ multiplicative steps. Table 5–4.1 compares the values of $(n-1)(n!)$ and $(n^3 + 2n - 3)/3$ for several values of n. For $n = 10$, 339 multiplicative steps is a lot of work, but 32,659,200 steps is something else.

TABLE 5-4.1

n	(n − 1)(n!)	$\frac{n^3 + 2n - 3}{3}$
2	2	3
3	12	10
4	72	23
5	480	44
6	3,600	75
10	32,659,200	339

EXERCISES 5-4

1. Evaluate the following determinants by using elementary operations to reduce the calculations to the point where only one product must be computed.

   a) $\begin{vmatrix} 3 & 7 \\ 2 & 5 \end{vmatrix}$

   b) $\begin{vmatrix} 1 & 6 \\ 2 & 5 \end{vmatrix}$

   c) $\begin{vmatrix} 2 & 4 & 1 \\ 3 & -2 & 4 \\ 6 & 3 & 1 \end{vmatrix}$

   d) $\begin{vmatrix} 2 & 3 & 4 \\ 5 & 6 & 7 \\ 8 & 9 & 10 \end{vmatrix}$

   e) $\begin{vmatrix} 2 & -2 & 3 \\ 5 & 3 & -2 \\ 7 & -4 & -5 \end{vmatrix}$

   f) $\begin{vmatrix} 1 & -1 & 2 & 1 \\ -1 & 3 & 2 & -2 \\ -3 & 0 & 1 & 0 \\ 1 & 4 & 3 & 5 \end{vmatrix}$

2. Verify Theorem 4.3 for the matrix products

   a) $\begin{bmatrix} 3 & 7 \\ 2 & 6 \end{bmatrix} \begin{bmatrix} 3 & -4 \\ 1 & 5 \end{bmatrix}$

   b) $\begin{bmatrix} 1 & 2 & 3 \\ 2 & 1 & -1 \\ 3 & -1 & 1 \end{bmatrix} \begin{bmatrix} 2 & 1 & -1 \\ 1 & 2 & 1 \\ -1 & 1 & 2 \end{bmatrix}$

3. Show that if L is a matrix in lower triangular form, then det L is just the product of the diagonal elements of L. Show that if U is a matrix in upper triangular form, then det U is just the product of the diagonal element of U.

5-5 COFACTORS

For the $n \times n$ matrix $A = [a_{ij}]$ consider any one element a_{ij}. The element a_{ij} appears as a factor in some of the terms in the formula for the determinant and does not appear in others. Thus, det A can be considered to be of the form

$$\det A = a_{ij}A_{ij} + \text{(terms which do not contain } a_{ij} \text{ as a factor)}. \quad (5.1)$$

The scalar A_{ij} is called the *cofactor* of a_{ij}.

The first problem we face is that of computing the cofactor A_{ij} directly from the matrix A. As a special case, consider the cofactor A_{11}. We have to decide which terms of (3.6) are included in $a_{11}A_{11}$ and which are not. Clearly, a term is included in $a_{11}A_{11}$ if and only if $\pi(1) = 1$. Thus, $a_{11}A_{11}$ is of the form

$$a_{11}A_{11} = \sum_{\substack{\pi \\ \pi(1)=1}} \operatorname{sgn} \pi \, a_{11} a_{\pi(2),2} \cdots a_{\pi(n),n}, \quad (5.2)$$

$$A_{11} = \sum_{\substack{\pi \\ \pi(1)=1}} \operatorname{sgn} \pi \, a_{\pi(2),2} \, a_{\pi(3),3} \cdots a_{\pi(n),n}, \quad (5.3)$$

where the sums are taken only over those permutations for which $\pi(1) = 1$. A permutation π for which $\pi(1) = 1$ permutes the elements $\{2, 3, \ldots, n\}$ among themselves. Thus, we can define a permutation π' on $\mathscr{S}' = \{2, 3, \ldots, n\}$ which coincides with π on $\mathscr{S}'$. No inversion of π involves the element 1. Hence sgn π' = sgn π. The significance of this observation is that formula (5.3) for the cofactor A_{11} is in the form of the determinant formula. It is the determinant of the matrix obtained from A by deleting the first column and the first row.

Suppose we start with an $n \times n$ matrix A and delete any r rows and any r columns of A. The remaining elements, in the same relative order in which they appeared in A, can be considered to be a matrix of order $n-r$, a *submatrix* of A. The determinant of the submatrix is called a *minor* of A of order $n-r$. In this terminology, A_{11} is a minor of A of order $n-1$. In fact, all cofactors are either minors of order $n-1$ or negatives of such minors. To see this, consider the process of moving the element a_{ij} to row 1 column 1 by performing a sequence of interchanges in which we interchange only adjacent rows or columns. By interchanging only adjacent rows or columns in this way we can preserve the relative order of all rows and columns other than row i and column j. It takes $i-1$ such interchanges to move the element a_{ij} into the first row and it takes $j-1$ such interchanges to move it into the first column, $i+j-2$ interchanges in all. The determinant of the new matrix is the determinant of A multiplied by $(-1)^{i+j-2} = (-1)^{i+j}$. As already observed, the cofactor of the element in the first row and first column is the minor obtained by deleting the first row and first column. This is the same as the minor obtained by deleting row i and column j in the original matrix A. Thus, A_{ij} is this minor multiplied by $(-1)^{i+j}$.

Each term in the formula for det A contains exactly one factor from each row and each column of A. Thus, for any given row r each term contains exactly one factor from that row. Hence, for that row r,

$$\det A = \sum_{j=1}^{n} a_{rj} A_{rj}. \tag{5.4}$$

Similarly, for any given column k each term contains exactly one factor from that column. Hence for that column k,

$$\det A = \sum_{i=1}^{n} a_{ik} A_{ik}. \tag{5.5}$$

Formula (5.4) is called an expansion of det A in terms of cofactors of row r, and (5.5) is called an expansion of det A in terms of cofactors of column k.

From a computational point of view, formulas (5.4) and (5.5) offer no advantage over direct use of the formula. Each cofactor is a minor of order $n-1$, and has $(n-1)!$ terms. There are n cofactors to compute, so there are still $n!$ terms in all. If any row or column has several zeros in it, the cofactors of those terms do not have to be computed. But these terms do not have to be computed in the formula either.

Supply the justification, either in terms of elementary

operations or cofactor expansions, for each step in the following calculation.

$$\det A = \begin{vmatrix} 2 & 3 & -2 & 5 \\ 6 & -2 & 1 & 4 \\ 5 & 10 & 3 & -2 \\ -1 & 2 & 2 & 3 \end{vmatrix} = \begin{vmatrix} 14 & -1 & 0 & 13 \\ 6 & -2 & 1 & 4 \\ -13 & 16 & 0 & -14 \\ -13 & 6 & 0 & -5 \end{vmatrix} \quad (5.6)$$

$$= -\begin{vmatrix} 14 & -1 & 13 \\ -13 & 16 & -14 \\ -13 & 6 & -5 \end{vmatrix} = -\begin{vmatrix} 14 & -1 & 13 \\ 0 & 10 & -9 \\ -13 & 6 & -5 \end{vmatrix}$$

$$= -\left\{ 14 \begin{vmatrix} 10 & -9 \\ 6 & -5 \end{vmatrix} - 13 \begin{vmatrix} -1 & 13 \\ 10 & -9 \end{vmatrix} \right\} = -1629.$$

In this example there was no attempt to choose the most efficient steps. The steps chosen merely illustrate the variety of choices available and the consequence of each choice. But it should illustrate that we can combine elementary operations and cofactor expansions to the same advantage that Gaussian elimination provides.

Suppose we multiply the elements in row i by the cofactors of the elements in row k, where $k \neq i$. The cofactor A_{kj} from row k is computed from a minor of the matrix A obtained by crossing out row k. Thus, the value of A_{kj} does not depend on the elements in row k. Consider replacing the elements in row k by the corresponding elements from row i, that is, replace a_{kj} by a_{ij}. This gives us a matrix with two identical rows, and the determinant of this new matrix is 0. Thus,

$$\sum_{j=1}^{n} a_{ij} A_{kj} = 0 \quad \text{for} \quad k \neq i, \quad (5.7)$$

since this is just the cofactor expansion of the new determinant. But this is what we set out to compute, the sum of the products of the elements from one row by the cofactors from another row. By using the Kronecker delta, the formulas (5.4) and (5.7) can be stated in one formula in the form

$$\sum_{j=1}^{n} a_{ij} A_{kj} = \delta_{ik} \det A. \quad (5.8)$$

A similar discussion applies to cofactor expansions in columns and we can obtain the formula

$$\sum_{i=1}^{n} a_{ij} A_{ik} = \delta_{jk} \det A. \tag{5.9}$$

The matrix obtained from A by taking the transpose of the matrix of cofactors of A is called the *adjunct* of A, and it is denoted by adj A. Thus, adj $A = [A_{ij}]^T$.

Theorem 5.1 $A(\text{adj } A) = (\text{adj } A)A = (\det A)I$.

Proof.

$$A(\text{adj } A) = [a_{ij}] \cdot [A_{ij}]^T = \left[\sum_{j=1}^{n} a_{ij} A_{kj} \right]$$

$$= [(\det A)\ \delta_{ik}] = (\det A)I$$

$$(\text{adj } A)A = [A_{ij}]^T \cdot [a_{ij}] = \left[\sum_{i=1}^{n} A_{ik} a_{ij} \right]$$

$$= [(\det A)\delta_{kj}] = (\det A)I. \qquad \square$$

Although we have introduced cofactors as a means of computing determinants, it is really Theorem 5.1 (or the formulas (5.8) and (5.9)) that gives cofactors rather wide utility in the theory of linear algebra.

One application of Theorem 5.1 is that it gives a formula for the inverse of a matrix, when the matrix has an inverse. We have already pointed out that a square matrix A has an inverse if and only if $\det A \neq 0$. In this case, Theorem 5.1 yields

$$A^{-1} = \frac{1}{\det A} \text{ adj } A. \tag{5.10}$$

This method of obtaining the inverse of a matrix is not particularly convenient. It involves computing an n-th order determinant (first, to see if an inverse exists), and then n^2 $(n-1)$-st order determinants. This involves computing $n! + n^2(n-1)! = (n+1)!$ terms. For n greater than 3 the amount of work is discouragingly large. But (5.10) is a formula that is expressed in "closed" form rather than just a method or algorithm. Such a formula can provide very useful information in the analysis of a variety of different problems. For example, suppose A is a square matrix with elements that are integers. Then all cofactors are integers since they are determinants of matrices with integral elements. If det $A = 1$, formula (5.10) tells us that the inverse of A has all integral elements. Even if det A is not 1, formula (5.10) can be used to obtain estimates on the kinds of denominators that can appear in A^{-1}. There are problems in number theory and ring theory, for example, in which this kind of information is quite important.

EXERCISES 5-5

1. Find all cofactors of the elements in the first column of
$$\begin{bmatrix} 2 & 1 & 4 \\ 0 & 3 & 5 \\ 3 & -2 & -3 \end{bmatrix}$$

2. Find the inverse of $\begin{bmatrix} 2 & 3 \\ 4 & 5 \end{bmatrix}$ by finding cofactors.

3. Verify formula (5.8) for the elements of row 2 and the cofactors of row 3 in the following determinant.
$$\begin{vmatrix} 1 & 2 & 1 \\ 3 & 4 & -1 \\ 2 & -1 & 3 \end{vmatrix}$$

4. Verify formula (5.9) for the elements of column 1 and cofactors of column 3 for the following determinant.
$$\begin{vmatrix} 3 & 0 & 1 \\ 2 & 6 & -2 \\ 3 & 9 & 6 \end{vmatrix}$$

5. Evaluate the determinant in Exercises 3 and 4 by means of formulas (5.8) and (5.9).

Summary

A *determinant* is a scalar-valued function of n vectors in an n-dimensional vector space (or a scalar-valued function of an $n \times n$ matrix) such that

  I (The unit condition). det $\{\delta_1, \delta_2, \ldots, \delta_n\} = 1$ for some specified and fixed basis, $\{\delta_1, \delta_2, \ldots, \delta_n\}$.
  II (The linearity condition). det $\{\alpha_1, \alpha_2, \ldots, \alpha_n\}$ is linear in each vector variable separately.

III (The dependency condition). $\det\{\alpha_1, \alpha_2, \ldots, \alpha_n\} = 0$ if two vectors in the set $\{\alpha_1, \alpha_2, \ldots, \alpha_n\}$ are identical.

The determinant function is uniquely characterized by these conditions. For an n-th order matrix $A = [a_{ij}]$, the determinant function takes the form

$$\det A = \Sigma_\pi \text{ sgn } \pi \, a_{\pi(1),1} \, a_{\pi(2),2} \cdots a_{\pi(n),n}. \tag{S.1}$$

Though the determinant has an interpretation as a volume, its main utility to us is that it has the value 0 if and only if the n vectors are linearly dependent.

6
EIGENVECTORS AND EIGENVALUES

In this chapter we develop the theory of eigenvectors and eigenvalues. In this connection we are concerned exclusively with linear transformations of a vector space $\mathscr{V}$ into itself. Given a linear transformation σ of $\mathscr{V}$ into itself, we wish to find subspaces $\mathscr{S}$ such that σ maps each such subspace $\mathscr{S}$ into itself. For such a subspace $\mathscr{S}$ we can consider σ as defining, or inducing, a linear transformation on $\mathscr{S}$. For $\alpha \in \mathscr{S}$, $\sigma(\alpha) \in \mathscr{S}$. Thus, this amounts merely to restricting one's attention to a smaller set than all of $\mathscr{V}$.

Generally, a problem in linear algebra is easier to analyze in spaces of lower dimension. Thus, finding subspaces in which the given linear transformation induces a linear transformation results in a simplification of the problem. Finding such subspaces is not particularly easy in full generality. A problem as important as this has been studied from many points of view and a number of different approaches have been taken, some of them quite sophisticated.

Since the problems we are considering are simplified the most when posed in spaces of least dimension, the ultimate in simplification occurs in spaces of dimension 1. If $\mathscr{S}$ is a subspace of dimension 1 which is mapped into itself by σ, the non-zero vectors in $\mathscr{S}$ are called eigenvectors of σ. This chapter is mainly concerned with eigenvectors and how to find them.

6-1 ALGEBRA OF SUBSPACES

Let $\mathscr{S}_1$ and $\mathscr{S}_2$ be any two subsets of a vector space $\mathscr{V}$. By $\mathscr{S}_1 + \mathscr{S}_2$ we mean the set of all $\alpha + \beta$ where $\alpha \in \mathscr{S}_1$ and $\beta \in \mathscr{S}_2$.

Theorem 1.1 If $\mathscr{S}_1$ and $\mathscr{S}_2$ are subspaces, then $\mathscr{S}_1 + \mathscr{S}_2$ is a subspace.

Proof. Let $\alpha_1 + \alpha_2 \in \mathscr{S}_1 + \mathscr{S}_2$ where $\alpha_i \in \mathscr{S}_i$ and let $\beta_1 + \beta_2 \in \mathscr{S}_1 + \mathscr{S}_2$ where $\beta_i \in \mathscr{S}_i$. Then $(\alpha_1 + \alpha_2) + (\beta_1 + \beta_2) = (\alpha_1 + \beta_1) + (\alpha_2 + \beta_2) \in \mathscr{S}_1 + \mathscr{S}_2$ since $\alpha_i + \beta_i \in \mathscr{S}_i$. Also, $a(\alpha_1 + \alpha_2) = a\alpha_1 + a\alpha_2 \in \mathscr{S}_1 + \mathscr{S}_2$ since $a\alpha_i \in \mathscr{S}_i$. Thus, $\mathscr{S}_1 + \mathscr{S}_2$ is a subspace. □

Clearly, the definition and the theorem can be extended to a sum of any finite number of subspaces.

Generally, the representation of an element of $\mathscr{S}_1 + \mathscr{S}_2$ as a sum is not unique. For example, consider $\mathscr{S} + \mathscr{S}$ where $\mathscr{S}$ is a subspace. Then $\mathscr{S} + \mathscr{S} = \mathscr{S}$ and any $\alpha \in \mathscr{S}$ can be written as $\alpha + 0 = 0 + \alpha = \alpha/2 + \alpha/2$, etc. However, if $\mathscr{S}_1 \cap \mathscr{S}_2 = \{0\}$ the representation of an element of $\mathscr{S}_1 + \mathscr{S}_2$ as a sum is unique. Suppose $\mathscr{S}_1 \cap \mathscr{S}_2 = \{0\}$ and $\alpha_1 + \alpha_2 = \beta_1 + \beta_2$ where $\alpha_i, \beta_i \in \mathscr{S}_i$. Then $\alpha_1 - \beta_1 = \beta_2 - \alpha_2$. The left side is in $\mathscr{S}_1$ and the right side is in $\mathscr{S}_2$. Thus, both sides are in $\mathscr{S}_1 \cap \mathscr{S}_2$. This means $\alpha_1 - \beta_1 = 0$ and $\beta_2 - \alpha_2 = 0$ and the representation of $\alpha_1 + \alpha_2$ as a sum is unique.

When the representation of a sum in $\mathscr{S}_1 + \mathscr{S}_2$ is unique we say the sum is *direct*. If we know a sum is direct we denote it by $\mathscr{S}_1 \oplus \mathscr{S}_2$. Similarly, the sum of more than two subspaces is said to be direct if the representation of any element as a sum is unique. When we write $\mathscr{S}_1 + \mathscr{S}_2$, that does not mean the sum is not direct. It means merely that we do not know whether the sum is direct.

We have already encountered sums of subspaces and direct sums in the following way. Let $\mathscr{S}_i$ be a subspace spanned by a single element, $\mathscr{S}_i = \langle \alpha_i \rangle$. Assuming $\alpha_i \neq 0$, $\langle \alpha_i \rangle$ is 1-dimensional. Any vector in $\langle \alpha_i \rangle$ is of the form $a_i \alpha_i$ where a_i is a scalar. Any vector in the sum $\langle \alpha_1 \rangle + \langle \alpha_2 \rangle + \cdots + \langle \alpha_n \rangle = \mathscr{S}$ can be written in the form $a_1 \alpha_1 + a_2 \alpha_2 + \cdots + a_n \alpha_n$. Thus, $\mathscr{S}$ is the subspace spanned by $\{\alpha_1, \ldots, \alpha_n\}$. If the sum is direct, the representation in the form $a_1 \alpha_1 + \cdots + a_n \alpha_n$ is unique. That is, the set $\{\alpha_1, \ldots, \alpha_n\}$ is then linearly independent. Conversely, if $\{\alpha_1, \ldots, \alpha_n\}$ is linearly independent, the sum $\langle \alpha_1 \rangle + \cdots + \langle \alpha_n \rangle$ is direct.

There is a familiar theorem in elementary solid geometry to the effect that two planes are either identical or parallel, or intersect in a common line. In this discussion, 2-dimensional subspaces are planes, but not all planes are 2-dimensional subspaces. All subspaces must contain the origin. In this context, then, the theorem from solid geometry would say that two 2-dimensional subspaces either are identical or intersect in a common line. In the second alternative, the common line would also have to contain the origin since both 2-dimensional subspaces do. This conclusion is a special case of the following theorem.

Theorem 1.2 *If $\mathscr{S}_1$ and $\mathscr{S}_2$ are subspaces, then $\dim(\mathscr{S}_1 + \mathscr{S}_2) + \dim(\mathscr{S}_1 \cap \mathscr{S}_2) = \dim \mathscr{S}_1 + \dim \mathscr{S}_2$.*

Proof. Let $\{\alpha_1, \ldots, \alpha_r\}$ be a basis of $\mathscr{S}_1 \cap \mathscr{S}_2$. Since $\mathscr{S}_1 \cap \mathscr{S}_2 \subset \mathscr{S}_1$, $\{\alpha_1, \ldots, \alpha_r\}$ can be extended to a basis $\{\alpha_1, \ldots, \alpha_r, \beta_1, \ldots, \beta_s\}$ of $\mathscr{S}_1$. Similarly, $\{\alpha_1, \ldots, \alpha_r\}$ can be extended to a basis $\{\alpha_1, \ldots, \alpha_r, \gamma_1, \ldots, \gamma_t\}$ of $\mathscr{S}_2$. Then $\{\alpha_1, \ldots, \alpha_r, \beta_1, \ldots, \beta_s, \gamma_1, \ldots, \gamma_t\}$ spans $\mathscr{S}_1 + \mathscr{S}_2$. We wish to show this set is linearly independent and a basis for $\mathscr{S}_1 + \mathscr{S}_2$. Suppose $\sum_{i=1}^{r} a_i \alpha_i + \sum_{j=1}^{s} b_j \beta_j + \sum_{k=1}^{t} c_k \gamma_k = 0$. Then $\sum_{i=1}^{r} a_i \alpha_i + \sum_{j=1}^{s} b_j \beta_j = -\sum_{k=1}^{t} c_k \gamma_k$. The left side of this expression is in $\mathscr{S}_1$ and the right side is in $\mathscr{S}_2$, so both sides are in $\mathscr{S}_1 \cap \mathscr{S}_2$. Hence it (both sides are just different representations of the same vector) can be expressed as a linear combination of $\{\alpha_1, \ldots, \alpha_r\}$. Thus, $-\sum_{k=1}^{t} c_k \gamma_k = \sum_{i=1}^{r} d_i \alpha_i$, or $\sum_{i=1}^{r} d_i \alpha_i + \sum_{k=1}^{t} c_k \gamma_k = 0$. Since $\{\alpha_1, \ldots, \alpha_r, \gamma_1, \ldots, \gamma_t\}$ is linearly independent, all $d_i = 0$ and all $c_k = 0$. Thus, $\sum_{i=1}^{r} a_i \alpha_i + \sum_{j=1}^{s} b_j \beta_j = 0$. Since $\{\alpha_1, \ldots, \alpha_r, \beta_1, \ldots, \beta_s\}$ is linearly independent, all $a_i = 0$ and all $b_j = 0$. This means $\{\alpha_1, \ldots, \alpha_r, \beta_1, \ldots, \beta_s, \gamma_1, \ldots, \gamma_t\}$ is linearly independent and, hence, a basis for $\mathscr{S}_1 + \mathscr{S}_2$. We have shown that $\dim(\mathscr{S}_1 + \mathscr{S}_2) = r + s + t$, $\dim(\mathscr{S}_1 \cap \mathscr{S}_2) = r$, $\dim \mathscr{S}_1 = r + s$, and $\dim \mathscr{S}_2 = r + t$. Thus, the equality asserted in the theorem holds. □

Let us be explicit about the connection between this theorem and the remarks made preceding the statement of the theorem. Suppose dim $\mathscr{S}_1 =$ dim $\mathscr{S}_2 = 2$, and dim $\mathscr{V} = 3$. Then dim$(\mathscr{S}_1 + \mathscr{S}_2)$ is either 2 (if $\mathscr{S}_1 = \mathscr{S}_2$) or 3 (since $\mathscr{S}_1 + \mathscr{S}_2 \subset \mathscr{V}$). In the first case $\mathscr{S}_1$ and $\mathscr{S}_2$ are identical. In the second case dim$(\mathscr{S}_1 \cap \mathscr{S}_2) = 1$ because of Theorem 1.2.

If the dimension of $\mathscr{V}$ is 4 or more, the conclusion that two non-identical planes through the origin have a line in common does not follow. It might happen that dim$(\mathscr{S}_1 + \mathscr{S}_2) = 4$ and dim$(\mathscr{S}_1 \cap \mathscr{S}_2) = 0$. If it defies your geometric intuition to visualize two planes with only a point in common, you must realize this is just a limitation of our human experience and training. An example of this possibility must be given algebraically, and this is done in Exercise 3.

Theorem 1.3 *If $\mathscr{S} = \mathscr{S}_1 + \mathscr{S}_2$, then dim $\mathscr{S} \leq$ dim $\mathscr{S}_1$ + dim $\mathscr{S}_2$. If $\mathscr{S}$ is the direct sum of $\mathscr{S}_1$ and $\mathscr{S}_2$, then dim $\mathscr{S} =$ dim $\mathscr{S}_1$ + dim $\mathscr{S}_2$.*

Proof. This theorem is an immediate corollary of Theorem 1.2, but an independent proof is very easy and gives additional insight into this theorem.

Let $\{\alpha_1, \ldots, \alpha_r\}$ be a basis of $\mathscr{S}_1$ and let $\{\beta_1, \ldots, \beta_s\}$ be a basis of $\mathscr{S}_2$. Then $\{\alpha_1, \ldots, \alpha_r, \beta_1, \ldots, \beta_s\}$ spans $\mathscr{S} = \mathscr{S}_1 + \mathscr{S}_2$. Thus, dim $\mathscr{S} \leq r + s =$ dim $\mathscr{S}_1$ + dim $\mathscr{S}_2$.

Assume $\mathscr{S} = \mathscr{S}_1 \oplus \mathscr{S}_2$, and suppose that $\sum_{i=1}^{r} a_i \alpha_i + \sum_{j=1}^{s} b_j \beta_j = 0$. Since the sum is direct, the representation of 0 as a sum is unique. One such representation is $0 = 0 + 0$. Thus, $\sum_{i=1}^{r} a_i \alpha_i = 0$ and $\sum_{j=1}^{s} b_j \beta_j = 0$. Because of the linear independence of the bases of $\mathscr{S}_1$ and $\mathscr{S}_2$, all $a_i = 0$ and all $b_j = 0$. Thus, $\{\alpha_1, \ldots, \alpha_r, \beta_1, \ldots, \beta_s\}$ is a basis of $\mathscr{S}$ and dim $\mathscr{S} = r + s =$ dim $\mathscr{S}_1$ + dim $\mathscr{S}_2$. □

Let $\{\alpha_1, \alpha_2, \ldots, \alpha_r\}$ be a set of vectors and define $\mathscr{S}_k = \langle \alpha_1, \ldots, \alpha_k \rangle$. Then, for each k, $\mathscr{S}_{k+1} = \mathscr{S}_k + \langle \alpha_{k+1} \rangle$. Since dim $\langle \alpha_{k+1} \rangle = 1$, there are just two possibilities. Either dim $\mathscr{S}_{k+1} =$ dim $\mathscr{S}_k$ or dim $\mathscr{S}_{k+1} =$ dim $\mathscr{S}_k + 1$. In the first case $\mathscr{S}_{k+1} = \mathscr{S}_k$, and this means that $\alpha_{k+1} \in \mathscr{S}_k$. This, in turn, means that α_{k+1} can be expressed as a linear combination of the vectors in $\{\alpha_1, \ldots, \alpha_k\}$. In the second case $\alpha_{k+1} \notin \mathscr{S}_k$, and this means α_{k+1} cannot be expressed as a linear combination of the vectors in $\{\alpha_1, \ldots, \alpha_k\}$. This observation gives us an alternative way of expressing Theorem 6.1 of Chapter 2.

To allow us to state this alternative in its simplest form, define $\mathscr{S}_0 = \langle \emptyset \rangle = \{0\}$. Then: the set $\{\alpha_1, \ldots, \alpha_r\}$ is linearly dependent if and only if there is a k, $0 \le k \le r-1$, such that $\dim \mathscr{S}_{k+1} = \dim \mathscr{S}_k$. An equivalent statement is: the set $\{\alpha_1, \ldots, \alpha_r\}$ is linearly independent if for all $k, 0 \le k \le r-1$, $\dim \mathscr{S}_{k+1} = \dim \mathscr{S}_k + 1$.

When the m-tuples representing the vectors $\{\alpha_1, \ldots, \alpha_r\}$ are arranged into columns in a matrix and the resulting matrix is reduced to row-echelon form, the considerations of the previous paragraph are demonstrated. Each column that has a non-zero element in a new row corresponds to an index where the dimension increases by 1. For the other columns the dimension does not increase.

Theorem 1.4 *Let $\mathscr{S}$ be any subspace of $\mathscr{V}$. There exists another subspace $\mathscr{S}'$ such that $\mathscr{V} = \mathscr{S} \oplus \mathscr{S}'$.*

Proof. Let $\{\alpha_1, \ldots, \alpha_r\}$ be a basis of $\mathscr{S}$. Extend this linearly independent set to a basis $\{\alpha_1, \ldots, \alpha_r, \beta_1, \ldots, \beta_s\}$ of $\mathscr{V}$. Let $\mathscr{S}'$ be the subspace spanned by $\{\beta_1, \ldots, \beta_s\}$. Then any $\alpha \in \mathscr{V}$ can be expressed in the form $\alpha = \sum_{i=1}^{r} a_i \alpha_i + \sum_{j=1}^{s} b_j \beta_j \in \mathscr{S} + \mathscr{S}'$. The fact that $\mathscr{S} + \mathscr{S}'$ is a direct sum follows from the uniqueness of the representation of any $\alpha \in \mathscr{V}$ as a linear combination of the basis elements. $\square$

Corollary 1.5 *If $\mathscr{S}$ and $\mathscr{T}$ are subspaces and $\mathscr{S} \subset \mathscr{T}$, there exists a subspace $\mathscr{S}'$ such that $\mathscr{S} \oplus \mathscr{S}' = \mathscr{T}$.* $\square$

Any subspace $\mathscr{S}'$ such that $\mathscr{V} = \mathscr{S} \oplus \mathscr{S}'$ is called a *complementary* subspace of $\mathscr{S}$. Since the basis of $\mathscr{S}$ can be extended to a basis of $\mathscr{V}$ in many different ways the complementary subspace is not unique. As an example of how this can occur, consider the real plane $\mathscr{R}^2 = \mathscr{V}$, and let $\mathscr{S}$ be the x_1-axis. Any 1-dimensional subspace, other than $\mathscr{S}$ itself, can serve as a complementary subspace.

EXERCISES 6–1

1. Let $\mathscr{S}_1$ and $\mathscr{S}_2$ be any subsets of a vector space $\mathscr{V}$. $\mathscr{S}_1$ generates the subspace of $\langle \mathscr{S}_1 \rangle$ and $\mathscr{S}_2$ generates the subspace $\langle \mathscr{S}_2 \rangle$. Find a set of generators for the subspace $\langle \mathscr{S}_1 \rangle + \langle \mathscr{S}_2 \rangle$.

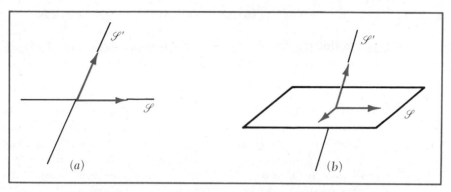

Figure 6-1.1 Complementary subspaces. In a 2-dimensional space, any two non-identical 1-dimensional subspaces are complementary. In a 3-dimensional space, any 2-dimensional subspace $\mathscr{S}$ and a 1-dimensional subspace $\mathscr{S}'$ not contained in $\mathscr{S}$ are complementary. In (a) and (b), basis vectors for $\mathscr{S}$ and basis vectors for $\mathscr{S}'$ are indicated in color. Together, they form a basis for the entire space $\mathscr{V}$.

2. Let $\mathscr{W}_1$ be a subspace spanned by $\{(1, 1, 10, 3), (2, 1, 16, 1)\}$ and let $\mathscr{W}_2$ be a subspace spanned by $\{(1, 1, 7, 9), (2, 1, 11, 11)\}$. Find a basis for $\mathscr{W}_1 \cap \mathscr{W}_2$. (The recommended way to do this is to find a system of equations for $\mathscr{W}_1$ in the spirit of Section 2–8. Similarly, find a system of equations for $\mathscr{W}_2$. $\mathscr{W}_1 \cap \mathscr{W}_2$ satisfies both systems.) Find a basis for $\mathscr{W}_1 + \mathscr{W}_2$.

3. Let $\mathscr{W}_1$ be a subspace spanned by $\{(1, 1, 2, 3), (2, 1, 0, 1)\}$ and let $\mathscr{W}_2$ be spanned by $\{(1, 2, 3, 4), (2, -4, 3, 1)\}$. Find $\mathscr{W}_1 + \mathscr{W}_2$. Use this information to find $\mathscr{W}_1 \cap \mathscr{W}_2$.

4. If $\mathscr{W}_1$ and $\mathscr{W}_2$ are subspaces, show that $\mathscr{W}_1 \cup \mathscr{W}_2$ is a subspace if and only if $\mathscr{W}_1 \subset \mathscr{W}_2$ or $\mathscr{W}_1 \supset \mathscr{W}_2$.

5. Let $\mathscr{W}_1$ be the subspace of $\mathscr{R}^4$ spanned by $\{(1, 2, 1, 2), (2, 1, -3, -2)\}$. Find a complementary subspace for $\mathscr{W}_1$.

6. Show that if $\mathscr{W}_1$ and $\mathscr{W}_2$ are subspaces of $\mathscr{V}$ such that $\dim \mathscr{W}_1 = \dim \mathscr{W}_2$ and $\mathscr{W}_1 \subset \mathscr{W}_2$, then $\mathscr{W}_1 = \mathscr{W}_2$.

7. Let $\mathscr{S}, \mathscr{W}_1$, and $\mathscr{W}_2$ be three subspaces of a finite dimensional vector space $\mathscr{V}$ for which $\mathscr{S} \cap \mathscr{W}_1 = \mathscr{S} \cap \mathscr{W}_2, \mathscr{S} + \mathscr{W}_1 = \mathscr{S} + \mathscr{W}_2$, and $\mathscr{W}_1 \subset \mathscr{W}_2$. Show that $\mathscr{W}_1 = \mathscr{W}_2$.

8. Show that it is possible to have three subspaces $\mathscr{S}, \mathscr{W}_1, \mathscr{W}_2$ satisfying any two of the three conditions without satisfying the conclusion that $\mathscr{W}_1 = \mathscr{W}_2$.

6-2 EIGENVECTORS AND INVARIANT SUBSPACES

Definition *Let σ be a linear transformation of $\mathscr{V}$ into itself. A subspace $\mathscr{S}$ such that $\sigma(\mathscr{S}) \subset \mathscr{S}$ is said to be **invariant** under σ.*

If we can find invariant subspaces for σ, the representation of σ by a matrix and the analysis of the properties of σ are considerably simplified. It is, generally, not particularly easy to find invariant subspaces for a given σ. In this chapter we will take up a method for finding one type of invariant subspace, a subspace consisting of eigenvectors of σ.

Let $\mathscr{S}$ be an invariant subspace for σ. Let $\{\alpha_1, \ldots, \alpha_r\}$ be a basis for $\mathscr{S}$, and extend this linearly independent set to a basis $\{\alpha_1, \ldots, \alpha_r, \alpha_{r+1}, \ldots, \alpha_n\}$. For $j \leq r$, $\alpha_j \in \mathscr{S}$ and $\sigma(\alpha_j) \in \mathscr{S}$. Thus, $\sigma(\alpha_j) = \sum_{i=1}^{r} a_{ij}\alpha_i$, where this sum uses only the elements of the basis of $\mathscr{S}$. The matrix representing σ has the form

$$\begin{bmatrix} a_{11} & \cdots & a_{1r} & a_{1,r+1} & \cdots & a_{1n} \\ \cdot & & \cdot & \cdot & & \cdot \\ \cdot & & \cdot & \cdot & & \cdot \\ \cdot & & \cdot & \cdot & & \cdot \\ a_{r1} & \cdots & a_{rr} & a_{r,r+1} & \cdots & a_{rn} \\ 0 & \cdots & 0 & a_{r+1,r+1} & \cdots & a_{r+1,n} \\ \cdot & & \cdot & \cdot & & \cdot \\ \cdot & & \cdot & \cdot & & \cdot \\ \cdot & & \cdot & \cdot & & \cdot \\ 0 & \cdots & 0 & a_{m,r+1} & \cdots & a_{mn} \end{bmatrix} \quad (2.1)$$

The important thing to notice is the block of zeros in the lower left-hand corner.

Even greater simplification can be obtained if we can find two invariant subspaces $\mathscr{S}_1$ and $\mathscr{S}_2$ such that $\mathscr{V} = \mathscr{S}_1 \oplus \mathscr{S}_2$. Let $\{\alpha_1, \ldots, \alpha_r\}$ be a basis for $\mathscr{S}_1$ and let $\{\beta_1, \ldots, \beta_s\}$ be a basis for $\mathscr{S}_2$. By the same argument used to prove Theorem 1.3 $\{\alpha_1, \ldots, \alpha_r, \beta_1, \ldots, \beta_s\}$ is a basis for $\mathscr{V}$. With respect to this basis the matrix representing σ would have zeros in both the lower left corner and the upper right corner.

As an example, consider a rotation in a 3-dimensional coordinate space. The axis is one invariant subspace. A plane

6-2 Eigenvectors and Invariant Subspaces

through the origin and perpendicular to the axis is another. If α_1 is a vector in the axis of rotation and if $\{\alpha_2, \alpha_3\}$ are mutually perpendicular vectors of length one in the plane of rotation, the matrix representing the rotation has the form.

$$\begin{bmatrix} 1 & 0 & 0 \\ 0 & \cos\theta & -\sin\theta \\ 0 & \sin\theta & \cos\theta \end{bmatrix}. \qquad (2.2)$$

It is very desirable to find invariant subspaces of dimension one. Let $\mathscr{S} = \langle \xi \rangle$ be a 1-dimensional invariant subspace. Since $\sigma(\xi) \in \langle \xi \rangle$, $\sigma(\xi) = d\xi$ for some scalar d.

Definition *Whenever a non-zero vector ξ and a scalar d are related to the linear transformation σ by*

$$\sigma(\xi) = d\xi, \qquad (2.3)$$

*we call d an **eigenvalue** of σ and ξ an **eigenvector** of σ. Also, we say that ξ is an eigenvector corresponding to d and d is the eigenvalue corresponding to ξ.*

Given an eigenvector ξ, there is just one eigenvalue associated with it since the value of d is determined by the rela-

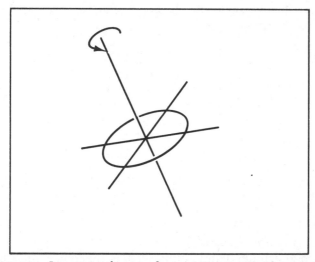

Figure 6–2.1 Invariant subspaces for a rotation in $\mathscr{R}^3$. The axis of the rotation is one invariant subspace. A plane through the origin and perpendicular to the axis is another.

tion $\sigma(\xi) = d\xi$. But, given an eigenvalue d there are many eigenvectors corresponding to d. In fact, if ξ is an eigenvector associated with d and a is any scalar, then $\sigma(a\xi) = a\sigma(\xi) = ad\xi = d(a\xi)$. That is, $a\xi$ is another eigenvector corresponding to d.

For any scalar d the set of all ξ satisfying the relation $\sigma(\xi) = d\xi$ is precisely the set of all ξ satisfying the relation $(\sigma - d)\xi = 0$, where here d represents the corresponding scalar transformation. This is the kernel of $\sigma - d$. The zero vector satisfies this condition, but the zero vector is not considered to be an eigenvector. Thus the non-zero vectors in the kernel of $\sigma - d$ are the eigenvectors corresponding to d. For most values of d the linear transformation $\sigma - d$ is non-singular and the kernel of $\sigma - d$ is just the zero subspace. In this case d is not an eigenvalue and there are no eigenvectors corresponding to d. There are eigenvectors corresponding to d if and only if $\sigma - d$ is singular and has a kernel of positive dimension. For an eigenvalue d the kernel of $\sigma - d$ is called the *eigenspace* corresponding to d, and we denote it by $\mathscr{S}(d)$. The dimension of $\mathscr{S}(d)$ is called the *geometric multiplicity* of d. Clearly $\mathscr{S}(d)$ is an invariant subspace under σ, and any subspace of $\mathscr{S}(d)$ is also invariant.

The ultimate in simplification occurs when $\mathscr{V}$ has a basis consisting of eigenvectors of σ. Suppose $\mathscr{X} = \{\xi_1, \ldots, \xi_n\}$ is a basis of $\mathscr{V}$, where each ξ_j is an eigenvector. Let d_j be the eigenvalue corresponding to ξ_j. Then $\sigma(\xi_j) = d_j\xi_j$ for each j. Thus, with respect to the basis $\mathscr{X}$, σ is represented by the diagonal matrix

$$\begin{bmatrix} d_1 & 0 & \cdots & 0 \\ 0 & d_2 & \cdots & 0 \\ \cdot & \cdot & & \cdot \\ \cdot & \cdot & & \cdot \\ \cdot & \cdot & & \cdot \\ 0 & 0 & \cdots & d_n \end{bmatrix} = D. \tag{2.4}$$

This means that if σ is represented by a matrix A with respect to a basis $\mathscr{A} = \{\alpha_1, \ldots, \alpha_n\}$, there is a matrix of transition P from the basis $\mathscr{A}$ to the basis $\mathscr{X}$ such that $P^{-1}AP = D$ is a diagonal matrix. We say that A is similar to a diagonal matrix, or that A is *diagonalizable*.

On the other hand, suppose A is diagonalizable. Let P be a non-singular matrix such that $P^{-1}AP = D$ is a diagonal matrix. We interpret A as representing a linear transformation σ of $\mathscr{V}$ into itself and P as a matrix of transition. Since D represents σ with respect to the new basis, $\mathscr{X} = \{\xi_1, \ldots, \xi_n\}$, we must have $\sigma(\xi_j) = d_j \xi_j$. That is, each ξ_j is an eigenvector for σ and the elements in the main diagonal of D are eigenvalues of σ. We shall refer to these eigenvalues as eigenvalues of A as well.

Not all linear transformations have enough eigenvectors to make up a basis. For example, a rotation of the plane about the origin through any angle, other than 0° and 180°, would move every line through the origin to a new position. Thus, there would be no invariant 1-dimensional subspaces. Thus, such a rotation has no eigenvectors.

As another example, consider the shear described in Section 4–2. There the x_1-axis is left fixed and any vector in that line is an eigenvector. But no other 1-dimensional subspace is invariant, so there are no other eigenvectors. There are numerous eigenvectors, all in one 1-dimensional subspace. But there are not two linearly independent eigenvectors. Thus, there is no basis of eigenvectors.

If the eigenvalues are known, there is no problem in finding the eigenvectors. If d is the eigenvalue we must find the kernel of $\sigma - d$, a problem that we have already examined in detail. For example, suppose σ is represented by the matrix

$$\begin{bmatrix} 3 & 2 & 2 \\ 1 & 4 & 1 \\ -2 & -4 & -1 \end{bmatrix}, \quad (2.5)$$

and we are given that 1, 2, and 3 are eigenvalues. Then $\sigma - d$ is represented by the matrix $A - dI = C(d)$, which is

$$C(d) = \begin{bmatrix} 3-d & 2 & 2 \\ 1 & 4-d & 1 \\ -2 & -4 & -1-d \end{bmatrix}. \quad (2.6)$$

For $d = 1$, we have

$$C(1) = \begin{bmatrix} 2 & 2 & 2 \\ 1 & 3 & 1 \\ -2 & -4 & -2 \end{bmatrix}. \quad (2.7)$$

Reducing to basic form, we get

$$\begin{bmatrix} 1 & 0 & 1 \\ 0 & 1 & 0 \\ 0 & 0 & 0 \end{bmatrix}. \quad (2.8)$$

Whatever question there might be as to how the values of the eigenvalues were obtained, there is no doubt now that 1 is an eigenvalue. This is assured because $C(1)$ has a positive nullity. From (2.8) we see that $(-1, 0, 1)$ represents an eigenvector corresponding to the eigenvalue 1. For $d = 2$ we have

$$C(2) = \begin{bmatrix} 1 & 2 & 2 \\ 1 & 2 & 1 \\ -2 & -4 & -3 \end{bmatrix}. \quad (2.9)$$

Reducing this to basic form, we obtain

$$\begin{bmatrix} 1 & 2 & 0 \\ 0 & 0 & 1 \\ 0 & 0 & 0 \end{bmatrix}. \quad (2.10)$$

This gives $(-2, 1, 0)$ as a representation of the eigenvector. Similarly, for $d = 3$ we get $(0, -1, 1)$ as a representation of the corresponding eigenvector.

To find a diagonal representation of σ, we use a basis of eigenvectors. Now, for this example we see that three eigenvectors are represented by $\{(-1, 0, 1), (-2, 1, 0), (0, -1, 1)\}$. If these three eigenvectors are linearly independent they will form a basis of eigenvectors. We could verify that they are linearly independent by direct calculations along familiar lines. But we shall shortly prove a theorem that will allow us to conclude that they are linearly independent without calculation. Then the matrix of transition to the basis of eigenvectors is

$$P = \begin{bmatrix} -1 & -2 & 0 \\ 0 & 1 & -1 \\ 1 & 0 & 1 \end{bmatrix}. \quad (2.11)$$

To determine the diagonal representation of σ we compute $P^{-1}AP$. When computing $P^{-1}AP$ we can compute AP first or we can compute $P^{-1}A$ first. The columns of P are eigenvectors of A. Thus, the columns of AP should be the columns of P multiplied by the corresponding eigenvalues. Look for this in the following calculation.

$$P^{-1}AP = P^{-1} \begin{bmatrix} 3 & 2 & 2 \\ 1 & 4 & 1 \\ -2 & -4 & -1 \end{bmatrix} \begin{bmatrix} -1 & -2 & 0 \\ 0 & 1 & -1 \\ 1 & 0 & 1 \end{bmatrix}$$

$$= \begin{bmatrix} 1 & 2 & 2 \\ -1 & -1 & -1 \\ -1 & -2 & -1 \end{bmatrix} \begin{bmatrix} -1 & -4 & 0 \\ 0 & 2 & -3 \\ 1 & 0 & 3 \end{bmatrix} \qquad (2.12)$$

$$= \begin{bmatrix} 1 & 0 & 0 \\ 0 & 2 & 0 \\ 0 & 0 & 3 \end{bmatrix}.$$

If we are confident of the accuracy of our work, we could just write down the diagonal matrix as the product $P^{-1}AP$ and skip all intermediate calculations. Computing AP has the value of providing a check that the columns of P are really eigenvectors of A. But that purpose can also be achieved by calculating $P^{-1}A$ first. Since $P^{-1}AP = D$ is a diagonal matrix, $D = (P^{-1}AP)^T = P^T A^T (P^T)^{-1}$. This means $(P^T)^{-1}$ is a matrix of transition which diagonalizes A^T. Thus, the columns of $(P^T)^{-1} = (P^{-1})^T$ are eigenvectors of A^T. This fact will show up in the calculation of $P^{-1}A$, which is the transpose of $A^T(P^T)^{-1}$. The rows of $P^{-1} A$ will be the rows of P^{-1} multiplied by the corresponding eigenvalues of A.

How much of this type of calculation one does in a particular application depends on what is needed and how necessary it is to check results. In some cases it is sufficient to know the eigenvalues. In such a case all calculations beyond determining the eigenvalues could be omitted unless it is necessary to check the result. In other applications one wants to know both the eigenvalues and the eigenvectors. In such a case one could determine them, write down the matrix of transition P, and conclude that $P^{-1}AP = D$ without calculating this product, provided one was sure his calculations were accurate.

In the example just discussed, there was a point where we assumed the three eigenvectors to be linearly independent so they formed a basis. The following theorem justifies that assumption.

Theorem 2.1 *If $\xi_1, \xi_2, \ldots, \xi_r$ are eigenvectors with distinct eigenvalues, then $\xi_1 + \xi_2 + \cdots + \xi_r \neq 0$.*

Proof. The theorem is true for $r = 1$ because an eigenvector is non-zero by definition.

Assume the theorem to be true for sums involving less than r eigenvectors, and suppose

$$\xi_1 + \xi_2 + \cdots + \xi_r = 0. \qquad (2.13)$$

Applying σ to the relation in (2.13) we have

$$d_1 \xi_1 + d_2 \xi_2 + \cdots + d_r \xi_r = 0. \qquad (2.14)$$

Multiply relation (2.13) by d_r and subtract the result from (2.14) to obtain

$$(d_1 - d_r)\xi_1 + (d_2 - d_r)\xi_2 + \cdots + (d_{r-1} - d_r)\xi_{r-1} = 0. \qquad (2.15)$$

Since $d_i - d_r \neq 0$ for $i \neq r$, each $(d_i - d_r)\xi_i \neq 0$ is an eigenvector for $i \neq r$. Since relation (2.15) is a contradiction to the induction assumption, the theorem is proved. □

Corollary 2.2 *If $\xi_1, \xi_2, \ldots, \xi_r$ are eigenvectors corresponding to distinct eigenvalues, they are linearly independent.*

Proof. Suppose there is a non-trivial relation of the form

$$a_1 \xi_1 + a_2 \xi_2 + \cdots + a_r \xi_r = 0. \qquad (2.16)$$

For each non-zero a_i, $a_i \xi_i$ is an eigenvector corresponding to d_i. If we drop the zero terms from the relation (2.16), the remaining sum is a sum of distinct eigenvectors. Since the relation (2.16) is non-trivial this sum would contain at least one eigenvector, which is a contradiction to Theorem 2.1. □

Corollary 2.3 *If $\mathscr{S}(d_1), \ldots, \mathscr{S}(d_r)$ are eigenspaces corresponding to distinct eigenvalues the sum $\mathscr{S}(d_1) + \cdots + \mathscr{S}(d_r)$ is direct.*

Proof. Suppose two different representations as a sum were possible. That is, suppose $\alpha_1 + \cdots + \alpha_r = \beta_1 + \cdots + \beta_r$ where $\alpha_i, \beta_i \in \mathscr{S}(d_i)$. Then $0 = (\alpha_1 - \beta_1) + \cdots + (\alpha_r - \beta_r)$. Since each $\alpha_i - \beta_i$ is an eigenvector or 0, if any $\alpha_i - \beta_i \neq 0$ we would have a contradiction to Theorem 2.1. Thus, the sum is direct. □

6-2 Eigenvectors and Invariant Subspaces

There are usually two ways in which Theorem 2.1 and its corollaries are used. Often, for a linear transformation in an n-dimensional vector space we can find n distinct eigenvalues. For each eigenvalue there is an eigenvector, which can be obtained by the methods illustrated in the example. Theorem 2.1 assures us these eigenvectors are linearly independent and they form a basis.

In more complicated situations there are not n distinct eigenvalues. But for some of the eigenvalues the corresponding eigenspace has dimension greater than 1. We find a basis for each eigenspace separately. Theorem 2.1 and its corollary assure us the collection of all the eigenvectors obtained this way is linearly independent. If the total number obtained is n we have a basis of eigenvectors.

The example already given illustrates the first situation. We will now give an example illustrating the second possibility.

Suppose we are given that the linear transformation represented by

$$\begin{bmatrix} -1 & -6 & 3 \\ 3 & 8 & -3 \\ 6 & 12 & -4 \end{bmatrix} \qquad (2.17)$$

has 2 and -1 as eigenvalues. Again, we find the eigenvectors by finding the kernel of $\sigma - d$. Taking $d = 2$ in $C(d) = A - dI$, we have

$$C(2) = \begin{bmatrix} -3 & -6 & 3 \\ 3 & 6 & -3 \\ 6 & 12 & -6 \end{bmatrix}. \qquad (2.18)$$

We reduce (2.18) to basic form and obtain

$$\begin{bmatrix} 1 & 2 & -1 \\ 0 & 0 & 0 \\ 0 & 0 & 0 \end{bmatrix}. \qquad (2.19)$$

Since the reduced form is of rank 1, the kernel of $\sigma - 2$ is of dimension 2. This is the dimension of the eigenspace $\mathscr{S}(2)$. From (2.19) we see that $\{(-2, 1, 0), (1, 0, 1)\}$ is a basis of $\mathscr{S}(2)$.

For the eigenvalue $d = -1$ we obtain an eigenvector represented by $(-1, 1, 2)$. Let us examine how Theorem 2.1 and its corollary allow us to conclude that the set $\{(-2, 1, 0),$

$(1, 0, 1)$, $(-1, 1, 2)\}$ is linearly independent. Suppose we have a linear relation of the form

$$a_1(-2, 1, 0) + a_2(1, 0, 1) + a_3(-1, 1, 2) = 0. \quad (2.20)$$

Then $a_1(-2, 1, 0) + a_2(1, 0, 1)$ belongs to the eigenspace $\mathscr{S}(2)$ and $a_3(-1, 1, 2)$ belongs to the eigenspace $\mathscr{S}(-1)$. Thus, $a_1(-2, 1, 0) + a_2(1, 0, 1) = 0$ and $a_3(-1, 1, 0) = 0$ because of Corollary 2.2. This shows $a_3 = 0$, and $a_1 = a_2 = 0$ because the set $\{(-2, 1, 0), (1, 0, 1)\}$ is linearly independent. Thus, $\{(-2, 1, 0), (1, 0, 1), (-1, 1, 2)\}$ represents a basis of eigenvectors of σ.

Continuing with this example, the matrix of transition to the basis of eigenvectors is

$$P = \begin{bmatrix} -2 & 1 & -1 \\ 1 & 0 & 1 \\ 0 & 1 & 2 \end{bmatrix}. \quad (2.21)$$

We should know that $P^{-1}AP$ is a diagonal matrix with the eigenvalues 2, 2, -1 in the main diagonal. To check we compute

$$P^{-1}AP = \begin{bmatrix} 1 & 3 & -1 \\ 2 & 4 & -1 \\ -1 & -2 & 1 \end{bmatrix} \begin{bmatrix} -1 & -6 & 3 \\ 3 & 8 & -3 \\ 6 & 12 & -4 \end{bmatrix} \begin{bmatrix} -2 & 1 & -1 \\ 1 & 0 & 1 \\ 0 & 1 & 2 \end{bmatrix}$$

$$= \begin{bmatrix} 1 & 3 & -1 \\ 2 & 4 & -1 \\ -1 & -2 & 1 \end{bmatrix} \begin{bmatrix} -4 & 2 & 1 \\ 2 & 0 & -1 \\ 0 & 2 & -2 \end{bmatrix} \quad (2.22)$$

$$= \begin{bmatrix} 2 & 0 & 0 \\ 0 & 2 & 0 \\ 0 & 0 & -1 \end{bmatrix}.$$

Notice that in the calculation of AP the columns of the product are the columns of P multiplied by the eigenvalues.

EXERCISES 6-2

1. Find a linearly independent set of eigenvectors and a matrix of transition to diagonalize the matrix

6-2 Eigenvectors and Invariant Subspaces

$$A = \begin{bmatrix} 7 & -2 \\ 15 & -4 \end{bmatrix}$$

given that 1 and 2 are eigenvalues of A.

2. Find a linearly independent set of eigenvectors and a matrix of transition to diagonalize the matrix

$$B = \begin{bmatrix} 27 & 15 \\ -50 & -28 \end{bmatrix}$$

given that 2 and -3 are eigenvalues of B.

3. Find a linearly independent set of eigenvectors and a matrix of transition to diagonalize the matrix

$$C = \begin{bmatrix} -5 & -8 & -12 \\ -6 & -10 & -12 \\ 6 & 10 & 13 \end{bmatrix}$$

given that 1, -1 and -2 are eigenvalues of C.

4. Find a linearly independent set of eigenvectors and a matrix of transition to diagonalize the matrix

$$D = \begin{bmatrix} 0 & -2 & -2 \\ -2 & -3 & -2 \\ 3 & 6 & 5 \end{bmatrix}$$

given that 1, -1, and 2 are eigenvalues of D.

5. Find a linearly independent set of eigenvectors and a matrix of transition to diagonalize the matrix

$$E = \begin{bmatrix} 9 & 8 & 4 \\ 4 & 5 & 2 \\ -22 & -22 & -10 \end{bmatrix}$$

given that 1 and 2 are eigenvalues of E.

6. Find a linearly independent set of eigenvectors for the matrix

$$F = \begin{bmatrix} 6 & 3 & 2 \\ 1 & 0 & 0 \\ -10 & -2 & -2 \end{bmatrix}.$$

F cannot be diagonalized, but use

$$\begin{bmatrix} -1 & -1 & 4 \\ -1 & 0 & 2 \\ 4 & 2 & -11 \end{bmatrix}$$

to find a matrix similar to F. Compare it with the diagonal matrix obtained in Exercise 5.

7. Show that similar matrices have the same eigenvalues.

8. Show that 0 is an eigenvalue of a matrix A if and only if A is singular.

9. Show that if d is an eigenvalue of the linear transformation σ, then d^2 is an eigenvalue of σ^2, and d^n is an eigenvalue of σ^n.

10. Let σ_1 and σ_2 be linear transformations such that ξ is an eigenvector for both. Let d_1 be an eigenvalue of σ_1 corresponding to ξ, and let d_2 be an eigenvalue of σ_2 corresponding to ξ. Show that $d_1 + d_2$ is an eigenvalue of $\sigma_1 + \sigma_2$.

11. Let $p(x) = a_n x^n + a_{n-1} x^n + \cdots + a_1 x + a_0$ be a polynomial with coefficients in F. Let $\mathscr{V}$ be a vector space over F, and let σ be a linear transformation of σ into itself. By $p(\sigma)$ we mean the linear transformation $a_n \sigma^n + a_{n-1} \sigma^{n-1} + \cdots + a_1 \sigma + a_0 \, 1$. (Note that the constant term must be replaced by a scalar transformation.) Show that if d is an eigenvalue of σ corresponding to the eigenvector ξ, then $p(d)$ is an eigenvalue of $p(\sigma)$ corresponding to ξ.

6-3 THE CHARACTERISTIC MATRIX AND CHARACTERISTIC EQUATION

By rearranging in order the ideas of the previous section, we can obtain an effective computational method for finding eigenvalues and eigenvectors. We observed that the kernel of $\sigma - d$ contains eigenvectors of σ if and only if $\sigma - d$ is singular. Thus, we wish to find those values of d for which $\sigma - d$ is singular. The determinant function provides a tool for doing this.

Let A be any matrix representing σ. Then $A - dI$ is a matrix representing $\sigma - d$. $\sigma - d$ is singular as a linear transformation if and only if $A - dI$ is singular as a matrix. And $A - dI$ is singular if and only if $\det(A - dI) = 0$.

Let x denote a scalar unknown and consider the matrix

$$C(x) = A - xI = \begin{bmatrix} a_{11} - x & a_{12} & \cdots & a_{1n} \\ a_{21} & a_{22} - x & \cdots & a_{2n} \\ \vdots & \vdots & & \vdots \\ a_{n1} & a_{n2} & \cdots & a_{nn} - x \end{bmatrix}. \quad (3.1)$$

6-3 The Characteristic Matrix and Characteristic Equation

Notice that $C(x)$ is a matrix whose elements are polynomials in x of degree at most 1. The $n \times n$ matrix $C(x)$ is called the *characteristic matrix* of A.

The determinant function $\det C(x)$ is a sum of $n!$ terms, each of which is a product of n elements from $C(x)$. Thus, each term in $\det C(x)$ is a polynomial in x of degree at most n. However, only one term in the expansion of $\det C(x)$ is a polynomial of degree n, the term $(a_{11}-x)(a_{22}-x)\cdots(a_{nn}-x)$. Thus, the expansion of $\det C(x)$ contains the term $(-1)^n x^n$, which is the only term of degree n. Hence, $\det C(x)$ is a polynomial of degree n and the coefficient of x^n is $(-1)^n$. This polynomial is called the *characteristic polynomial* of A.

All the coefficients of $\det C(x)$ can be expressed explicitly in terms of the elements of A, but we shall compute only two more coefficients of the characteristic polynomial. Since $\det C(x)$ is a formal polynomial expression in terms of the formula for the determinant, we can give x scalar values. Clearly, $C(0) = A$ so $\det C(0) = \det A$. When x is given the value 0 in a polynomial, the value of the polynomial is the term of degree zero. Thus, $\det A$ is the constant term of the characteristic polynomial.

The coefficient of x^{n-1} is also important and quite easy to obtain. Any term of the determinant formula that contains a_{ij}, $i \neq j$, as a factor contains neither $(a_{ii}-x)$ nor $(a_{jj}-x)$ as a factor, and must be of degree $n-2$ or less. Hence, no term in the expansion of the determinant contributes to the $(n-1)$-degree term of the characteristic polynomial except the product $(a_{11}-x)(a_{22}-x)\cdots(a_{nn}-x)$. The coefficient of x^{n-1} in this product is $(-1)^{n-1}(a_{11}+a_{22}+\cdots+a_{nn}) = (-1)^{n-1}\sum_{i=1}^{n} a_{ii}$. The number $\sum_{i=1}^{n} a_{ii} = \text{Tr}(A)$ is called the *trace* of A. The trace is just the sum of the elements in the main diagonal of A.

The polynomial equation $\det C(x) = 0$ is called the *characteristic equation* of A. Since $C(d)$ is singular if d is an eigenvalue, an eigenvalue is a solution of the characteristic equation. On the other hand, a solution of the characteristic equation is not necessarily an eigenvalue. An additional condition is needed. In the eigenvector equation $\sigma(\xi) = d\xi$, ξ and $\sigma(\xi)$ are vectors in the vector space, and d is a scalar in the field over which the vector space is defined. Thus, an eigenvalue must be an element of the field, a scalar. If we are considering a vector space over the rational numbers, the elements in the matrix A representing σ will be rational numbers and the coefficients appearing in $\det C(x)$ will be rational numbers. The characteristic equation will be a polynomial equation with rational coefficients. This equation may well have non-rational solutions, and these cannot be eigenvalues. If we are considering a vector space over the real numbers, the representing matrix A and the characteristic polynomial det $C(x)$ will have real elements and coefficients. Again, the characteristic equation may have non-real solutions, which cannot be eigenvalues.

On the other hand, a solution of the characteristic equation that is an element of the field of scalars is an eigenvalue. If d is a scalar solution of the characteristic equation, $A - dI$ and $\sigma - d$ are singular. Thus, the kernel of $\sigma - d$ has positive dimension and any non-zero vector in the kernel is an eigenvector corresponding to d. Despite the fact that non-scalar solutions of the characteristic equation are not eigenvalues, there is no purpose served by discarding them. They often provide useful information on the basis of their sign, size, or interrelationships. We shall call solutions of the characteristic equation *characteristic values*. An eigenvalue is always a characteristic value; a characteristic value is an eigenvalue if and only if it is in the field of scalars.

An eigenvalue problem is seldom given abstractly. We are usually given the matrix A representing the linear transformation. If the problem is given abstractly, we are expected to select a basis and determine the representing matrix A. Thus, we consider the eigenvalue problem as starting with a given $n \times n$ matrix A. We are expected to find the eigenvalues of A and to find the corresponding eigenvectors. This means we are expected to find representations of the eigenvectors using the same basis as was used to represent the linear transformation by A. Starting with A the procedure is as follows:

1. Determine the characteristic matrix $C(x) = A - xI$.
2. Determine the characteristic polynomial det $C(x)$.
3. Find the eigenvalues of A by solving the characteristic equation det $C(x) = 0$ and selecting those solutions that are in the given field of scalars.
4. For each eigenvalue d, solve the system of linear homogeneous equations $C(d)X = 0$, as illustrated in (2.6), (2.7), and (2.8) of the previous section. Find a basis of the solution subspace.
5. Collect in one set the eigenvectors obtained in step 4 for all eigenvalues. Because of Corollary 2.2, this set is linearly independent.
6. If the eigenvectors obtained in step 5 form a basis

(which they will if there are as many as the order of A) write them in the columns of a matrix of transition P and compute $P^{-1}AP = D$ as a check.
7. If insufficiently many eigenvectors are obtained in step 5 to form a basis, terminate the work. In this case the work can be continued in several different ways; for example, to obtain what is known as a Jordan normal form. But these further possibilities are not included in this book.

These computational steps are illustrated in the following examples:

(1) We are given

$$A = \begin{bmatrix} 4 & 2 & 2 \\ -7 & -11 & -7 \\ 6 & 12 & 8 \end{bmatrix},$$

a matrix over the real numbers, and we are asked to find its eigenvalues and eigenvectors. The characteristic matrix for A is

$$C(x) = \begin{bmatrix} 4-x & 2 & 2 \\ -7 & -11-x & -7 \\ 6 & 12 & 8-x \end{bmatrix}.$$

From $C(x)$ we can calculate $\det C(x)$ to obtain the characteristic polynomial, $f(x) = -x^3 + x^2 + 14x - 24$. The characteristic polynomial can be factored in the form $f(x) = -(x-2)(x-3)(x+4)$. Thus, 2, 3, and -4 are eigenvalues for A. Substituting the eigenvalue 2 in the characteristic matrix, we obtain

$$C(2) = \begin{bmatrix} 2 & 2 & 2 \\ -7 & -13 & -7 \\ 6 & 12 & 6 \end{bmatrix}.$$

We reduce $C(2)$ to basic form to obtain a basis for the kernel. The reduced form is

$$\begin{bmatrix} 1 & 0 & 1 \\ 0 & 1 & 0 \\ 0 & 0 & 0 \end{bmatrix}$$

and a representation of a vector in the kernel is $(-1, 0, 1)$. $(-1, 0, 1)$ represents an eigenvector of the linear transformation represented by A.

Eigenvectors and Eigenvalues

In the same way, we substitute the eigenvalues 3 and -4 for x in the characteristic matrix $C(x)$, reduce to basic form, and determine the representations $(-2, 1, 0)$ and $(0, -1, 1)$.

Now arrange the coordinates of the eigenvectors obtained in the columns of a matrix of transition P,

$$P = \begin{bmatrix} -1 & -2 & 0 \\ 0 & 1 & -1 \\ 1 & 0 & 1 \end{bmatrix}.$$

By a straightforward calculation we can obtain the inverse of P,

$$P^{-1} = \begin{bmatrix} 1 & 2 & 2 \\ -1 & -1 & -1 \\ -1 & -2 & -1 \end{bmatrix}.$$

Now calculate AP to obtain

$$AP = \begin{bmatrix} -2 & -6 & 0 \\ 0 & 3 & 4 \\ 2 & 0 & -4 \end{bmatrix}.$$

By examining AP we can check our work to this point. The columns of AP should be multiples of the columns of P. Now we calculate $P^{-1}AP = P^{-1}(AP) = D$ to obtain

$$D = \begin{bmatrix} 2 & 0 & 0 \\ 0 & 3 & 0 \\ 0 & 0 & -4 \end{bmatrix}.$$

According to formula 6.14 of Chapter 4, D represents the same linear transformation represented by A, but with respect to a basis consisting of eigenvectors. The matrix D is a diagonal matrix, which is the simple form of the representing matrix that was sought at the beginning of this chapter.

The steps outlined to find the eigenvalues, eigenvectors, and a diagonal representation are effective, but only in a theoretical sense. To the student working through the example above, it may seem that computing a determinant to find the characteristic polynomial is the biggest drawback in the method. Certainly, that step is tedious, but it does not present the most serious difficulty. Solving the characteristic equation is not trivial. In the example given above, an exact solution of the characteristic equation was easy to obtain.

The example was contrived so that solving it would be easy. Generally, exact formulas are available for solving polynomial equations of degrees two, three, and four. There is no formula (involving a finite number of elementary algebraic operations) for solving an arbitrary polynomial equation of degree five or higher. Furthermore, it has been proved that there cannot exist any such formula for polynomial equations of degree five or higher.

The trouble is that an approximate value for the eigenvalue does not work in step 4, the step in which we find the corresponding eigenvector. When we solve a polynomial equation, an approximation to a solution, when substituted for the unknown, causes the value of the polynomial to be small. When the objective is to solve the polynomial equation this is a result that usually has a satisfying interpretation. But in step 4, when the characteristic polynomial does not have a value exactly zero, the characteristic matrix $C(d)$ is not singular and the solution subspace will have dimension zero. This means an approximate solution to the characteristic equation will not yield an approximate eigenvector by this method. It will yield no eigenvector at all. This difficulty causes problems even when dealing with matrices of order less than five. The formulas that exist for polynomial equations of degrees two, three, and four involve extractions of roots. If any of these radicals must be replaced by a real number approximation, the method described will not yield an eigenvector. One must retain the radicals in the expression for the eigenvalue in the characteristic matrix and the resulting homogeneous equations, and solve these equations in symbolic form.

If one must solve a practical eigenvalue-eigenvector problem and come up with numerical results, one must use other methods (as usual, "beyond the scope of this text"). These methods are studied in numerical analysis and are based on the fundamental eigenvalue-eigenvector equation $\sigma(\xi) = d\xi$, or its matrix equivalent $AX = dX$. By iteration one finds successively better approximations for both the eigenvalue and corresponding eigenvector at the same time.

In this context there is little point in trying to solve numerically involved problems with the methods given here. Accordingly, the examples given here and the problems to follow are contrived to avoid this difficulty. Just don't expect the problems to be representative of the kind of thing that can occur in a realistic applied situation.

With this said, let us look at a more complicated example with a repeated eigenvalue (a multiple solution of the characteristic equation).

(2) Let
$$A = \begin{bmatrix} 5 & -6 & 6 \\ 6 & -7 & 6 \\ 3 & -3 & 2 \end{bmatrix}$$

be a matrix given over the real numbers. We are to find its eigenvalues and eigenvectors. The corresponding characteristic matrix is

$$C(x) = \begin{bmatrix} 5-x & -6 & 6 \\ 6 & -7-x & 6 \\ 3 & -3 & 2-x \end{bmatrix}.$$

From the characteristic matrix we obtain the characteristic polynomial $-x^3 + 3x + 2 = -(x+1)^2(x-2)$. In this case -1 is a root of the characteristic equation of multiplicity two. The *algebraic multiplicity* of this eigenvalue is two. The other eigenvalue is 2. Substituting -1 in the characteristic matrix, we obtain

$$C(-1) = \begin{bmatrix} 6 & -6 & 6 \\ 6 & -6 & 6 \\ 3 & -3 & 3 \end{bmatrix}.$$

We reduce $C(-1)$ to basic form to obtain a basis for the kernel. The reduced form is

$$\begin{bmatrix} 1 & -1 & 1 \\ 0 & 0 & 0 \\ 0 & 0 & 0 \end{bmatrix}.$$

In this case the reduced form is of rank 1, the kernel is of dimension 2. Hence, we can find two linearly independent eigenvectors corresponding to the eigenvalue -1. There are many ways two linearly independent eigenvectors can be selected. Unless conditions are given that are irrelevant at this point of the discussion, there does not seem to be any way to make a selection that is obviously natural or optimal. Hence, the reader cannot expect his answer to conform to an answer given in the text. But any selection must yield a basis for the same subspace. We select $\{(1, 1, 0), (-1, 0, 1)\}$.

6-3 The Characteristic Matrix and Characteristic Equation

In the same way, we substitute 2 in the characteristic matrix and obtain $(2, 2, 1)$ as a representation of an eigenvector corresponding to the eigenvalue 2. Corollary 2.2 assures us that the set $\{(1, 1, 0), (-1, 0, 1), (2, 2, 1)\}$ is linearly independent, and we have obtained a basis of eigenvectors.

In the same way as in example (1), we obtain the matrix of transition P and its inverse,

$$P = \begin{bmatrix} 1 & -1 & 2 \\ 1 & 0 & 2 \\ 0 & 1 & 1 \end{bmatrix}, \quad P^{-1} = \begin{bmatrix} -2 & 3 & -2 \\ -1 & 1 & 0 \\ 1 & -1 & 1 \end{bmatrix}.$$

As a check, we calculate $P^{-1}AP$ and obtain a diagonal matrix with the eigenvalues in the main diagonal.

(3) Let

$$A = \begin{bmatrix} 3 & -4 & 1 \\ 1 & -2 & 1 \\ -3 & 3 & -1 \end{bmatrix}.$$

This matrix has the same characteristic polynomial as the matrix given in example (2). Hence, again, -1 is an eigenvalue of algebraic multiplicity two. When -1 is substituted in the characteristic matrix, we obtain

$$C(-1) = \begin{bmatrix} 4 & -4 & 1 \\ 1 & -1 & 1 \\ -3 & 3 & 0 \end{bmatrix}.$$

When this is reduced to basic form, we obtain

$$\begin{bmatrix} 1 & -1 & 0 \\ 0 & 0 & 1 \\ 0 & 0 & 0 \end{bmatrix}.$$

The observation to be made here is that this matrix is of rank 2, so only one linearly independent eigenvector corresponding to the eigenvalue -1 can be obtained. Another eigenvector will be obtained corresponding to the eigenvalue 2, and no other independent eigenvectors can be obtained. Thus, in this case a basis of eigenvectors cannot be obtained and we terminate our work on this example.

EXERCISES 6-3

As pointed out, there is little value in trying to develop the method given here to solve realistic eigenvalue problems. Accordingly, the problems given below are contrived to avoid all numerical difficulties. All the characteristic equations that you will meet have integral solutions. (Hence all characteristic values obtained are also eigenvalues.)

1. Find the characteristic equation and eigenvalues for both
$$A = \begin{bmatrix} 1 & 0 & 0 \\ 0 & 2 & 0 \\ 0 & 0 & 2 \end{bmatrix} \text{ and } B = \begin{bmatrix} 1 & 0 & 0 \\ 0 & 2 & 1 \\ 0 & 0 & 2 \end{bmatrix}$$

2. Find the characteristic equation and eigenvalues for both
$$C = \begin{bmatrix} -1 & 6 & -3 \\ -4 & 10 & -4 \\ -6 & 12 & -4 \end{bmatrix} \text{ and } D = \begin{bmatrix} -1 & 9 & -5 \\ -4 & 16 & -8 \\ -6 & 21 & -10 \end{bmatrix}$$

3. Find the matrix of transition to diagonalize the matrix C in Exercise 2. Use the same matrix of transition to obtain a matrix similar to D.

4. Show that D in Exercise 2 cannot be diagonalized.

5. Show that if two matrices, A and B, are similar then they have the same characteristic equation.

6. Show that C and D in Exercise 2 are not similar to each other though they have the same characteristic equation.

6-4 EQUIVALENCE AND SIMILARITY

In Section 4-6 we defined two matrices A and A' to be *similar* if there was a non-singular matrix P such that $A' = P^{-1}AP$. In the previous section we have shown how, under certain circumstances, a matrix might be similar to a diagonal matrix. Similarity is a particular case of what is known in mathematics as an equivalence relation.

Definition Let $\mathscr{S}$ be a set of objects. A **relation exists** in the set $\mathscr{S}$ whenever for every pair of elements (a, b) one and only one of two statements holds unequivocally: either a is related to b, or a is not related to b. If a is related to b we write, aRb. If a is not related to b we write aR̸b. A relation is an **equivalence relation** if the relation satisfies the following three properties.

1. For all $a \in \mathscr{S}$, aRa. (reflexivity)
2. If aRb, then bRa. (symmetry)
3. If aRb and bRc, then aRc. (transitivity)

A definition as formal as this certainly looks artificial, but in fact the idea involved is very common and most readers have already encountered several different kinds of equivalence relations early in their studies. In elementary geometry, congruence, similarity, and parallelism are examples of equivalence relations. The importance of the three properties of an equivalence relation is that it permits us to subdivide $\mathscr{S}$ into disjoint classes such that two elements are equivalent if and only if they are in the same class. These classes are called *equivalence classes*, and they are determined in the following way. For each element $a \in \mathscr{S}$, let $\mathscr{C}(a)$ be the set containing all elements equivalent to a. That is, $x \in \mathscr{C}(a)$ if and only if aRx. In the first place, every element is in some class since $a \in \mathscr{C}(a)$. This follows from reflexivity. Next, we wish to show that classes defined by two different elements are either disjoint or identical. Thus, suppose $c \in \mathscr{C}(a) \cap \mathscr{C}(b)$. Then aRc and bRc. By symmetry, cRb. By transitivity, aRb. Thus, $b \in \mathscr{C}(a)$. For any $x \in \mathscr{C}(b)$, $x \in \mathscr{C}(a)$, by transitivity again. Thus, $\mathscr{C}(b) \subset \mathscr{C}(a)$. Because of the symmetry between a and b in this argument, it follows in the same way that $\mathscr{C}(a) \subset \mathscr{C}(b)$. Thus, $\mathscr{C}(a) = \mathscr{C}(b)$.

Conversely, if $\mathscr{S}$ is subdivided into disjoint classes, an equivalence relation can be defined which gives rise to the given classes. All that is necessary is to say that two elements are equivalent if and only if they are in the same class. It is quite easy to see that such a relation satisfies the three properties required for an equivalence relation.

The idea of an equivalence relation is one of the most frequently encountered concepts in mathematics. In its various occurrences it serves many different purposes, and raises and resolves many different kinds of problems. At this point we wish to discuss two instances of this idea in linear algebra. In the first chapter of this text we were concerned with equivalent systems of linear equations. When these systems were represented by matrices, two matrices were equivalent when one could be transformed into the other by a sequence of elementary row operations. Later, we showed that an elementary operation could be represented by an elementary matrix. We also showed that every non-singular matrix was representable as a product of elementary matrices. Thus, two matrices A and A' are *row-equivalent* if and only if there is a non-singular matrix P such that $A' = PA$. On this basis it is quite easy to show that row-equivalence satisfies the three conditions for an equivalence relation. (1) Reflexivity: $A = IA$. (2) Symmetry: If $A = PB$ where P is non-singular, then $Q = P^{-1}$ is non-singular and $B = QA$. (3) Transitivity: If $A = PB$ and $B = QC$ where P and Q are non-singular, then $A = (PQ)C$ where PQ is non-singular.

In spite of the fact that the definition of row-equivalence is unequivocal, given two matrices A and B of the same order, it is not very convenient to determine whether they are row-equivalent by direct application of the definition of row-equivalence. As a matter of fact, it is easier to reduce both A and B to row-echelon form and compare the two resulting row-echelon forms with each other. This is an illustration of the value of knowing that the row-echelon form to which a matrix is row-equivalent is unique. That is, two different matrices in row-echelon form are not row-equivalent.

For reasons such as this, it is desirable to choose a representative from each equivalence class. This chosen representative is often called a *normal form* or a *canonical form*. Given any object in $\mathscr{S}$, it should be possible, by direct and effective methods, to find the normal form to which it is equivalent. It is by this means that the equivalence class to which the element belongs is determined. For row-equivalence, every matrix is row-equivalent to a row-echelon form. Every equivalence class contains one and only one matrix in row-echelon form. Thus, the row-echelon form defines the equivalence class.

There is a similar concept of column-equivalence, about which similar statements can be made.

In Chapter 4 we considered linear transformations of one vector space into another. We showed, in formula (6.12) of that chapter, that if A and B were matrices representing the same linear transformation, there were non-singular matrices P and Q such that $B = Q^{-1}AP$. The relation between A and B established in this way is also an equivalence relation. We also showed that, with this notion of equivalence, every matrix was equivalent to a matrix of the form given in formula

(6.13) of Chapter 4. That is the normal form for the kind of equivalence discussed there.

In this chapter we are considering linear transformations of a vector space into itself. In this situation, if A and B are matrices representing the same linear transformation, there exists a non-singular matrix P, the matrix of transition, such that $B = P^{-1}AP$. This equivalence relation we have referred to as *similarity*; we say that A and B are *similar*. We have seen that under some circumstances a square matrix is similar to a diagonal matrix. When this occurs, the diagonal form is a suitable normal form. Unfortunately, as example (3) of the previous section shows, there are classes of similar matrices that do not contain a diagonal matrix. If the characteristic equation has solutions that are not scalars, that is, if there are characteristic values that are not eigenvalues, not much more can be said. If the characteristic polynomial factors into linear factors in the scalar field, that is, if all characteristic values are eigenvalues, then a normal form known as the *Jordan normal form* can be obtained. Though of interest and utility, this is a topic that will not be taken up in this book. For a reference, see Linear Algebra and Matrix Theory, E. D. Nering, published by John Wiley & Sons, Inc.

We summarize what can be said about the normal form under similarity on the basis of what we have done.

For an $n \times n$ matrix A, if the characteristic polynomial for A factors into n distinct linear factors in the scalar field, then A is similar to a diagonal matrix. This follows because there are then n distinct eigenvalues, and the eigenvectors corresponding to these eigenvalues are linearly independent. If the characteristic polynomial factors into linear factors in the scalar field, but the factors are not necessarily distinct, the matrix A is similar to a diagonal matrix if n linearly independent eigenvectors can be obtained. Otherwise, A is not similar to a diagonal matrix.

EXERCISES 6-4

1. Show that the matrix relation, $A \, R \, B$ if and only if there exists a non-singular matrix P such that $AP = B$, is an equivalence relation.

2. Let m be any positive integer. We say that two integers, a and b, are *congruent modulo m*, written $a \equiv b \pmod{m}$, if and only if $a - b$ is divisible by m. Show that congruence

modulo m is an equivalence relation. How many equivalence classes are there? Find a representative element in each equivalence class.

3. Two real valued functions f and g of a real variable are said to be related if and only if $f(1) = g(1)$. Show this is an equivalence relation. Show there is one equivalence class for each real number.

4. Two people are said to be hair-equivalent if they have the same number of hairs in their heads. Show there is at least one equivalence class containing two people. (Hint: Make a reasonable estimate of the maximum number of hairs on any person's head.)

5. Let σ be a linear transformation of a vector space $\mathscr{U}$ into a vector space $\mathscr{V}$. Two vectors, α and $\beta \in \mathscr{U}$, are said to be congruent modulo σ if and only if $\sigma(\alpha) = \sigma(\beta)$. Show this is an equivalence relation in $\mathscr{U}$.

6. In Exercise 5, show that each equivalence class is a coset of the kernel of σ.

7. Let $\mathscr{W}$ be any subspace of a vector space $\mathscr{V}$ over a field F. We say that two vectors α and $\beta \in \mathscr{V}$, are congruent modulo $\mathscr{W}$ if and only if $\alpha - \beta \in \mathscr{W}$. Show this is an equivalence relation.

8. In Exercise 7, let $\mathscr{C}(\alpha)$ denote the equivalence class containing α. Define $\mathscr{C}(\alpha) + \mathscr{C}(\beta)$ to be equivalence class containing $\alpha + \beta$. That is, $\mathscr{C}(\alpha) + \mathscr{C}(\beta) = \mathscr{C}(\alpha + \beta)$. Define $a\mathscr{C}(\alpha)$, for $a \in F$, to be the equivalence class containing $a\alpha$. That is, $a\mathscr{C}(\alpha) = \mathscr{C}(a\alpha)$. Show that, with these definitions, the set of equivalence classes is a vector space over F.

9. The set of equivalence classes is called a *factor set*. The factor set defined in Exercise 8, with the operations defined there, is denoted by $\mathscr{V}/\mathscr{W}$ and is called a *factor space*. Show that the mapping of $\alpha \in \mathscr{V}$ onto $\mathscr{C}(\alpha) \in \mathscr{V}/\mathscr{W}$ is a linear transformation. Find the kernel of this linear transformation.

6-5 SPECTRAL RESOLUTION

Let $\mathscr{X} = \{\xi_1, \xi_2, \ldots, \xi_n\}$ be any basis of $\mathscr{V}$. We wish to construct projections π_k with image $\langle \xi_k \rangle$ and kernel $\langle \xi_1, \ldots, \xi_{k-1}, \xi_{k+1}, \ldots, \xi_n \rangle$. A set $\{\pi_1, \pi_2, \ldots, \pi_n\}$ of projections of this type will have some useful properties, which will be shown shortly.

Let $\mathscr{A} = \{\alpha_1, \ldots, \alpha_n\}$ be the given basis and let

$$\xi_j = \sum_{i=1}^{n} p_{ij}\alpha_i, \qquad (5.1)$$

so that $P = [p_{ij}]$ is the matrix of transition from the basis $\mathscr{A}$ to the basis $\mathscr{X}$. Let $Q = P^{-1} = [q_{ij}]$. Consider the linear transformation π_k represented by the matrix

$$E_k = [p_{ik} q_{kj}] \qquad (5.2)$$

with respect to the basis $\mathscr{A}$. This means

$$\pi_k(\alpha_j) = \sum_{i=1}^{n} p_{ik} q_{kj} \alpha_i. \qquad (5.3)$$

Then

$$\pi_k(\xi_r) = \pi_k \left(\sum_{j=1}^{n} p_{jr} \alpha_j \right)$$

$$= \sum_{j=1}^{n} p_{jr} \pi_k(\alpha_j)$$

$$= \sum_{j=1}^{n} p_{jr} \left(\sum_{i=1}^{n} p_{ik} q_{kj} \alpha_i \right)$$

$$= \sum_{i=1}^{n} p_{ik} \left(\sum_{j=1}^{n} q_{kj} p_{jr} \right) \alpha_i \qquad (5.4)$$

$$= \sum_{i=1}^{n} p_{ik} \delta_{kr} \alpha_i$$

$$= \delta_{kr} \sum_{i=1}^{n} p_{ik} \alpha_i$$

$$= \delta_{kr} \xi_k.$$

Formula (5.4) tells us that $\pi_k(\xi_k) = \xi_k$ and $\pi_k(\xi_r) = 0$ for $r \neq k$. Thus, the kernel of π_k is $\langle \xi_1, \ldots, \xi_{k-1}, \xi_{k+1}, \ldots, \xi_n \rangle$, and the image of π_k is $\langle \xi_k \rangle$. Furthermore, π_k is a projection since $\pi_k^2 = \pi_k$. Also, $\pi_k \pi_r = 0$ for $k \neq r$ since the image of π_r is in the kernel of π_k.

Since $\mathscr{X}$ is a basis, any vector $\alpha \in \mathscr{V}$ can be written in the form

$$\alpha = \sum_{j=1}^{n} x_j \xi_j. \tag{5.5}$$

Because of formula (5.4),

$$\pi_k(\alpha) = \pi_k \left(\sum_{j=1}^{n} x_j \xi_j \right)$$

$$= \sum_{j=1}^{n} x_j \pi_k(\xi_j) \tag{5.6}$$

$$= x_k \xi_k.$$

Hence,

$$\alpha = \sum_{k=1}^{n} x_k \xi_k = \sum_{k=1}^{n} \pi_k(\alpha) = \left(\sum_{k=1}^{n} \pi_k \right)(\alpha). \tag{5.7}$$

The significance of formula (5.7) is that it implies that

$$\sum_{k=1}^{n} \pi_k = 1, \tag{5.8}$$

the identity transformation. We say we have *resolved* the identity into a sum of projections.

As an example, let $\mathscr{X} = \{\xi_1, \xi_2, \xi_3\}$ be represented by $\{(2, -1, 0), (1, 0, -1), (0, 1, -1)\}$. Then

$$P = \begin{bmatrix} 2 & 1 & 0 \\ -1 & 0 & 1 \\ 0 & -1 & -1 \end{bmatrix} \tag{5.9}$$

is the matrix of transition from $\mathscr{A}$ to $\mathscr{X}$. Then $Q = P^{-1}$ is

$$Q = \begin{bmatrix} 1 & 1 & 1 \\ -1 & -2 & -2 \\ 1 & 2 & 1 \end{bmatrix}. \tag{5.10}$$

Using formula (5.2) we compute

$$E_1 = \begin{bmatrix} 2 \\ -1 \\ 0 \end{bmatrix} \begin{bmatrix} 1 & 1 & 1 \end{bmatrix} = \begin{bmatrix} 2 & 2 & 2 \\ -1 & -1 & -1 \\ 0 & 0 & 0 \end{bmatrix}. \tag{5.11}$$

Notice that each row of E_1 is a multiple of the first row of Q, and each column of E_1 is a multiple of the first column of P. In a similar way we compute

$$E_2 = \begin{bmatrix} 1 \\ 0 \\ -1 \end{bmatrix} \begin{bmatrix} -1 & -2 & -2 \end{bmatrix} = \begin{bmatrix} -1 & -2 & -2 \\ 0 & 0 & 0 \\ 1 & 2 & 2 \end{bmatrix} \quad (5.12)$$

$$E_3 = \begin{bmatrix} 0 \\ 1 \\ -1 \end{bmatrix} \begin{bmatrix} 1 & 2 & 1 \end{bmatrix} = \begin{bmatrix} 0 & 0 & 0 \\ 1 & 2 & 1 \\ -1 & -2 & -1 \end{bmatrix}. \quad (5.13)$$

By direct calculation we can show that

$$E_1 + E_2 + E_3 = I, \quad (5.14)$$

which is a verification of formula (5.8). Also, by direct calculation we can show that

$$E_k E_r = \delta_{rk} E_k. \quad (5.15)$$

Now suppose that σ is a linear transformation of $\mathscr{V}$ into itself and that $\mathscr{X} = \{\xi_1, \ldots, \xi_n\}$ is a basis of eigenvectors of σ. Let d_i be the eigenvalue corresponding to ξ_i. Suppose this basis is used to construct the projections $\{\pi_1, \ldots, \pi_n\}$, as shown above. For each ξ_j,

$$\left(\sum_{k=1}^{n} d_k \pi_k \right)(\xi_j) = \sum_{k=1}^{n} d_k \pi_k(\xi_j) = d_j \xi_j = \sigma(\xi_j). \quad (5.16)$$

Since formula (5.16) holds for every vector in a basis,

$$\sigma = \sum_{k=1}^{n} d_k \pi_k \quad (5.17)$$

as mappings. The set of eigenvalues of a linear transformation σ is called the *spectrum* of σ, and formula (5.17) is called a *spectral resolution* of σ.

For example, consider the linear transformation represented by

$$A = \begin{bmatrix} -1 & 2 & 2 \\ 2 & 2 & 2 \\ -3 & -6 & -6 \end{bmatrix}. \quad (5.18)$$

It is an easy calculation to see that $f(x) = -x(x+2)(x+3)$ is the characteristic polynomial, and $\{-2, -3, 0\}$ is the set of eigenvalues, the spectrum of A. It is also a direct calculation to find that $\{(2, -1, 0), (1, 0, -1), (0, 1, -1)\}$ is the set of eigenvectors corresponding to the given eigenvalues. This is the basis used in the numerical example given above, from which we constructed the representations E_1, E_2, and E_3. We can then verify that

$$(-2)E_1 + (-3)E_2 + 0E_3 = A, \tag{5.19}$$

which is the matrix form of formula (5.17).

Formula (5.17) allows one to calculate some rather complicated expressions in σ with ease. For example,

$$\sigma^2 = \sum_{k=1}^{n} d_k^2 \pi_k, \tag{5.20}$$

$$\sigma^n = \sum_{k=1}^{n} d_k^n \pi_k. \tag{5.21}$$

If $f(x)$ is any polynomial in x with scalar coefficients, then

$$f(\sigma) = \sum_{k=1}^{n} f(d_k) \pi_k. \tag{5.22}$$

In the above numerical example, we have

$$A^{10} = (-2)^{10} E_1 + (-3)^{10} E_2$$

$$= 2^{10} \begin{bmatrix} 2 & 2 & 2 \\ -1 & -1 & -1 \\ 0 & 0 & 0 \end{bmatrix} + 3^{10} \begin{bmatrix} -1 & -2 & -2 \\ 0 & 0 & 0 \\ 1 & 2 & 2 \end{bmatrix} \tag{5.23}$$

$$= \begin{bmatrix} 2^{11} - 3^{10} & 2^{11} - 2 \cdot 3^{10} & 2^{11} - 2 \cdot 3^{10} \\ -2^{10} & -2^{10} & -2^{10} \\ 3^{10} & 2 \cdot 3^{10} & 2 \cdot 3^{10} \end{bmatrix}.$$

Quite obviously, this calculation is much easier than trying to compute A^{10} by raising it to 10th power by direct multiplication.

6-6 THE HAMILTON-CAYLEY THEOREM

Consider the characteristic matrix $C(x) = A - xI$. We have suggested that x is a variable for which we can substitute a scalar value. But what about substituting a matrix for x?

6-6 The Hamilton-Cayley Theorem

If we consider $A - xI$ as a polynomial of first degree with matrix coefficients, there is no problem. We obtain merely $A - AI = 0$. But in formula (3.1) we have expressed the characteristic matrix in the form

$$C(x) = \begin{bmatrix} a_{11}-x & a_{12} & \cdots & a_{1n} \\ a_{21} & a_{22}-x & \cdots & a_{2n} \\ \cdot & \cdot & & \cdot \\ \cdot & \cdot & & \cdot \\ \cdot & \cdot & & \cdot \\ a_{n1} & a_{n2} & \cdots & a_{nn}-x \end{bmatrix}. \qquad (6.1)$$

In this form it makes no sense to substitute a matrix for x. With this as an illustration, it should be clear that one must be careful how one handles and interprets polynomial expressions involving matrices.

If a polynomial $p(x) = a_m x^m + \cdots + a_1 x + a_0$ is given, we can substitute a square matrix A for x in the form $p(A) = a_m A^m + \cdots + a_1 A + a_0 I$. Notice, in particular, that the constant term a_0 must be replaced by $a_0 I$ so that every term in the sum is a matrix of the same order. In this situation, no problems are encountered and such polynomials can be treated in the usual manner. We must also consider polynomials where the coefficients are matrices. $A - xI$ considered at the beginning of this section is an example of such a polynomial. More generally, a polynomial of this type would look like $A_m x^m + \cdots + A_1 x + A_0$. All coefficients must be matrices of the same order so that the sum makes sense when a scalar is substituted for x. Polynomials with matrix coefficients are more difficult to handle because the coefficients do not commute under multiplication. But our use of such polynomials is limited. We will never attempt to substitute a matrix for x and we will not try to factor such a polynomial. We need only accept the fact that we can multiply such polynomials in the obvious way, and that two polynomials with matrix coefficients are equal if and only if their coefficients are identical.

A matrix whose elements are scalar polynomials can always be rewritten in the form of a polynomial with matrix coefficients. The idea is really rather simple, and an example should suffice to show what is involved.

$$\begin{bmatrix} x^2+x+1 & x-2 \\ -x^2+1 & 2x^2+x+1 \end{bmatrix} = \begin{bmatrix} 1 & 0 \\ -1 & 2 \end{bmatrix} x^2 + \begin{bmatrix} 1 & 1 \\ 0 & 1 \end{bmatrix} x \qquad (6.2)$$

$$+ \begin{bmatrix} 1 & -2 \\ 1 & 1 \end{bmatrix}$$

If the intention is to substitute a scalar for x it doesn't matter which representation is used. But if the intention is to substitute a matrix for x it is necessary to use the representation in the form of a polynomial with matrix coefficients.

Theorem 6.1 (Hamilton-Cayley Theorem) *If A is a square matrix and $f(x)$ is its characteristic polynomial, then $f(A) = 0$.*

Proof. The characteristic matrix $C(x)$ for A is a matrix of order n in which the diagonal elements are first degree polynomials. Thus, adj $C(x)$ will contain elements that are polynomials of degree $n-1$ at most. Then adj $C(x)$ can be rewritten as a polynomial with matrix coefficients of degree at most $n-1$;

$$\text{adj } C(x) = C_{n-1}x^{n-1} + C_{n-2}x^{n-2} + \cdots + C_1 x + C_0, \quad (6.3)$$

where each C_i is an n-th order matrix with scalar elements.

By Theorem 5.1 of Chapter 5, we have

$$\begin{aligned}
(\text{adj } C(x)) &\cdot C(x) \\
&= (\det C(x))I = f(x)I \\
&= (\text{adj } C(x))(A - xI) = (\text{adj } C(x))A - (\text{adj } C(x))x. \\
&= -C_{n-1}x^n - C_{n-2}x^{n-1} - \cdots - C_0 x \\
&\quad + C_{n-1}Ax^{n-1} + \cdots + C_1 Ax + C_0 A \\
&= -C_{n-1}x^n + (C_{n-1}A - C_{n-2})x^{n-1} \\
&\quad + (C_{n-2}A - C_{n-3})x^{n-2} + \cdots \\
&\quad + (C_1 A - C_0)x + C_0 A.
\end{aligned} \quad (6.4)$$

From this it follows by substitution that

$$f(A) = f(A)I = -C_{n-1}A^n + C_{n-1}A^n - C_{n-2}A^{n-1} + \cdots \\ - C_0 A + C_0 A = 0. \quad \square \quad (6.5)$$

Theorem 6.2 *Similar matrices have the same characteristic polynomial and satisfy the same polynomial equations.*

Proof. Let A and $B = P^{-1}AP$ be similar matrices. Then the characteristic polynomial for B is

$$\begin{aligned}
\det(B - xI) &= \det(P^{-1}AP - xI) = \det P^{-1}(A - xI)P \\
&= \det P^{-1} \det(A - xI) \det P \\
&= \det P^{-1} \det P \det(A - xI) = \det(A - xI).
\end{aligned} \quad (6.6)$$

Let $g(x) = b_m x^m + b_{m-1} x^{m-1} + \cdots + b_1 x + b_0$ be any polynomial for which $g(A) = 0$. Then

$$\begin{aligned} g(B) &= b_m B^m + b_{m-1} B^{m-1} + \cdots + b_1 B + b_0 I \\ &= b_m (P^{-1}AP)^m + b_{m-1}(P^{-1}AP) + \cdots + b_1(P^{-1}AP) \\ &\quad + b_0 I \\ &= b_m P^{-1} A^m P + b_{m-1} P^{-1} A^{m-1} P + \cdots + b_1 P^{-1} AP \quad (6.7) \\ &\quad + b_0 I \\ &= P^{-1}(b_m A^m + b_{m-1} A^{m-1} + \cdots + b_1 A + b_0 I) P \\ &= P^{-1} g(A) P = 0. \end{aligned}$$

By the symmetry of the relation between A and B, any polynomial equation which B satisfies is also satisfied by A. ☐

For any square matrix A, the Hamilton-Cayley theorem establishes one polynomial equation satisfied by A. Among all polynomials which A satisfies, there is one of lowest degree. The polynomial of lowest degree and leading coefficient 1 which A satisfies is called the *minimum polynomial* for A. Let it de denoted by $m(x)$. If $g(x)$ is any other polynomial for which $g(A) = 0$, then $g(x)$ is divisible by $m(x)$. By the division algorithm, we can write $g(x)$ in the form

$$g(x) = m(x) \cdot q(x) + r(x), \quad (6.8)$$

where $r(x)$ is a polynomial of degree less than the degree of $m(x)$ or $r(x)$ is zero. Upon substituting A for x we have $0 = g(A) = m(A) \cdot q(A) + r(A) = r(A)$. This would imply that $r(A) = 0$. Since $m(x)$ is the minimum polynomial for A we must have $r(x) = 0$, that is, $g(x)$ is divisible by $m(x)$.

In view of this observation, the second part of Theorem 6.2 is equivalent to the statement that similar matrices have the same minimum polynomial. The converse of neither part of Theorem 6.2 is true. Examples 2 and 3 of Section 6-3 give matrices with the same characteristic polynomial. One has a basis of eigenvectors and the other does not. Thus, one is similar to a diagonal matrix and the other is not. Hence they cannot be similar. To show that the converse of the second part of the theorem does not hold is somewhat more involved.

If A and $P^{-1}AP = B$ are similar, they may be considered to be different representations of the same linear transformation with respect to different bases where P is the matrix of transition from one basis to the other. The various eigenspaces of the linear transformation are not affected by the particular matrix that is used to represent the linear transformation.

Thus, whether we compute the eigenvalues and eigenvectors using A or B we will get the same eigenvalues, and the eigenvectors will correspond to each other. We will just get different coordinates for them since the two representations involve two different coordinate systems.

Now consider the two matrices

$$\begin{bmatrix} 1 & 0 & 0 & 0 \\ 0 & 1 & 0 & 0 \\ 0 & 0 & 1 & 1 \\ 0 & 0 & 0 & 1 \end{bmatrix}, \begin{bmatrix} 1 & 1 & 0 & 0 \\ 0 & 1 & 0 & 0 \\ 0 & 0 & 1 & 1 \\ 0 & 0 & 0 & 1 \end{bmatrix}. \quad (6.9)$$

By direct calculation we can show that both these matrices satisfy the polynomial equation $(x-1)^2 = 0$, and neither satisfies the equation $(x-1) = 0$. Since the minimum polynomial for both matrices must divide $(x-1)^2$, $(x-1)^2$ is the minimum polynomial for both. The only eigenvalue for both matrices is 1. By using the characteristic matrix for each and finding the corresponding eigenspace, we find that the first has an eigenspace of dimension 3 and the second has an eigenspace of dimension 2. Thus, these two matrices cannot be similar.

EXERCISES 6-6

1. In Section 2-7, Exercise 14, it was shown that the set of all $n \times n$ matrices is a vector space of dimension n^2 over the scalar field F. Show that if A is any $n \times n$ matrix, there is an equation of degree n^2 or less that A satisfies. (Do not use the Hamilton-Cayley theorem to prove this.)

2. The polynomial equation of lowest degree that an $n \times n$ matrix A satisfies is called the *minimum polynomial equation* for A. Show that the minimum polynomial is of degree n or less.

3. For polynomial with coefficients in a field, the division algorithm says that for any two polynomials f and g, there exist polynomials q and r such that $f(x) = g(x)q(x) + r(x)$, where either $r=0$ or degree of $r <$ degree of g. Use the division algorithm to show that the minimum polynomial for A divides the characteristic polynomial for A.

4. In the theory of Markov processes, a stochastic matrix is an $n \times n$ matrix $P = [p_{ij}]$ with the properties $p_{ij} \geq 0$ and $\sum_{j=1}^{n} p_{ij} = 1$. The interpretation is that there is a system with n states and, at a given time, the system can change states and p_{ij} is the probability of a change from state i to state j. If the system can change states at a succession of times, and if P is the stochastic matrix for all possible times of changes, the situation described is called a stationary Markov process. The element in row i column j of P^2 is the probability of moving from state i to state j in two steps. Similarly, the element in row i column j of P^m is the probability of passing from state i to state j in m steps. If that element is zero, it is impossible to pass from state i to state j in exactly m steps. Show that if it is impossible to pass from state i to state j in n steps or less, it is never possible to pass from state i to state j. ("Impossible" means the probability is zero.)

Summary

A subspace $\mathscr{S}$ is *invariant* under a linear transformation σ if $\sigma(\mathscr{S}) \subset \mathscr{S}$. The analysis of a linear transformation σ can be simplified if we can find invariant subspaces $\mathscr{S}_1$ and $\mathscr{S}_2$ such that $\mathscr{S}_1 \oplus \mathscr{S}_2 = \mathscr{V}$ (a direct sum).

This chapter concentrates on finding a particular kind of invariant subspace, an eigenspace. A non-zero vector ξ is an *eigenvector* for σ if there is a scalar d, an *eigenvalue* for σ such that

$$\sigma(\xi) = d\xi. \tag{S.1}$$

For a particular eigenvalue d, the kernel of $\sigma - d$ is an *eigenspace* $\mathscr{S}(d)$. The non-zero vectors in $\mathscr{S}(d)$ are eigenvectors for σ.

The ultimate in simplification occurs when there exists a basis of eigenvectors for $\mathscr{V}$. In this case σ is represented by a diagonal matrix with respect to the basis of eigenvectors.

The method for finding eigenvalues and eigenvectors given in this chapter is the following. Start with any matrix A representing σ.
1. Determine the characteristic matrix $C(x) = A - xI$.
2. Determine the characteristic polynomial $\det C(x)$.
3. Find the eigenvalues of A (and σ) by solving the char-

acteristic equation det $C(x) = 0$ and selecting solutions that are scalars.

4. For each eigenvalue d solve the system of homogeneous linear equations $C(d)X = 0$ and obtain a basis of the eigenspace.
5. Collect the eigenvectors obtained for all eigenvalues and obtain a linearly independent set of eigenvectors.
6. If the eigenvectors obtained in step 5 form a basis of $\mathscr{V}$, let P be the matrix of transition to this basis. Compute $P^{-1}AP$.
7. If the eigenvectors obtained in step 5 do not form a basis for $\mathscr{V}$, terminate the work. A basis of eigenvectors does not exist.

Equivalence relations are important throughout mathematics. Row-equivalence, similarity, and unitary similarity are three equivalence relations that are studied in this book. This chapter is concerned with similarity.

The chapter closes with a proof of the Hamilton-Cayley theorem, which says that a matrix satisfies its characteristic equation. Outside of a few exercises, no application of the Hamilton-Cayley theorem occurs in this book.

7

ORTHOGONAL AND UNITARY SPACES

We have appealed to the reader's previous experience with the concepts of distance, angle, and area for motivation and illustration. But, in a strictly logical sense, these concepts have not been used in the development of linear algebra up to this chapter. For example, while the concept of area was used to motivate the definition of the determinant function, the definition itself and the consequences that followed could have been made without any reference to area.

There is considerable advantage in developing as much of linear algebra as possible without reference to distance and angle. Many applications of linear algebra do not involve such concepts. The same is true for calculus. Calculus is usually developed with reference to functions plotted on a cartesian coordinate system. But many of the functions discussed in applications involve variables with different physical meanings; for example, we often have a function expressing distance as a function of time.

A space in which a distance function is defined is called a metric space, whether the underlying space is a vector space or not. We are interested in a more refined concept in which the distance function and the algebraic structure are tied together. What is usually done in this context is to make the distance function translation invariant. This means that if α and β are two points (represented by vectors) and γ is any other point, then the distance between α and β is the same as the distance between $\alpha + \gamma$ and $\beta + \gamma$. As a result, the distance structure at one point is the same as it is at any other point.

It is sufficient, then, to define the distance structure at just one point, usually at the origin of a vector space. A vector space with the distance function defined at the origin is called a normed vector space.

A still more refined concept is that of an inner product. An inner product can be used to define a norm, and a norm can be used to define a distance function. We will introduce the concepts in that order. In general, a distance function does not define a norm and a norm does not define an inner product.

Almost all computational methods have already been developed. The only new method is the Gram-Schmidt orthonormalization procedure. Even that will appear in the exercises in only a modest way. The most important thing in this chapter is an understanding of the situation, what can and what cannot be done. Thus, the chapter will have a rather theoretical flavor. Nevertheless, these ideas are important in many applications of linear algebra, particularly to physical problems.

Basically, the idea is to select coordinate systems that are most suitable for use in a vector space in which distance is defined. Essentially, these are coordinate systems in which the coordinate axes are mutually orthogonal and the same distance scale is used on each coordinate axis. These are the orthonormal or cartesian coordinate systems.

This means that coordinate changes will be restricted to changes from one orthonormal coordinate system to another. Except for the restriction in what is permitted, the formulas used in carrying out a change of coordinates are the same as those we have been using all along.

Since fewer coordinate changes are allowed, it is more difficult to make a simplifying change of coordinates. However, the matrices that appear in many applications where orthonormal coordinates are used are also restricted to a few types. The problem is to show that, for these types, the desired simplification can be effected. Once it is known that these steps can be carried out, the methods involved are the same as in earlier chapters. The occasional use of the Gram-Schmidt procedure is the most important exception.

7-1 INNER PRODUCTS

Let $\mathscr{V}$ be a vector space of dimension n over the real numbers. Let a basis $\mathscr{A} = \{\alpha_1, \alpha_2, \ldots, \alpha_n\}$ be chosen. The definition we are going to give depends on the basis chosen. It is necessary to keep this in mind and consider later how the choice of a different basis would affect things. With the chosen basis we have a one-to-one correspondence between vectors and n-tuples representing them.

Definition. Let $\alpha = a_1 \alpha_1 + \cdots + a_n \alpha_n$ be a vector represented by the n-tuple $A = (a_1, a_2, \ldots, a_n)$, and $\beta = b_1 \alpha_1 + \cdots + b_n \alpha_n$ be represented by the n-tuple $B = (b_1, b_2, \ldots, b_n)$. We define an **inner product**, or **dot product**, of α and β to be

$$(\alpha, \beta) = a_1 b_1 + a_2 b_2 + \cdots + a_n b_n. \tag{1.1}$$

Remembering the convention introduced in Section 2 of Chapter 2 that the n-tuple A is also represented by a 1-column matrix, the product in (1.1) can also be written in the form

$$(\alpha, \beta) = A^T B. \tag{1.2}$$

The product $A^T B$ in (1.2) is sometimes also denoted by $A \cdot B$ (hence the name "dot product").

The product defined here is sometimes also called the *scalar product*. By careful use of words we could distinguish between scalar multiplication (in which a vector is multiplied by a scalar to obtain a vector) and the scalar product (in which two vectors are multiplied to obtain a scalar), but it seems better to avoid using the term "scalar product" altogether, particularly since the term "inner product" is available and widely used.

For all practical purposes, we will use the terms "inner product" and "dot product" as if they were equivalent. But we shall usually use the term "inner product" when referring to the product of two vectors, (α, β), and we shall usually use the term "dot product" when referring to the product of two n-tuples, $X^T Y = X \cdot Y$. This will allow us to talk of two different inner products, depending on the coordinate system chosen. For each choice of a coordinate system the representing dot product would be calculated the same way.

Theorem 1.1 *The inner product has the following properties.*
  (1) $(\alpha, \alpha) \geq 0$, and $(\alpha, \alpha) > 0$ if $\alpha \neq 0$.
  (2) $(\alpha, \beta) = (\beta, \alpha)$.
  (3) $(\alpha, b\beta + c\gamma) = b(\alpha, \beta) + c(\alpha, \gamma)$,
      $(a\alpha + b\beta, \gamma) = a(\alpha, \gamma) + b(\beta, \gamma)$.

Proof. (1) If $\alpha = a_1\alpha_1 + \cdots + a_n\alpha_n$, then $(\alpha, \alpha) = a_1^2 + a_2^2 + \cdots + a_n^2$. As a sum of squares, $(\alpha, \alpha) \geq 0$. If $\alpha \neq 0$, at least one $a_i \neq 0$ and $(\alpha, \alpha) > 0$.

(2) If $\beta = b_1\alpha_1 + \cdots + b_n\alpha_n$, then $(\alpha, \beta) = a_1b_1 + \cdots + a_nb_n = b_1a_1 + \cdots + b_na_n = (\beta, \alpha)$.

(3) Let $\gamma = c_1\alpha_1 + \cdots + c_n\alpha_n$. Then $b\beta + c\gamma = (bb_1 + cc_1)\alpha_1 + \cdots + (bb_n + cc_n)\alpha_n$, and

$$
\begin{aligned}
(\alpha, b\beta + c\gamma) &= a_1(bb_1 + cc_1) + \cdots + a_n(bb_n + cc_n) \\
&= b(a_1b_1 + \cdots + a_nb_n) + c(a_1c_1 + \cdots + a_nc_n) \\
&= b(\alpha, \beta) + c(\alpha, \gamma).
\end{aligned}
$$

The second part of (3) follows by using (2). Namely, $(a\alpha + b\beta, \gamma) = (\gamma, a\alpha + b\beta) = a(\gamma, \alpha) + b(\gamma, \beta) = a(\alpha, \gamma) + b(\beta, \gamma)$. □

We refer to property (1) of the theorem by saying that the inner product is *positive definite*. If the product had the property that $(\alpha, \alpha) \leq 0$, and $(\alpha, \alpha) < 0$ if $\alpha \neq 0$, we would say the product was *negative definite*. If we drop the requirement that $(\alpha, \alpha) > 0$ for $\alpha \neq 0$, we would call the product *non-negative semi-definite*.

We refer to property (2) by saying that the inner product is *symmetric*.

Property (3) says the inner product is linear in each variable separately, and we refer to this property by saying that the inner product is *bilinear*.

A function of several variables that has scalar values is sometimes called a form. In this terminology, the inner product is a positive definite, symmetric, bilinear form.

A *length*, or *norm*, of a vector α is defined to be

$$\|\alpha\| = \sqrt{(\alpha, \alpha)}. \tag{1.3}$$

The fact that the inner product is positive definite means that the number under the radical sign is never negative. Furthermore, $\|\alpha\| \geq 0$ and $\|\alpha\| > 0$ if $\alpha \neq 0$.

There is a similar definition of an inner product for vector spaces over the complex numbers. The consequences of the definitions are similar in both the real case and the complex case, and the proofs differ very little in the two cases. We have to make a choice between developing the theory in the real case and then repeating much of the material to obtain the same results in the complex case, or developing both theories at the same time. We choose the second alternative. So we will give a definition of an inner product in the complex case. However, it turns out that we cannot define an inner product that is positive definite, symmetric and bilinear. If the definition were given precisely as it was for a vector space over the real numbers, the inner product would be symmetric and bilinear, but it would not be positive definite since $(i\alpha_1, i\alpha_1) = i^2 = -1$ would not be positive. In the complex case we can make the inner product positive definite by defining the inner product of α and β to be

$$(\alpha, \beta) = \bar{a}_1 b_1 + \bar{a}_2 b_2 + \cdots + \bar{a}_n b_n, \quad (1.4)$$

where $\bar{a}_i$ is the conjugate complex of a_i. The complex form of Theorem 1.1 then becomes

Theorem 1.2 *For the inner product defined over the complex number we have*

(1) $(\alpha, \alpha) \geq 0$, and $(\alpha, \alpha) > 0$ if $\alpha \neq 0$.
(2) $(\alpha, \beta) = \overline{(\beta, \alpha)}$.
(3) $(\alpha, b\beta + c\gamma) = b(\alpha, \beta) + c(\alpha, \gamma)$,
    $(a\alpha + b\beta, \gamma) = \bar{a}(\alpha, \gamma) + \bar{b}(\beta, \gamma)$.

Proof. The details are similar to those in the proof of Theorem 1.1. □

Property (1) is the reason for defining the inner product in the complex case as we did. It allows us to define the norm of a vector with precisely the same formula as in (1.3). The norm is so useful that it is better to give up the symmetry and bilinearity that the real inner product has. The complex inner product is said to be *conjugate symmetric,* or *Hermitian symmetric,* and *conjugate bilinear,* or *Hermitian bilinear.*

Notice that the definition of the inner product in the complex case reduces to the real inner product if all numbers involved are real numbers. This means that unless a proof depends on the existence of a complex number with certain properties, a proof that is valid for the complex inner product will also be valid for the real inner product. The proof of Schwarz's inequality is an illustration of this principle.

Theorem 1.3 (*Schwarz's inequality*) *For all $\alpha, \beta \in \mathscr{V}$,*

$$|(\alpha, \beta)| \leq \|\alpha\| \cdot \|\beta\|. \quad (1.5)$$

Proof. For t any real number we have

$$0 \leq \|(\alpha, \beta)t\alpha - \beta\|^2 = |(\alpha, \beta)|^2 t^2 \|\alpha\|^2 - 2t|(\alpha, \beta)|^2 + \|\beta\|^2. \quad (1.6)$$

If $\|\alpha\| = 0$, $|(\alpha, \beta)| = 0$ since the inequality must hold for arbitrary t. In this case Schwarz's inequality holds since the left side of (1.5) is 0. If $\|\alpha\| \neq 0$, take $t = 1/\|\alpha\|^2$. Then (1.6) is equivalent to Schwarz's inequality. □

Schwarz's inequality is a fundamental inequality, and it has many important consequences. For example, it implies that the inner product of two small vectors is small. This kind of condition is important in vector calculus because it means that the inner product is continuous. For a vector space over the real numbers, (α, β) is a real number and Schwarz's inequality implies that for $\|\alpha\| \neq 0$ and $\|\beta\| \neq 0$,

$$-1 \leq \frac{(\alpha, \beta)}{\|\alpha\| \cdot \|\beta\|} \leq 1. \quad (1.7)$$

In two and three dimensions, the ratio in (1.7) is the cosine of the angle between the vectors α and β. Because of Schwarz's inequality, this ratio can be interpreted as a cosine for a real vector space of any dimension. This is a reasonable interpretation. Actually, the subspace spanned by $\{\alpha, \beta\}$ is at most 2-dimensional. Thus, even though the whole vector space $\mathscr{V}$ may be of high dimension, the relation between α and β is 2-dimensional.

Theorem 1.4 *The norm defined in (1.3) satisfies the following three conditions.*

(1) $\|\alpha\| \geq 0$ and $\|\alpha\| > 0$ if $\alpha \neq 0$.
(2) $\|a\alpha\| = |a| \cdot \|\alpha\|$ for a scalar a.
(3) $\|\alpha + \beta\| \leq \|\alpha\| + \|\beta\|$.

Proof. We observed at the time that the norm was defined that condition (1) is satisfied.

Condition (2) is quite easy to show.

$$\|a\alpha\| = \sqrt{(a\alpha, a\alpha)} = \sqrt{a\bar{a}(\alpha, \alpha)} = |a| \cdot \|\alpha\|. \quad (1.8)$$

Condition (3) is deeper. It depends on Schwarz's inequality.

$$\begin{aligned}\|\alpha+\beta\|^2 &= (\alpha+\beta, \alpha+\beta) \\ &= (\alpha, \alpha) + (\alpha, \beta) + \underline{(\beta, \alpha)} + (\beta, \beta) \\ &= \|\alpha\|^2 + (\alpha, \beta) + \overline{(\alpha, \beta)} + \|\beta\|^2 \\ &= \|\alpha\|^2 + 2 \operatorname{Re}\{(\alpha, \beta)\} + \|\beta\|^2\end{aligned}$$

Here, $\operatorname{Re}\{(\alpha, \beta)\}$ denotes the real part of the complex number (α, β).

$$\leq \|\alpha\|^2 + 2|(\alpha, \beta)| + \|\beta\|^2 \tag{1.9}$$

This follows because the real or imaginary part of a complex number is no larger than the absolute value of the number.

$$\leq \|\alpha\|^2 + 2\|\alpha\| \cdot \|\beta\| + \|\beta\|^2$$

This is the step that uses Schwarz's inequality.

$$= (\|\alpha\| + \|\beta\|)^2.$$

Thus, $\|\alpha + \beta\|^2 \leq (\|\alpha\| + \|\beta\|)^2$. Taking the square root of both sides of this inequality yields the desired result. □

There is another useful inequality that follows from condition (3) of Theorem 1.4, and is equivalent to it. From condition (3) we have

$$\|\alpha\| = \|\alpha - \beta + \beta\| \leq \|\alpha - \beta\| + \|\beta\|,$$
$$\text{or } \|\alpha\| - \|\beta\| \leq \|\alpha - \beta\|. \tag{1.10}$$

By interchanging the roles of α and β, we have

$$\|\beta\| - \|\alpha\| \leq \|\alpha - \beta\|. \tag{1.11}$$

Since both (1.10) and (1.11) hold, they are equivalent to

$$|\|\alpha\| - \|\beta\|| \leq \|\alpha - \beta\|. \tag{1.12}$$

The norm, giving the lengths of vectors, can be interpreted as giving the distance from the origin to the end point of the vector. Thus, the norm defines distance only radially from one point. The algebraic structure of the vector space is used to define distances from other points. If α and β

are two vectors, the distance from the end point of α to the end point of β is defined to be the norm of the difference. Thus, we define the distance function $d(\alpha, \beta)$ by the formula

$$d(\alpha, \beta) = \|\beta - \alpha\|. \tag{1.13}$$

Theorem 1.5 *The distance function satisfies the following three conditions.*

(1) $d(\alpha, \beta) \geq 0$ *and* $d(\alpha, \beta) > 0$ *if* $\alpha \neq \beta$.
(2) $d(\alpha, \beta) = d(\beta, \alpha)$.
(3) $d(\alpha, \beta) \leq d(\alpha, \gamma) + d(\gamma, \beta)$.

Proof. These properties follow directly from the corresponding properties of the norm by substituting in the defining formula. □

Condition (3) of Theorem 1.5 is called the *triangle inequality*. Because of the close connection between this condition and condition (3) of Theorem 1.4, that condition is also called the triangle inequality, despite the fact that the vectors appearing in condition (3) of Theorem 1.4 do not form a triangle (all vectors have their tails at the origin).

As far as we are concerned, the distance function gives an interpretation of the norm. We will not use the distance function as a separate idea. When helpful, we will refer to the distance between α and β in terms of the norm; that is, $\|\beta - \alpha\|$ is the distance between α and β.

EXERCISES 7-1

1. Let $\mathscr{V}$ be a space of continuous real valued functions defined on the interval $[0, 1]$. For $f, g \in \mathscr{V}$ we can define

$$(f, g) = \int_0^1 f(t)\, g(t)\, dt. \tag{1.14}$$

Show that this defined as an inner product in $\mathscr{V}$. (The most sophisticated step is to show that the inner product is positive definite. This requires a theorem from calculus that if the integral of a non-negative continuous function over an interval is zero, then the function is zero on that interval.)

2. Use Schwarz's inequality to prove that if α is a vector for which $\|\alpha\| = 0$, then $(\alpha, \beta) = 0$ for all $\beta \in \mathscr{V}$. In particular, show that $(0, \beta) = (\beta, 0) = 0$ for all $\beta \in \mathscr{V}$.

3. For α represented by $(1, 3, 2, -1)$ and β represented by $(2, 0, 1, 2)$, determine the value of (α, β).

4. For α represented by $(1, i, 2)$ and β represented by $(1, i, -2)$, determine the value of (α, β).

5. Show that for a real vector space
$$\|\alpha - \beta\|^2 = \|\alpha\|^2 + \|\beta\|^2 - 2(\alpha, \beta). \tag{1.15}$$
Set $\cos \theta = \dfrac{(\alpha, \beta)}{\|\alpha\| \cdot \|\beta\|}$, as suggested in (1.7), and show that (1.15) is equivalent to the cosine law from trigonometry.

6. Let $\mathscr{R}^n$ be an n-dimensional real coordinate space. For an n-tuple $A = (a_1, a_2, \ldots, a_n)$, define
$$\|A\| = \max\{|a_1|, |a_2|, \ldots, |a_n|\}, \tag{1.16}$$
where $|a|$ is the ordinary absolute value of a. Show that (1.16) defines a norm. That is, the function defined by (1.16) satisfies the three conditions of Theorem 1.4. (Incidentally, there is no inner product that will give this norm.)

7. Show that $\|\alpha + \beta\|^2 + \|\alpha - \beta\|^2 = 2\|\alpha\|^2 + 2\|\beta\|^2$. This is interpreted as saying that for a parallelogram, the sum of the squares of the diagonals is equal to the sum of the squares of the sides. See Figure 7.1.1.

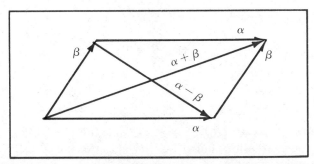

Figure 7-1.1 The parallelogram formula, $\|\alpha + \beta\|^2 + \|\alpha - \beta\|^2 = 2\|\alpha\|^2 + 2\|\beta\|^2$.

7-2. ORTHOGONALITY AND ORTHONORMAL BASES

Two vectors, α and β, are said to be *orthogonal* if $(\alpha, \beta) = 0$. As pointed out in the previous section, for a real vector space we can define the angle θ between α and β by the formula $\cos \theta = (\alpha, \beta)/\|\alpha\| \cdot \|\beta\|$. It would then be consistent to take $\theta = \pi/2$ as the angle between two orthogonal vectors. However, there are important applications that do not have a simple geometric interpretation. An example is the inner product defined in Exercise 1 of the previous section. It is better to connect the word "orthogonal" with the condition "$(\alpha, \beta) = 0$" and to rely as little as possible on thinking in terms of the vectors being at "right angles."

A set of vectors is said to be orthogonal if each vector in the set is orthogonal to every other vector in the set. A set of vectors is said to be *orthonormal* if it is orthogonal and every vector in the set is of length 1.

Let $\mathscr{A} = \{\alpha_1, \alpha_2, \ldots, \alpha_n\}$ be the basis through which the inner product was defined. Then for each i, $(\alpha_i, \alpha_i) = 1$, and for $i \neq j$, $(\alpha_i, \alpha_j) = 0$. In other words, the defining basis is an orthonormal basis. By using the Kronecker delta, the condition for orthonormality can be written somewhat more concisely in the form $(\alpha_i, \alpha_j) = \delta_{ij}$.

Theorem 2.1 *An orthogonal set of non-zero vectors is linearly independent.*

Proof. Let $\mathscr{B} = \{\beta_1, \beta_2, \ldots\}$ be an orthogonal set of non-zero vectors. Suppose there is a linear relation among these vectors of the form

$$\sum_{i=1}^{m} b_i \beta_i = 0. \tag{2.1}$$

Then, for each β_k, $k = 1, 2, \ldots, m$, we have

$$\left(\beta_k, \sum_{i=1}^{m} b_i \beta_i\right) = (\beta_k, 0) = 0$$
$$= \sum_{i=1}^{m} b_i (\beta_k, \beta_i) = b_k (\beta_k, \beta_k). \tag{2.2}$$

Since $\beta_k \neq 0$, $(\beta_k, \beta_k) \neq 0$. Hence, $b_k = 0$. This shows every linear relation must be trivial, and the set β is linearly independent. □

Among other things, this shows that an orthonormal set is linearly independent. Also, an orthonormal set can contain no more vectors than the dimension of the space, or subspace, that contains the set. The basis used to define the inner product is one orthonormal basis. We shall show shortly, by the Gram-Schmidt orthonormalization procedure, that there are many others.

One advantage of an orthonormal basis is that there is a simple and useful relation expressing the coordinates of a vector in terms of the inner product. Let $\mathscr{X} = \{\xi_1, \xi_2, \ldots, \xi_n\}$ be an orthonormal basis for $\mathscr{V}$ and let $\alpha = a_1\xi_1 + \cdots + a_n\xi_n$ be a representation of a vector α in $\mathscr{V}$ as a linear combination of the basis vectors. Then

$$(\xi_i, \alpha) = (\xi_i, \sum_{j=1}^{n} a_j\xi_j)$$

$$= \sum_{j=1}^{n} a_j(\xi_i, \xi_j) \qquad (2.3)$$

$$= \sum_{j=1}^{n} a_j\delta_{ij} = a_i.$$

Thus,

$$\alpha = \sum_{i=1}^{n} (\xi_i, \alpha)\xi_i. \qquad (2.4)$$

Let us illustrate this situation with an example. Let $\mathscr{A} = \{\alpha_1, \alpha_2, \alpha_3\}$ be the basis used to define the inner product. Let $\{\xi_1, \xi_2, \xi_3\}$ be an orthonormal basis of a 3-dimensional space over the real numbers. Let the ξ_i be represented by $\left\{\frac{1}{3}(2, 2, 1), \frac{1}{3}(2, -1, -2), \frac{1}{3}(1, -2, 2)\right\}$ with respect to the defining basis. It is easy to check that these vectors are, in fact, orthonormal. Now let α be represented by $(1, 2, 3)$, also with respect to the defining basis. Then, we can compute

$$(\xi_1, \alpha) = 3, \quad (\xi_2, \alpha) = -2, \quad (\xi_3, \alpha) = 1. \qquad (2.5)$$

We can check that

$$3 \cdot \frac{1}{3}(2, 2, 1) - 2 \cdot \frac{1}{3}(2, -1, -2) + 1 \cdot \frac{1}{3}(1, -2, 2)$$
$$= (1, 2, 3). \qquad (2.6)$$

The calculations can also be represented in matrix form.

$$\begin{bmatrix} \frac{2}{3} & \frac{2}{3} & \frac{1}{3} \\ \frac{2}{3} & -\frac{1}{3} & -\frac{2}{3} \\ \frac{1}{3} & -\frac{2}{3} & \frac{2}{3} \end{bmatrix} \begin{bmatrix} 1 \\ 2 \\ 3 \end{bmatrix} = \begin{bmatrix} 3 \\ -2 \\ 1 \end{bmatrix}. \tag{2.7}$$

Formula (2.8) of Chapter 3 gives a method for calculating the coordinates of a vector with respect to a new basis. In this respect, formula (2.3) is no different. But formula (2.3) has two important advantages. First, one does not have to compute the inverse of the matrix of transition. The reason why this is not necessary will be discussed in some detail in the next section. Second, the calculation of a_i, the coefficient of ξ_i, does not require knowledge of any vector in the basis other than ξ_i.

The observation concerning the independence of the computation in formula (2.3) from the other elements of an orthonormal basis has a geometric interpretation. The situation is general in spaces of all dimensions, but is most easily illustrated in three dimensions. Let ξ_i be a vector in an orthonormal basis. In Figure 7-2.1, the straight line shown contains the vector ξ_i and the plane orthogonal to the line contains the other basis vectors. To simplify the picture, none of the other basis vectors is shown. An arbitrary vector α is shown. α is expressed as a sum of two vectors. One is $a_i \xi_i$, which lies in the line. The other is $\alpha - a_i \xi_i$, which is orthogonal to ξ_i and

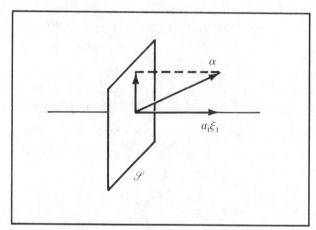

Figure 7-2.1 Let $\mathscr{X} = \{\xi_1, \ldots, \xi_n\}$ be an orthonormal basis, and let $\alpha \in \mathscr{V}$ be written in the form $\alpha = a_1 \xi_1 + \ldots + a_n \xi_n$. Then $a_1 \xi_1$ is orthogonal to the plane containing $\{\alpha_2, \ldots, \alpha_n\}$, and $\alpha - a_1 \xi_1$ is in this plane. The decomposition of α into $\alpha = a_1 \xi_1 + (\alpha - a_1 \xi_1)$ where $a_1 \xi_1$ and $(\alpha - a_1 \xi_1)$ are orthogonal depends on ξ_1 and not on the other vectors in $\mathscr{H}$.

lies in the plane. This breakdown is independent of the particular vectors that make up the rest of the orthonormal basis. All that is important is that they all lie in the same plane regardless of how they are selected. Thus, $a_i \xi_i$ is independent of how the rest of the orthonormal basis is chosen.

It turns out that every subspace has an orthonormal basis—even very many orthonormal bases. An important method for finding convenient orthonormal bases is the method contained in the proof of the following theorem.

Theorem 2.2 *Let $\{\beta_1, \beta_2, \ldots, \beta_m\}$ be a given linearly independent set. There exists an orthonormal set $\{\xi_1, \xi_2, \ldots, \xi_m\}$ such that for each k, $\{\beta_1, \ldots, \beta_k\}$ and $\{\xi_1, \ldots, \xi_k\}$ span the same subspace. Or, equivalently, each ξ_k is a linear combination of the vectors in the set $\{\beta_1, \ldots, \beta_k\}$.*

Proof. The proof involves a step-by-step construction of the desired orthonormal basis. The procedure is known as the *Gram-Schmidt orthonormalization procedure*. The name is a mouthful and the formal description leaves most students cold. But the idea behind it is quite simple, and it is more helpful to understand the idea than it is to memorize a formula or a method.

Since $\{\beta_1, \ldots, \beta_m\}$ is linearly independent, $\beta_1 \neq 0$. Thus, $\|\beta_1\| > 0$. Let

$$\xi_1 = \frac{\beta_1}{\|\beta_1\|}. \tag{2.8}$$

Notice that ξ_1 is of length 1. The step by which we obtain the vector ξ_1 of length 1 from the non-zero vector β_1 is called *normalization*. This is where the "normal" part of orthonormal comes from. We will perform this step several times.

As pointed out in the discussion in connection with Figure 7-2.1, $(\xi_1, \beta_2)\xi_1$ is the component of β_2 in the direction of ξ_1, and $\beta_2' = \beta_2 - (\xi_1, \beta_2)\xi_1$ is in a subspace orthogonal to ξ_1. We need to show that β_2' is non-zero so that it can be normalized. If β_2' were 0, β_2 would be a multiple of ξ_1, which is, in turn, a multiple of β_1. This would contradict the linear independendence of $\{\beta_1, \beta_2\}$. Hence $\beta_2' \neq 0$. Let

$$\xi_2 = \frac{\beta_2'}{\|\beta_2'\|}. \tag{2.9}$$

The subsequent steps are similar (just longer) and the reasoning is the same. Suppose we have arrived at the point where $\{\xi_1, \xi_2, \ldots, \xi_k\}$ are orthonormal and each ξ_i is a linear combination of $\{\beta_1, \ldots, \beta_i\}$. Then set

$$\beta_{k+1}' = \beta_{k+1} - (\xi_1, \beta_{k+1})\xi_1 - \cdots - (\xi_k, \beta_{k+1})\xi_k$$
$$= \beta_{k+1} - \sum_{i=1}^{k} (\xi_i, \beta_{k+1})\xi_i. \tag{2.10}$$

In (2.10) we have substracted from β_{k+1} the components of β_{k+1} in the direction of the orthonormal vectors already obtained. The difference β_k' is orthogonal to all ξ_i, for $i \leq k$. Namely,

$$\begin{aligned}(\xi_j, \beta_{k+1}') &= \left(\xi_j, \beta_{k+1} - \sum_{i=1}^{k} (\xi_i, \beta_{k+1})\xi_i\right) \\ &= (\xi_j, \beta_{k+1}) - \sum_{i=1}^{k} (\xi_i, \beta_{k+1})(\xi_j, \xi_i) \\ &= (\xi_j, \beta_{k+1}) - \sum_{i=1}^{k} (\xi_i, \beta_{k+1})\delta_{ij} \\ &= (\xi_j, \beta_{k+1}) - (\xi_j, \beta_{k+1}) = 0.\end{aligned} \tag{2.11}$$

Again, $\beta_{k+1}' \neq 0$ since otherwise β_{k+1} would be a linear combination of the $\{\xi_1, \ldots, \xi_k\}$ which are, in turn, linear combinations of the $\{\beta_1, \ldots, \beta_k\}$. Then normalize β_{k+1}' to obtain

$$\xi_{k+1} = \frac{\beta_{k+1}'}{\|\beta_{k+1}'\|}. \tag{2.12}$$

In this way we can proceed step-by-step until we obtain the orthonormal set $\{\xi_1, \ldots, \xi_m\}$, where each ξ_i is a linear combination of $\{\beta_1, \ldots, \beta_i\}$. □

The numerical work involved in the Gram-Schmidt procedure is not difficult, and not even particularly tedious. But one must be careful about details which might be confusing. Follow through the details of the following example.

In a 3-dimensional vector space $\mathscr{V}$, let $\{\beta_1, \beta_2, \beta_3\}$ be represented by $\{(1, -1, 1), (4, -1, 4), (3, 2, 1)\}$. As a first step we should verify that this set is linearly independent, because that is a hypothesis of the theorem. However, if we just proceed with the Gram-Schmidt process this question will settle itself. If one of the β_i is a linear combination of the preceding elements of the set, β_i' will be 0 in the calculation. If no β_i is a linear combination of the preceding elements, all $\beta_i' \neq 0$ and the procedure can be continued to termination.

The orthonormal set obtained will be linearly independent by Theorem 2.1.

We normalize first to obtain $\frac{1}{\sqrt{3}}(1, -1, 1)$ as the representation of ξ_1. Then we compute

$$(4, -1, 4) - \frac{1}{\sqrt{3}} \begin{bmatrix} 1 & -1 & 1 \end{bmatrix} \begin{bmatrix} 4 \\ -1 \\ 4 \end{bmatrix} \frac{1}{\sqrt{3}}(1, -1, 1) \quad (2.13)$$

$$= (4, -1, 4) - \frac{9}{3}(1, -1, 1) = (1, 2, 1).$$

This is the representation of β_2', so we normalize to obtain $\frac{1}{\sqrt{6}}(1, 2, 1)$. Then we compute

$$(3, 2, 1) - \frac{2}{\sqrt{3}} \cdot \frac{1}{\sqrt{3}}(1, -1, 1) - \frac{8}{\sqrt{6}} \cdot \frac{1}{\sqrt{6}}(1, 2, 1)$$

$$= (3, 2, 1) - \frac{2}{3}(1, -1, 1) - \frac{4}{3}(1, 2, 1) \quad (2.14)$$

$$= (1, 0, -1).$$

Normalizing, we obtain $\frac{1}{\sqrt{2}}(1, 0, -1)$ as the representation of ξ_3. Thus, $\left\{ \frac{1}{\sqrt{3}}(1, -1, 1), \frac{1}{\sqrt{6}}(1, 2, 1), \frac{1}{\sqrt{2}}(1, 0, -1) \right\}$ is the representation of the required orthonormal set. Since $\mathscr{V}$ is 3-dimensional, the set obtained is actually an orthonormal basis.

The numbers do not always work out so conveniently as they do in this contrived example, even if the coordinates of the given vectors are integers. For working examples of low dimension with small numbers, a few suggestions might be helpful. First, to normalize a vector, it is sufficient to normalize a multiple of it. For example, to normalize $1/9(13, 10, 1)$ it is sufficient to normalize $(13, 10, 1)$. In either case, you must end up with a vector of length 1, and it is a multiple of either vector.

Second, when finding an orthonormal basis, the Gram-Schmidt procedure places no effective conditions on the last vector obtained. It is just a normalized vector orthogonal to all those already obtained. For vector spaces of small dimension, this can often be obtained by inspection. In the example given, the third vector must be orthogonal to $(1, -1, 1)$ and

(1, 2, 1). It is not too difficult to see that $(1, 0, -1)$ has that property. All that remains is to normalize it.

Take heart. Numerical examples involving high dimensions are rare, even outside a textbook situation. Even where they do occur, the Gram-Schmidt procedure is an algorithm that is easily programmed. The most important applications are theoretical; for example, to show that every subspace has an orthonormal basis. Numerical examples in the text following this section involve only very low dimensions.

EXERCISES 7-2

1. Show that $\{(1, 2, 2), (2, 1, -2), (2, -2, 1)\}$ is an orthogonal set. Show that $\left\{\frac{1}{3}(1, 2, 2), \frac{1}{3}(2, 1, -2), \frac{1}{3}(2, -2, 1)\right\}$ is an orthonormal set.

2. Show that $\{(1, i), (1, -i)\}$ is an orthogonal set.

3. We know from elementary geometry that the diagonals of a rhombus are orthogonal. There is an analogue of this conclusion for inner product spaces. Show that if $\|\alpha\| = \|\beta\|$, then $\alpha + \beta$ and $\alpha - \beta$ are orthogonal.

4. Given the linear independent set $\{(1, 1, 1, 1), (2, 0, 1, 1), (-1, 1, 2, -1)\}$, apply the Gram-Schmidt procedure to obtain an orthonormal set.

5. Let $\mathscr{V}$ be a finite dimensional real vector space with an inner product. Let $\mathscr{A} = \{\alpha_1, \alpha_2, \ldots, \alpha_n\}$ be a basis of $\mathscr{V}$ over $\mathscr{R}$ which is not necessarily orthonormal. Define g_{ij} to the value of (α_i, α_j), and let $G = [g_{ij}]$. Show that G is *symmetric* (that is, $G^T = G$) and non-singular. If $\xi = \sum_{i=1}^{n} x_i \alpha_i$ and $\eta = \sum_{i=1}^{n} y_i \alpha_i$ are vectors in $\mathscr{V}$, show that

$$(\xi, \eta) = X^T G Y, \tag{3.15}$$

where $X = (x_1, x_2, \ldots, x_n)$ and $Y = (y_1, y_2, \ldots, y_n)$.

6. Let $\mathscr{S}$ be any subset of an inner product space $\mathscr{V}$. Let $\mathscr{S}^\perp$ denote the set of all $\alpha \in \mathscr{V}$ such that $(\alpha, \beta) = 0$ for all $\beta \in \mathscr{S}$. All vectors in $\mathscr{S}^\perp$ are orthogonal to all vectors in $\mathscr{S}$. Show that $\mathscr{S}^\perp$ is a subspace.

7. In the context of Exercise 6, show that $\mathscr{S}^{\perp\perp} = (\mathscr{S}^\perp)^\perp \supset \mathscr{S}$. Thus, $\mathscr{S}^{\perp\perp}$ is a subspace containing $\mathscr{S}$

8. Show that if $\mathscr{S} \subset \mathscr{T}$, then $\mathscr{S}^\perp \supset \mathscr{T}^\perp$.

9. Show that $\mathscr{S}^{\perp\perp\perp} = \mathscr{S}^\perp$. (This can be proved using the conclusions of Exercises 7 and 8 and nothing else.)

10. Show that $\mathscr{S} \cap \mathscr{S}^\perp \subset \{0\}$. Show that if $\mathscr{S}$ is a subspace, then $\mathscr{S} \cap \mathscr{S}^\perp = \{0\}$.

11. We know from Theorem 1.4 of Chapter 6 that if $\mathscr{S}$ is a subspace of $\mathscr{V}$, there is a subspace $\mathscr{T}$ such that $\mathscr{V} = \mathscr{S} \oplus \mathscr{T}$. Show that if $\mathscr{V}$ is a finite dimensional inner product space and $\mathscr{S}$ is a subspace of $\mathscr{V}$, there is a subspace $\mathscr{T}$ such that $\mathscr{V} = \mathscr{S} \oplus \mathscr{T}$ and $\mathscr{T} \subset \mathscr{S}^\perp$.

12. Show that $\mathscr{V} = \mathscr{S} \oplus \mathscr{S}^\perp$.

13. Show that $(\mathscr{S} + \mathscr{T})^\perp = \mathscr{S}^\perp \cap \mathscr{T}^\perp$ and $(\mathscr{S} \cap \mathscr{T})^\perp = \mathscr{S}^\perp + \mathscr{T}^\perp$.

7-3 UNITARY AND ORTHOGONAL SIMILARITY

Just as in situations discussed previously in this book, we wish to consider changing the coordinate system—changing from one basis to another. The new factor is that we are restricting our attention to orthonormal bases, and therefore the changes that we can consider are those that change from one orthonormal basis to another. Since this is a special case of what we have already done, once a basis is selected the formulas for making the change of coordinates are the same as those we have already developed. However, the specialization also brings about some simplifications. The main problem is to determine how the desired bases can be selected and what can be achieved with the limited choices available.

Let $\mathscr{A} = \{\alpha_1, \ldots, \alpha_n\}$ be a given orthonormal basis and let $\mathscr{B} = \{\beta_1, \ldots, \beta_n\}$ be a new orthonormal basis. The matrix of transition $P = [p_{ij}]$ is defined exactly as it was in Section 2 of Chapter 3,

$$\beta_j = \sum_{i=1}^n p_{ij}\alpha_i. \qquad (3.1)$$

The formulas for changing coordinates of a vector and changing a matrix representing a linear transformation are the same as before.

Before going into further detail in that direction, let us see what consequences follow directly from the restriction to orthonormal bases. This means

$$(\beta_i, \beta_j) = \delta_{ij}, \text{ and} \tag{3.2}$$

$$\begin{aligned}(\beta_i, \beta_j) &= \left(\sum_{k=1}^{n} p_{ki}\alpha_k, \sum_{r=1}^{n} p_{rj}\alpha_r\right) \\ &= \sum_{k=1}^{n} \bar{p}_{ki}\left(\sum_{r=1}^{n} p_{rj}(\alpha_k, \alpha_r)\right) \\ &= \sum_{k=1}^{n} \bar{p}_{ki}\left(\sum_{r=1}^{n} p_{rj}\delta_{kr}\right) \\ &= \sum_{k=1}^{n} \bar{p}_{ki}p_{kj}.\end{aligned} \tag{3.3}$$

The right side of (3.3) is the dot product of column i and column j of the matrix P. Combined with the orthonormality condition in (3.2), this says that the columns of P are orthonormal. Since the columns of P are the n-tuples representing the vectors in an orthonormal basis, this conclusion should have no element of mystery. The conclusion is precisely the same as the condition imposed.

Formula $(3.2) = (3.3)$ allows us to form the inverse of P in a simple way. Let $Q = \bar{P}^T = P^*$, the conjugate transpose of P. That is, $q_{ij} = \bar{p}_{ji}$. Then

$$\sum_{k=1}^{n} q_{ik}p_{kj} = \sum_{k=1}^{n} \bar{p}_{ki}p_{kj} = \delta_{ij}. \tag{3.4}$$

Formula (3.4) says that Q is the inverse of P. Thus,

$$P^*P = I. \tag{3.5}$$

But an inverse on the left side is also an inverse on the right. That is, $PP^* = I$. Written out in terms of the elements of P, this says

$$\sum_{k=1}^{n} p_{ik}\bar{p}_{jk} = \delta_{ij}. \tag{3.6}$$

In (3.6), i and j are row indices, and formula (3.6) says that the rows of P are orthonormal.

A matrix over the complex numbers whose conjugate transpose is its inverse is called a *unitary matrix*. For a matrix over the real numbers, the conjugate transpose is just the transpose. In that case the matrix is called an *orthogonal matrix*. A unitary matrix represents a change of orthonormal bases when the underlying field is complex, and an orthogonal matrix represents a change of orthonormal bases when the underlying field is real. Necessarily, the elements in an orthogonal matrix are real numbers. The elements of a unitary matrix might, in some cases, also be real numbers. In the terminology used here such a matrix is not orthogonal, but most current usage of these terms would call a real unitary matrix an orthogonal matrix. Actually, the terminology is not important. What is important is to keep in mind the coordinate changes that are permissible in each of the two cases, the complex case and the real case.

$P^*P = I$ is equivalent to the orthonormality of the columns of P, and $PP^* = I$ is equivalent to the orthonormality of the rows. Since either condition is sufficient to make P^* the inverse of P, either the orthonormality of the columns or the orthonormality of the rows is sufficient to make a matrix unitary (or orthogonal). Thus, either condition is sufficient to make P a matrix of transition from one orthonormal basis to another.

The fact that the inverse of a unitary (or orthogonal) matrix is so easy to obtain makes orthonormal coordinate changes particularly convenient. Let P be the unitary matrix of transition defined by (3.1), and let $\xi = \sum_{i=1}^{n} x_i \alpha_i = \sum_{j=1}^{n} x_j' \beta_j$ be a vector with $X = (x_1, \ldots, x_n)$ and $X' = (x_1', \ldots, x_n')$ the representations with respect to the two bases. Then (2.6) and (2.8) of Chapter 3 become

$$X = PX', \qquad (3.7)$$

$$X' = P^*X. \qquad (3.8)$$

Next, consider a linear transformation σ of $\mathscr{V}$ into itself. Let A be the matrix representing σ with respect to the basis $\mathscr{A}$, and let A' be the matrix representing σ with respect to the basis $\mathscr{B}$. Then, according to formula (6.14) of Chapter 4

$$A' = P^{-1}AP = P^*AP. \qquad (3.9)$$

If P is unitary we say that A and A' are *unitary similar*. If P is orthogonal, we say they are *orthogonal similar*. Unitary similarity and orthogonal similarity are special cases of similarity in the sense that matrices that are unitary or orthogonal similar are similar, but some matrices that are similar are not unitary similar or orthogonal similar.

In Chapter 4 we saw that it was not necessarily true that an arbitrary matrix was similar to a diagonal matrix. Now we have an even more restricted situation. The set of matrices unitary or orthogonal similar to a diagonal matrix is smaller than the set of matrices similar to a diagonal matrix. However, the class of matrices that is unitary similar to a diagonal matrix and the class of real matrices that is orthogonal similar to a diagonal matrix are comparatively easy to identify. Even more important, these classes of matrices turn out to be the kind that come up in a number of significant applied problems.

Theorem 3.1 *If a real matrix A is orthogonal similar to a diagonal matrix, then it is symmetric. That is, $A^T = A$.*

Proof. Let A be orthogonal similar to a diagonal matrix D. That is, there is an orthogonal matrix P such that $P^T A P = D$. Then $A = PDP^T$, and $A^T = (PDP^T)^T = (P^T)^T D^T P^T = PDP^T = A$. $\square$

Theorem 3.2 *If a complex matrix A is unitary similar to a diagonal matrix, then it commutes with its conjugate transpose. That is, $A^*A = AA^*$.*

Proof. Let A be unitary similar to a diagonal matrix D. That is, there is a unitary matrix P such that $P^*AP = D$. Then $A = PDP^*$, and $A^*A = (PDP^*)^*(PDP^*) = (P^*)^*D^*P^*PDP^* = P\bar{D}DP^* = PD\bar{D}P^* = PDP^*PD^*P^* = (PDP^*)(PDP^*)^* = AA^*$. $\square$

The point of these theorems is that the converse of each of them is also true. A real symmetric matrix is orthogonal similar to a diagonal matrix. A complex matrix that commutes with its conjugate transpose is unitary similar to a diagonal matrix. Both these converses are quite a bit deeper than the theorems given above.

7-3 Unitary and Orthogonal Similarity

If separate proofs of these converses are given, the proof in the real case is quite a bit easier than the proof in the complex case. However, if a proof is given for the complex case, it is rather easy to obtain the desired implications for the real case. Since we are going through the discussion for the complex case, our major consideration of the real case is postponed until Section 7-7. For those who are interested only in the real case, we conclude this section with a sketch of the situation. Those who intend to go through the discussion for the complex case should skip the remainder of this section.

Theorem 3.3 *If A is a real symmetric matrix, the solutions of its characteristic equation are all real.*

Proof. Though A is a real matrix, let us carry out all computations in the field of complex numbers. The characteristic polynomial, $f(x) = \det(A - xI)$, has real coefficients. In the field of complex numbers, $f(x)$ can be factored entirely into linear factors in the form $f(x) = (x - d_1)(x - d_2) \cdots (x - d_n)$. We wish to show that all d_i are real.

Let d be any of the solutions of the equation $f(x) = 0$. Since $A - dI$ is singular, there is a non-zero (possibly complex) n-tuple X such that $(A - dI)X = 0$. That is, $AX = dX$. Let $B = X^*AX$. B is a 1×1 matrix. Furthermore, B is real since $\bar{B} = B^* = (X^*AX)^* = X^*A^*X = X^*AX = B$. But also $B = X^*(AX) = X^*dX = dX^*X$. Since X^*X is real and positive, d is real. □

Theorem 3.4 *If A is a real symmetric matrix, there is an orthogonal matrix P such that P^TAP is a diagonal matrix.*

Proof. The proof is by induction on n, the order of the matrices under consideration. The theorem is obviously true for $n = 1$, since all matrices of order 1 are symmetric and diagonal.

Assume that the theorem is true for real symmetric matrices of order $n - 1$ or less. Let A be a real symmetric matrix of order n. By Theorem 3.3, there is a real eigenvalue d_1 for A. There is a non-zero eigenvector corresponding to d_1. This non-zero vector can be extended to a basis. The Gram-Schmidt orthonormalization procedure, applied to this basis, will yield an orthonormal basis. The first element of this basis will be a multiple of the eigenvector of A and will also be an eigenvector corresponding to d_1. Let Q be the orthogonal matrix of transition to this orthonormal basis. Then consider the matrix $A' = Q^T A Q$. Since A is symmetric, A' is symmetric. The argument that this is so should be familiar by now. Namely, $(A')^T = (Q^T A Q)^T = Q^T A^T Q = Q^T A Q = A'$. Since the first column of Q represents an eigenvector of A, A' has the form

$$A' = \begin{bmatrix} d_1 & \cdot & \cdot & \cdot & \cdot \\ 0 & \cdot & \cdot & \cdot & \cdot \\ \cdot & \cdot & \cdot & \cdot & \cdot \\ \cdot & \cdot & \cdot & \cdot & \cdot \\ \cdot & \cdot & \cdot & \cdot & \cdot \\ 0 & \cdot & \cdot & \cdot & \cdot \end{bmatrix}. \qquad (3.10)$$

The last $n-1$ elements in the first column are zero. By the symmetry of A' the last $n-1$ elements in the first row are also zero. That is, A' has the form

$$A' = \begin{bmatrix} d_1 & 0 \\ 0 & B \end{bmatrix}. \qquad (3.11)$$

Here, B denotes an $(n-1) \times (n-1)$ matrix made up of the last $n-1$ rows and last $n-1$ columns of A'. The zeros stand for $n-1$ zeros in the first column and $n-1$ zeros in the first row.

Since A' is symmetric, B is also symmetric. The induction hypothesis applies to B, so there is an orthogonal matrix R of order $n-1$ such that $R^T B R = B'$ is a diagonal matrix. We then construct the matrix

$$S = \begin{bmatrix} 1 & 0 \\ 0 & R \end{bmatrix}. \qquad (3.12)$$

It is easy to check that S is an orthogonal matrix, and that $S^T A' S$ is a diagonal matrix. Thus, $S^T(Q^T A Q) S = (QS)^T A (QS)$ is a diagonal matrix. Finally, since Q and S both represent changes from one orthonormal basis to another, $QS = P$ represents a change from one orthonormal basis to another and is an orthogonal matrix. □

The proof of Theorem 3.4 does not provide a very practical way of obtaining the orthogonal matrix P which diagonalizes A. The step described in the proof would have to be repeated $n-1$ times. For each step the Gram-Schmidt orthonormalization procedure would be required and, finally, all the orthogonal matrices obtained would have to be multiplied to obtain P.

7-3 Unitary and Orthogonal Similarity

The important thing about Theorem 3.4 is that it says an orthogonal matrix P exists. Knowing that it can be obtained we can proceed to find it by more direct means. For that purpose we establish the following theorem.

Theorem 3.5 *Let d_1 and $d_2 \neq d_1$ be eigenvalues of the real symmetric matrix A. Eigenvectors corresponding to these eigenvalues are orthogonal.*

Proof. Let X_1 represent an eigenvector corresponding to d_1, and let X_2 represent an eigenvector corresponding to d_2. That is, $AX_1 = d_1 X_1$ and $AX_2 = d_2 X_2$. Then $d_1(X_1^T X_2) = (d_1 X_1)^T X_2 = (AX_1)^T X_2 = X_1^T A^T X_2 = X_1^T A X_2 = X_1^T (d_2 X_2) = d_2(X_1^T X_2)$. Therefore, $(d_1 - d_2) X_1^T X_2 = 0$, since $d_1 - d_2 \neq 0$, $X_1^T X_2 = 0$. □

The following procedure is now practical. For the real symmetric matrix A, determine the characteristic polynomial $f(x)$, and find the characteristic values. Theorem 3.3 assures us they will all be real and, therefore, eigenvalues of A. For each eigenvalue find a basis for the eigenspace. Apply the Gram-Schmidt orthonormalization procedure within each eigenspace to find an orthonormal basis for the eigenspace. Theorem 3.5 assures us that the vectors in different eigenspaces are orthogonal. Then the union of the orthonormal bases from each eigenspace is an orthonormal set. Theorem 3.4 assures us that an orthonormal basis of eigenvectors exists. Thus, since every eigenvector is in some eigenspace, we can obtain a basis of eigenvectors in this way.

EXERCISES 7-3

1. Show that the product of two orthogonal matrices is an orthogonal matrix.

2. Show that the product of two unitary matrices is a unitary matrix.

3. Show that orthogonal similarity is an equivalence relation.

4. Show that unitary similarity is an equivalence relation.

5. Show, by finding an example, that the product of two symmetric matrices might not be symmetric.

6. Test the following matrices for orthogonality. If a matrix is orthogonal, find its inverse.

(a) $\begin{bmatrix} 1/3 & -2\sqrt{2}/3 \\ 2\sqrt{2}/3 & 1/3 \end{bmatrix}$

(b) $\begin{bmatrix} \cos\theta & -\sin\theta \\ \sin\theta & \cos\theta \end{bmatrix}$

(c) $\begin{bmatrix} 0.8 & 0.6 \\ -0.6 & 0.8 \end{bmatrix}$

(d) $\begin{bmatrix} 1 & 0 & 0 \\ 0 & 1 & 0 \\ 0 & 0 & 1 \end{bmatrix}$

(e) $\begin{bmatrix} 1 & 0 & 0 \\ 0 & 1 & 0 \\ 0 & 0 & -1 \end{bmatrix}$

(f) $\begin{bmatrix} 1 & 0 & 0 \\ 0 & -1 & 0 \\ 0 & 0 & -1 \end{bmatrix}$

(g) $\begin{bmatrix} -1 & 0 & 0 \\ 0 & -1 & 0 \\ 0 & 0 & -1 \end{bmatrix}$

(h) $\frac{1}{3}\begin{bmatrix} 1 & 2 & 2 \\ 2 & 1 & -2 \\ 2 & -2 & 1 \end{bmatrix}$

(i) $\begin{bmatrix} \cos\theta & -\sin\theta & 0 \\ \sin\theta & \cos\theta & 0 \\ 0 & 0 & 1 \end{bmatrix}$

(j) $\begin{bmatrix} \cos\theta & -\sin\theta & 0 \\ \sin\theta & \cos\theta & 0 \\ 0 & 0 & -1 \end{bmatrix}$

7. Test to determine which of the following matrices are unitary. If a matrix is unitary, find its inverse.

(a) $\frac{1}{\sqrt{2}}\begin{bmatrix} 1 & 1 \\ i & -i \end{bmatrix}$

(b) $\frac{1}{\sqrt{2}}\begin{bmatrix} 1 & i \\ i & 1 \end{bmatrix}$

(c) $\frac{1}{2}\begin{bmatrix} 1+i & 1-i \\ 1-i & 1+i \end{bmatrix}$

7-4 ORTHOGONAL AND UNITARY TRANSFORMATIONS

There is another important way to look at orthogonal and unitary matrices. In Chapter 4 we used matrices to represent linear transformations. If the matrices representing linear transformations are unitary or orthogonal, this implies special properties for the linear transformations they represent.

Let σ be any linear transformation of $\mathscr{V}$ into itself which is represented by the matrix $A = [a_{ij}]$ with respect to the orthonormal basis $\mathscr{A} = \{\alpha_1, \ldots, \alpha_n\}$. This means

$$\sigma(\alpha_j) = \sum_{i=1}^{n} a_{ij} \alpha_i. \tag{4.1}$$

Then

$$(\alpha_i, \sigma(\alpha_j)) = \left(\alpha_i, \sum_{k=1}^{n} a_{kj} \alpha_k\right)$$

$$= \sum_{k=1}^{n} a_{kj} (\alpha_i, \alpha_k) \tag{4.2}$$

$$= \sum_{k=1}^{n} a_{kj} \delta_{ik} = a_{ij}.$$

Now let $\xi = \sum_{i=1}^{n} x_i \alpha_i$ and $\eta = \sum_{j=1}^{n} y_j \alpha_j$ be two vectors in $\mathscr{V}$. Then

$$(\xi, \sigma(\eta)) = \left(\sum_{i=1}^{n} x_i \alpha_i, \sigma\left(\sum_{j=1}^{n} y_j \alpha_j\right)\right)$$

$$= \sum_{i=1}^{n} \bar{x}_i \left(\alpha_i, \sum_{j=1}^{n} y_j \sigma(\alpha_j)\right)$$

$$= \sum_{i=1}^{n} \bar{x}_i \sum_{j=1}^{n} y_j (\alpha_i, \sigma(\alpha_j)) \tag{4.3}$$

$$= \sum_{i=1}^{n} \sum_{j=1}^{n} \bar{x}_i y_j a_{ij}.$$

The right side of (4.3) can be written as a matrix product. Let $X = (x_1, \ldots, x_n)$ be the n-tuple representing ξ, and let $Y = (y_1, \ldots, y_n)$ be the n-tuple representing η. Then

$$(\xi, \sigma(\eta)) = \sum_{i=1}^{n} \sum_{j=1}^{n} \bar{x}_i a_{ij} y_j = X^* A Y. \tag{4.4}$$

In other words, the inner product $(\xi, \sigma(\eta))$ is the dot product of X and AY.

Let σ^* denote the linear transformation represented by A^*. The right side of (4.4) can be written in the form

$$X^*(AY) = (X^*A)Y = (A^*X)^*Y = (\sigma^*(\xi), \eta). \quad (4.5)$$

Combining (4.4) and (4.5) we have

$$(\xi, \sigma(\eta)) = (\sigma^*(\xi), \eta). \quad (4.6)$$

In more sophisticated treatments of linear algebra, formula (4.6) is used to define σ^*. But to use (4.6) as a basis for a definition requires a more careful examination of the properties of the inner product than we are prepared to undertake. For our purposes it is sufficient to observe that (4.5) and (4.6) are essentially equivalent, and (4.5) depends on the associative law for matrix multiplication.

Now let us assume that A is unitary. Since $A^*A = I$, $\sigma^*\sigma$ is the identity transformation. Then, for any $\xi, \eta \in \mathcal{V}$, we have

$$(\sigma(\xi), \sigma(\eta)) = (\sigma^*\sigma(\xi), \eta) = (\xi, \eta). \quad (4.7)$$

In other words, if σ is represented by a unitary matrix it preserves the inner product. In particular, this implies that σ preserves length.

$$\|\sigma(\xi)\|^2 = (\sigma(\xi), \sigma(\xi)) = (\xi, \xi) = \|\xi\|^2. \quad (4.8)$$

Also, if α_i and α_j are elements of an orthonormal set, then

$$(\sigma(\alpha_i), \sigma(\alpha_j)) = (\alpha_i, \alpha_j) = \delta_{ij}. \quad (4.9)$$

In other words, σ maps an orthonormal set onto an orthonormal set. A unitary matrix was defined by the property that it was a matrix of transition that mapped one orthonormal basis onto another. It mapped a particular orthonormal set onto an orthonormal set. Now we see that the linear transformation that is defined by a unitary matrix maps any orthonormal set onto an orthonormal set. This property of the linear transformation is usually taken as the defining property for a *unitary transformation*. Similarly, a linear transformation in a real vector space that maps orthonormal sets onto orthonormal sets is called an *orthogonal transformation*.

Theorem 4.1 *The eigenvalues of a unitary or orthogonal linear transformation are of absolute value 1.*

Proof. Let σ be unitary (or orthogonal). Let d be an eigenvalue of σ, and let ξ be a corresponding eigenvector. Then, by (4.1) $\|\xi\| = \|\sigma(\xi)\| = \|d\xi\| = |d|\,\|\xi\|$. Since $\|\xi\| \neq 0$, $|d| = 1$. □

If σ is a linear transformation represented by the matrix A with respect to an orthonormal system, we have defined σ^* to be the linear transformation represented by A^*. The problem with a definition given in this way is that it depends on the matrix A and, therefore, on the selection of an orthonormal basis. As a matter of fact, the linear transformation σ^* determined in this way is the same regardless of which orthonormal basis and representing matrix A is chosen.

To see this, let U be the unitary matrix of transition from the given orthonormal basis to another. Then σ would be represented by $U^{-1}AU = U^*AU = A'$, and σ^* would be represented by U^*A^*U. But since $(U^*AU)^* = U^*A^*U = (A')^*$, we see that defining σ^* through A' would give the same result.

The linear transformation σ^* defined in this way is called the *adjoint* of σ. The point of defining σ^* is that it is really related to σ through the inner product. Let ξ be a vector represented by X, and let η be a vector represented by Y. Then

$$(\eta, \sigma(\xi)) = Y^*(AX) = (Y^*A)X = (A^*Y)^*X = (\sigma^*(\eta), \xi). \quad (4.10)$$

The middle terms in (4.10) involved nothing deeper than the associative law of matrix multiplication. The adjoint allows us to write $(\eta, \sigma(\xi)) = (\sigma^*(\eta), \xi)$, which is, therefore, the form of the associative law for the inner product.

EXERCISES 7-4

1. Show that the composition of two orthogonal transformations is an orthogonal transformation.

2. Show that the composition of two unitary transformations is a unitary transformation.

3. Show that the inverse of an orthogonal transformation is an orthogonal transformation.

4. Show that the inverse of a unitary transformation is a unitary transformation.

5. Show that $(\sigma\tau)^* = \tau^*\sigma^*$.

6. Show that $K(\sigma^*) = \text{Im}(\sigma)^\perp$.

7. Show that if σ is a scalar transformation (that is, $\sigma(\alpha) = a\alpha$), then $\sigma^*(\alpha) = \bar{a}\alpha$.

8. Show that if $\mathscr{W}$ is an invariant subspace under σ, then $\mathscr{W}^\perp$ is invariant under σ^*.

9. Show that if σ is orthogonal and $\mathscr{W}$ is invariant under σ, then $\mathscr{W}^\perp$ is invariant under σ.

10. Show that if σ is orthogonal and $\mathscr{W}$ is invariant under σ, then $\mathscr{W}$ is invariant under σ^*.

11. Show that if $(\alpha, \sigma(\alpha)) = 0$ for all $\alpha \in \mathscr{V}$, then $(\beta, \sigma(\alpha)) + (\alpha, \sigma(\beta)) = 0$ for all $\alpha, \beta \in \mathscr{V}$.

12. Let $\mathscr{V}$ be an inner product space over the complex numbers. Show that if $(\alpha, \sigma(\alpha)) = 0$ for all $\alpha \in \mathscr{V}$, then $(\beta, \sigma(\alpha)) - (\alpha, \sigma(\beta)) = 0$ for all $\alpha, \beta \in \mathscr{V}$. (Hint: consider $(\alpha + i\beta, \sigma(\alpha + i\beta))$.)

13. Let $\mathscr{V}$ be an inner product space over the complex numbers. Show that if $(\alpha, \sigma(\alpha)) = 0$ for all $\alpha \in \mathscr{V}$, then $(\alpha, \sigma(\beta)) = 0$ for all $\alpha, \beta \in \mathscr{V}$.

14. Show that if $(\alpha, \sigma(\beta)) = 0$ for all $\alpha, \beta \in \mathscr{V}$, then $\sigma(\beta) = 0$ for all $\beta \in \mathscr{V}$. (Hint: consider $\alpha = \sigma(\beta)$.)

15. Exercises 13 and 14 allow us to conclude that for a complex inner product space, $(\alpha, \sigma(\alpha)) = 0$ for all α implies $\sigma = 0$. Such a conclusion for a real vector space is not possible. Let a linear transformation σ on a 2-dimensional real inner product space be represented by $\begin{bmatrix} 0 & -1 \\ 1 & 0 \end{bmatrix}$ with respect to an orthonormal basis. Show that $\sigma^* = -\sigma$. Then show that $(\alpha, \sigma(\alpha)) = 0$ for all $\alpha \in \mathscr{V}$.

7-5 UPPER TRIANGULAR FORM

In this section we consider a relatively simple form to which every matrix is unitary similar. A matrix is said to be in upper triangular form, or superdiagonal form, if every element below the main diagonal is zero. That is, $A = [a_{ij}]$ is in upper triangular form if $a_{ij} = 0$ for $i > j$. Similarly, a matrix is in lower triangular form if every element above the main diagonal is zero.

Theorem 5.1 *If $A = [a_{ij}]$ is in upper (or lower) triangular form, the elements in the main diagonal are the eigenvalues of A.*

Proof. Let $A = [a_{ij}]$ be in upper (or lower) triangular form. Then $\det(A - xI) = (a_{11} - x)(a_{22} - x) \cdots (a_{nn} - x)$. Thus, the a_{ii} are the zeros of the characteristic polynomial. Since the a_{ii} are in the field over which A is defined, they are eigenvalues of A. □

Theorem 5.2 *Let $\mathscr{V}$ be a vector space over the complex numbers. Let σ be any linear transformation of $\mathscr{V}$ into itself, and let A be a matrix representing σ with respect to an orthonormal basis $\mathscr{A} = \{\alpha_1, \ldots, \alpha_n\}$. Then there is an orthonormal basis $\mathscr{B} = \{\beta_1, \ldots, \beta_n\}$ such that the matrix representing σ is in upper triangular form.*

Proof. Let $f(x) = \det(A - xI)$ be the characteristic polynomial for A (and σ). $f(x)$ factors into linear factors over the field of complex numbers. That is $f(x) = (x - d_1)(x - d_2) \cdots (x - d_n)$. Corresponding to d_1 there is an eigenvector ξ_1 of length 1. Since $\xi_1 \neq 0$, there is a basis of $\mathscr{V}$ containing ξ_1 as its first element. By the Gram-Schmidt procedure, an orthonormal basis can be obtained. Since ξ_1 was selected of length 1, the orthonormal basis obtained will contain ξ_1 as its first element. Let this orthonormal basis be $\mathscr{X} = \{\xi_1, \xi_2, \ldots, \xi_n\}$.

Since the matrix of transition from one orthonormal basis to another is unitary, we have shown that for any complex matrix A, there is a unitary matrix U such that U^*AU has an eigenvalue of A in the upper left-hand corner, and the first column otherwise is zero.

We prove the theorem by induction. The theorem is certainly true for $n = 1$, that is, for a 1×1 matrix. Assume that the theorem is true for matrices of order $< n$. Let A be a matrix of order n. By the remarks of the previous paragraph, there is a unitary matrix U_1 such that $U_1^*AU_1$ has the form

$$U_1^*AU_1 = \begin{bmatrix} d_1 & C \\ 0 & D \end{bmatrix}, \tag{5.1}$$

where C is a $1 \times (n-1)$ submatrix, and D is an $(n-1) \times (n-1)$ submatrix.

By the induction assumption, there is an $(n-1) \times (n-1)$ unitary matrix U_2 such that $U_2^*DU_2$ is in upper triangular form. Then

$$\begin{bmatrix} 1 & 0 \\ 0 & U_2^* \end{bmatrix} \begin{bmatrix} d_1 & C \\ 0 & D \end{bmatrix} \begin{bmatrix} 1 & 0 \\ 0 & U_2 \end{bmatrix} = \begin{bmatrix} d_1 & CU_2 \\ 0 & U_2^*DU_2 \end{bmatrix} \tag{5.2}$$

is in upper triangular form. Furthermore, $\begin{bmatrix} 1 & 0 \\ 0 & U_2 \end{bmatrix} = U_3$ is an $n \times n$ unitary matrix. Finally, $U_1 U_3 = U$ is a unitary matrix, and $U^*AU = (U_1U_3)^*A(U_1U_3) = U_3^*(U_1^*AU_1)U_3$ is in upper triangular form. □

Since A and U^*AU are similar, they have the same eigenvalues. By Theorem 5.1, the eigenvalues are in the main diagonal of U^*AU.

The steps in the proof of Theorem 5.2 can be carried out sequentially to find an upper triangular form to which a given matrix is similar. However, such a procedure is very inconvenient, and we shall not attempt to carry out such a calculation. The importance to us is the existence of such a form. There is an important class of matrices for which the upper triangular form is actually a diagonal form, and in this case the calculations can be carried out by methods already developed. It is the purpose of the next section to identify this class of matrices.

EXERCISES 7-5

1. Given any linear transformation σ of an inner product space $\mathscr{V}$ over the complex numbers into itself, show there is an orthonormal basis such that the matrix representing σ is in lower triangular form.

2. Show that over the field of complex numbers, every

square matrix is unitary similar to a matrix in lower triangular form.

3. Show that the product of two upper triangular matrices is an upper triangular matrix.

4. Show that if $\sigma^* = \sigma$, then any matrix A representing σ with respect to an orthonormal basis has the property that $A^* = A$.

5. Show that if $\sigma^* = \sigma$, then any matrix A in upper triangular form representing σ with respect to an orthonormal basis is a diagonal matrix.

7-6 NORMAL MATRICES

Theorem 6.1 *A matrix A in upper triangular form is a diagonal matrix if and only if $A^*A = AA^*$.*

Proof. If A is a diagonal matrix, it is in upper triangular form and $A^*A = AA^*$.

Now suppose A is in upper triangular form and that $A^*A = AA^*$. The assumption that $A^*A = AA^*$ means that

$$\sum_{k=1}^{n} \bar{a}_{ki} a_{kj} = \sum_{r=1}^{n} a_{ir} \bar{a}_{jr}. \tag{6.1}$$

Consider only those sums for which $i = j$. Then, since $\bar{a}_{ki} = 0$ for $k > i$, and $a_{ir} = 0$ for $i > r$, (6.1) becomes

$$\sum_{k=1}^{i} \bar{a}_{ki} a_{ki} = \sum_{r=i}^{n} a_{ir} \bar{a}_{ir}. \tag{6.2}$$

If A has any non-zero elements above the main diagonal, there would be a first row in which there is a non-zero element to the right of the diagonal. Let s be the index of this first row. Then $a_{ks} = 0$ for $k < s$; these elements are above the element a_{ss}. And there exists a non-zero element a_{sr} for $r > s$. For $i = s$, (6.2) reduces to

$$|a_{ss}|^2 = \sum_{r=s}^{n} |a_{sr}|^2. \tag{6.3}$$

All the terms in (6.3) are non-negative and the right side has at least one non-zero term besides the term $|a_{ss}|^2$, which appears on both sides. This is a contradiction to the supposition that there is a non-zero term above the main diagonal. Thus, A is a diagonal matrix. □

A matrix A for which $A^*A = AA^*$ is called a *normal matrix*.

Theorem 6.2 *If a matrix A is normal, every matrix unitary similar to A is normal.*

Proof. Assume A is normal and let $A' = U^*AU$ where U is unitary. Then

$$\begin{aligned}(A')^*A' &= (U^*AU)^*(U^*AU) \\ &= (U^*A^*U)U^*AU \\ &= U^*A^*AU \\ &= U^*AA^*U \\ &= (U^*AU)(U^*A^*U) \\ &= (U^*AU)(U^*AU)^* \\ &= A'(A')^*.\end{aligned} \qquad (6.4)$$
□

Theorem 6.3 *A matrix A is unitary similar to a diagonal matrix if and only if it is normal.*

Proof. By Theorem 5.2, A is unitary similar to an upper triangular matrix. Let U^*AU be in upper triangular form where U is unitary. By Theorem 6.1, U^*AU is a diagonal matrix if and only if U^*AU is normal. By Theorem 6.2, U^*AU is normal if and only if A is normal. □

While the class of normal matrices is restricted, it is large enough to include some rather important types of matrices. A matrix A is said to be *Hermitian* if $A^* = A$. It is *symmetric* if $A^T = A$. Clearly, a real Hermitian matrix is symmetric. A matrix A is said to be *skew-Hermitian* if $A^* = -A$. It is *skew-symmetric* if $A^T = -A$. A real skew-Hermitian matrix is skew-symmetric.

Theorem 6.4 *Unitary matrices, orthogonal matrices, Hermitian matrices, real symmetric matrices, skew-Hermitian matrices, and real skew-symmetric matrices are normal.*

Proof. If U is unitary, $U^*U = I = UU^*$. Since an orthogonal matrix is a real unitary matrix, the same condition holds for orthogonal matrices.

For a Hermitian matrix, $A^*A = AA = AA^*$. Since a real symmetric matrix is a real Hermitian matrix, the same condition holds for a real symmetric matrix.

For a skew-Hermitian matrix, $A^*A = -AA = AA^*$. The same condition holds for a real skew-symmetric matrix. □

A *normal linear transformation* is a linear transformation for which the matrix representing it with respect to an orthonormal basis is a normal matrix. In particular, this means that σ is normal if and only if $\sigma^*\sigma = \sigma\sigma^*$.

Theorem 6.5 *For a normal linear transformation σ, $\|\sigma(\xi)\| = \|\sigma^*(\xi)\|$ for all vectors ξ.*

Proof. $\|\sigma(\xi)\|^2 = (\sigma(\xi), \sigma(\xi))$
$= (\sigma^*\sigma(\xi), \xi)$
$= (\sigma\sigma^*(\xi), \xi)$
$= (\sigma^*(\xi), \sigma^*(\xi))$
$= \|\sigma^*(\xi)\|^2.$ □

Theorem 6.6 *For a normal linear transformation σ, if ξ is an eigenvector of σ with eigenvalue d, then ξ is also an eigenvector of σ^* with eigenvalue $\bar{d}$.*

Proof. Suppose $\sigma(\xi) = d\xi$. Then $(\sigma - d)(\xi) = 0$. Then $0 = \|(\sigma - d)(\xi)\| = \|(\sigma - d)^*(\xi)\| = \|(\sigma^* - \bar{d})(\xi)\|$. Hence $(\sigma^* - \bar{d})(\xi) = 0$, or $\sigma^*(\xi) = \bar{d}\xi$. □

Theorem 6.7 *Let σ be a normal linear transformation. If ξ_1 and ξ_2 are eigenvectors of σ with distinct eigenvalues d_1 and d_2, respectively, then ξ_1 and ξ_2 are orthogonal.*

Proof. $d_2(\xi_1, \xi_2) = (\xi_1, d_2\xi_2) = (\xi_1, \sigma(\xi_2)) = (\sigma^*(\xi_1), \xi_2) = (\bar{d_1}\xi_1, \xi_2) = d_1(\xi_1, \xi_2)$. Hence, $(d_1 - d_2)(\xi_1, \xi_2) = 0$. Since $d_1 - d_2 \neq 0$, we have $(\xi_1, \xi_2) = 0$. □

This is the result we really want and need. Knowledge of this theorem greatly simplifies the problem of finding the unitary matrix of transition that reduces a normal matrix to a diagonal form. Starting with a normal matrix, we proceed as we did with a general matrix in Chapter 6.

Given a normal matrix, find the eigenvalues and corresponding eigenvectors. Theorem 6.7 assures us that eigenvectors corresponding to distinct eigenvalues are orthogonal.

If any eigenvalue is of multiplicity one, all that is necessary is to normalize the corresponding eigenvector. If an eigenvalue is of multiplicity greater than one, we find as many linearly independent eigenvectors as we can. For each such eigenvalue, we apply the Gram-Schmidt orthonormalization procedure to produce an orthonormal set. Theorem 6.3 assures us that a basis of orthonormal vectors exists for the given normal matrix. The procedure described will yield as large a set of linearly independent eigenvectors as is available. Hence it will yield an orthonormal basis of eigenvectors. In other words, we do not have to go through the tedious sequence of steps outlined in the proof of Theorem 5.2. Follow the work in the following examples.

Example 1: Consider the matrix

$$A = \begin{bmatrix} \frac{5}{9} & \frac{14}{9} & \frac{16}{9} \\ \frac{14}{9} & \frac{14}{9} & -\frac{2}{9} \\ \frac{16}{9} & -\frac{2}{9} & -\frac{1}{9} \end{bmatrix} \quad (6.5)$$

A is real symmetric. Hence, by Theorem 6.4, A is normal. The characteristic polynomial is $f(x) = -x^3 + 2x^2 + 5x - 6 = -(x-1)(x+2)(x-3)$. The eigenvalues are 1, -2, and 3. The characteristic matrix corresponding to the eigenvalue 1 is

$$\begin{bmatrix} -\frac{4}{9} & \frac{14}{9} & \frac{16}{9} \\ \frac{14}{9} & \frac{5}{9} & -\frac{2}{9} \\ \frac{16}{9} & -\frac{2}{9} & -\frac{10}{9} \end{bmatrix}$$

We reduce this matrix to basic form and determine that the corresponding eigenvector is represented by $(1, -2, 2)$.

In a similar way, we determine that the eigenvector corresponding to -2 is represented by $(-2, 1, 2)$, and the eigenvector corresponding to 3 is represented by $(2, 2, 1)$.

According to Theorem 6.7, the eigenvectors obtained are mutually orthogonal. Thus, we can obtain an orthonormal basis merely by normalizing each eigenvector. The resulting orthonormal basis of eigenvectors is then represented by

$\left\{\frac{1}{3}(1,-2,2), \frac{1}{3}(-2,1,2), \frac{1}{3}(2,2,1)\right\}$. The orthogonal matrix of transition is then

$$P = \frac{1}{3}\begin{bmatrix} 1 & -2 & 2 \\ -2 & 1 & 2 \\ 2 & 2 & 1 \end{bmatrix}. \tag{6.6}$$

Check that $P^T A P$ is a diagonal matrix with the eigenvalues of A in the main diagonal.

Example 2: Consider the matrix

$$A = \begin{bmatrix} 2 & -2 & 1 \\ 1 & 2 & 2 \\ 2 & 1 & -2 \end{bmatrix}. \tag{6.7}$$

First, we should check that A is normal. Actually, A is a scalar multiple of an orthogonal matrix. The characteristic polynomial is $f(x) = -x^3 + 2x^2 + 6x - 27 = -(x+3)(x^2 - 5x + 9)$. The characteristic values are -3, $\frac{5 + \sqrt{11}\,i}{2}$, and $\frac{5 - \sqrt{11}\,i}{2}$.

If A is considered to be a real matrix, there is only one eigenvalue. Let us consider A to be a complex normal matrix. The eigenvector corresponding to -3 is represented by $(1, 1, -3)$. The characteristic matrix corresponding to $\frac{5 + \sqrt{11}\,i}{2}$ is

$$\begin{bmatrix} \frac{-1 - \sqrt{11}\,i}{2} & -2 & 1 \\ 1 & \frac{-1 - \sqrt{11}\,i}{2} & 2 \\ 2 & 1 & \frac{-9 - \sqrt{11}\,i}{2} \end{bmatrix}.$$

Reduce this matrix to basic form and determine that an eigenvector is $\left(\frac{3 + \sqrt{11}\,i}{2}, \frac{3 - \sqrt{11}\,i}{2}, 1\right)$. Since A is real and the eigenvalue $\frac{5 - \sqrt{11}\,i}{2}$ is the conjugate complex of $\frac{5 + \sqrt{11}\,i}{2}$, the work done to obtain the eigenvector in the two cases differs only in the taking of conjugate complexes. Thus, the eigenvector corresponding to $\frac{5 - \sqrt{11}\,i}{2}$ is represented by $\left(\frac{3 - \sqrt{11}\,i}{2}, \frac{3 + \sqrt{11}\,i}{2}, 1\right)$.

The eigenvalues are distinct, so Theorem 6.7 assures us that the corresponding eigenvectors are mutually orthogonal. That this is so is easily checked. Thus, to obtain the orthonormal basis of eigenvectors we have merely to normalize the eigenvectors obtained. Then the unitary matrix of transition is

$$U = \frac{1}{2\sqrt{11}} \begin{bmatrix} 2 & 3+\sqrt{11}\,i & 3-\sqrt{11}\,i \\ 2 & 3-\sqrt{11}\,i & 3+\sqrt{11}\,i \\ -6 & 2 & 2 \end{bmatrix}. \qquad (6.8)$$

Check that U^*AU is a diagonal matrix with the eigenvalues of A in the main diagonal.

Example 3: This is an example with a multiple eigenvalue. Consider the matrix

$$A = \begin{bmatrix} 1/2 & 3\sqrt{2}/4 & 3\sqrt{2}/4 \\ 3\sqrt{2}/4 & 5/4 & -3/4 \\ 3\sqrt{2}/4 & -3/4 & 5/4 \end{bmatrix}. \qquad (6.9)$$

A is real symmetric, so it is normal. The characteristic polynomial is $-x^3 + 3x^2 - 4 = -(x+1)(x-2)^2$. An eigenvector corresponding to the eigenvalue -1 is represented by $(-2, 1, 1)$. If we write down the characteristic matrix corresponding to the eigenvalue 2 and reduce to basic form to obtain

$$\begin{bmatrix} -2 & 1 & 1 \\ 0 & 0 & 0 \\ 0 & 0 & 0 \end{bmatrix}. \qquad (6.10)$$

This matrix has rank 1, and the eigenspace has dimension 2. Actually, for this eigenvalue we do not have to go through the work of reducing the characteristic matrix to basic form. Since A is normal, we know that a basis of eigenvectors exists. Thus, there must be two linearly independent eigenvectors corresponding to the eigenvalue 2. We also know that these eigenvectors are orthogonal to the eigenvector corresponding to the eigenvalue -1. Knowing the eigenvector corresponding to -1 we could proceed directly to find the eigenvectors corresponding to 2 by finding the vectors orthogonal to the known eigenvector. In fact, the basic form obtained in (6.10) has this eigenvector in its only non-zero row. There are many choices for a basis for this eigenspace. Some involve much simpler numbers than others, but there seems to be no simple way to ensure making the simplest choice. One choice for a basis of eigenvectors is $\{(1, 2, 0), (1, 0, 2)\}$. We use the Gram-Schmidt orthonormalization procedure to obtain the orthonormal set $\left\{\frac{1}{\sqrt{3}}(1, 2, 0), \frac{1}{2\sqrt{3}}(2, -1, 3)\right\}$. Normalizing the eigenvector previously obtained, we obtain the orthogonal matrix of transition

$$P = \begin{bmatrix} -\sqrt{2}/2 & \sqrt{3}/3 & \sqrt{6}/6 \\ 1/2 & \sqrt{6}/3 & -\sqrt{3}/6 \\ 1/2 & 0 & \sqrt{3}/2 \end{bmatrix}. \tag{6.11}$$

Check that P is orthogonal and that $P^T A P$ is a diagonal matrix.

EXERCISES 7-6

1. For each of the following matrices, determine which are real symmetric, Hermitian, orthogonal, unitary, or normal but not of the other types. For each, find an orthonormal basis of eigenvectors and a diagonalizing matrix of transition.

   (a) $\begin{bmatrix} 1 & 2 \\ 2 & 1 \end{bmatrix}$

   (b) $\frac{1}{25} \begin{bmatrix} 66 & 12 \\ 12 & 59 \end{bmatrix}$

   (c) $\begin{bmatrix} 8 & 4 & 1 \\ 4 & -7 & -4 \\ 1 & -4 & 8 \end{bmatrix}$

   (d) $\begin{bmatrix} -1 & 4 & -6 \\ 4 & 3 & 2 \\ -6 & 2 & -2 \end{bmatrix}$

   (e) $\begin{bmatrix} 1 & 2 & 2 \\ 2 & 1 & -2 \\ 2 & -2 & 1 \end{bmatrix}$

   (f) $\begin{bmatrix} 1 & -i \\ i & 1 \end{bmatrix}$

   (g) $\begin{bmatrix} 2 & 1-i \\ 1+i & 3 \end{bmatrix}$

   (h) $\begin{bmatrix} 1 & -i & 0 \\ i & 1 & -i \\ 0 & i & 1 \end{bmatrix}$

2. Show that if σ is normal, then $K(\sigma^*) = K(\sigma)$.

3. Show that if σ is normal, then $K(\sigma) = \text{Im}(\sigma)^\perp$.

4. Show that if σ is normal, then $\text{Im}(\sigma^*) = \text{Im}(\sigma)$.

5. Show that if σ is normal and $\mathscr{S}$ is a set of eigenvectors of σ, then $\mathscr{S}^\perp$ is an invariant subspace under σ.

6. Show that if σ is normal and $\mathscr{W}$ is an invariant subspace under σ, then $\mathscr{W}^\perp$ is also invariant under σ, and $\mathscr{V} = \mathscr{W} \oplus \mathscr{W}^\perp$.

7-7 THE SITUATION OVER THE REAL NUMBERS

The discussion so far has assumed (explicitly or implicity) that the underlying field is the field of complex numbers. Although many applications of matrices use, or permit the use of, complex numbers, there are also many applications that require a restriction to real numbers. The results given so far have implications for the case in which the underlying field is the field of real numbers. That is why we considered the complex case first. But some care must be exercised in interpreting the results for the real case.

Restricting our attention to real numbers means that we are considering finite dimensional vector spaces over the real numbers. A linear transformation of such a vector space into itself is represented by a matrix with real elements. The definition of the inner product is the same except, of course, that the taking of a conjugate complex has no effect. The definitions of orthogonality and orthonormality are the same. Finally, changing from one orthonormal basis to another means that the matrix of transition must be real. Thus the matrices of transition must be orthogonal matrices.

Interpreting the results for the complex field in the real case runs into difficulties wherever there is an assertion of existence. Specifically, Theorem 5.2 suffers from this difficulty. The proof uses the fact that a polynomial with complex coefficients can be factored into linear factors. This kind of conclusion cannot be made over the field of real numbers, and Theorem 5.2, in the form stated, is not true over the real numbers. This means that everything that follows from Theorem 5.2 must be examined separately for its implications in the real case.

7-7 The Situation Over the Real Numbers

Even though we must consider a real matrix as representing something (a linear transformation, a change of basis, etc.) over a real vector space, it is possible to consider the matrix as a complex matrix that happens to have real elements. However, any conclusions based on such a viewpoint must be justified in terms of the interpretation of the matrix in the real setting.

We have already shown in Theorem 3.1 that if a matrix A is orthogonal similar to a diagonal matrix it is symmetric. We will now prove the converse of this theorem; that a symmetric real matrix is orthogonal similar to a diagonal matrix. Suppose A is a real symmetric matrix. Then A is normal and there is a unitary matrix U such that $U^*AU = D$ is a diagonal matrix. Since D is a diagonal matrix $\bar{D} = D^*$. Thus, $\bar{D} = (U^*AU)^* = U^*A^*U = U^*AU = D$. We see that D is actually real. By theorem 5.1 the elements in the diagonal of D are eigenvalues of A.

For each eigenvalue d, $C(d) = A - dI$ is a real matrix. Theorem 6.3 assures us that when we solve (in the field of complex numbers) the linear problem $(A - dI)X = 0$ we will obtain enough linearly independent solutions for each eigenvalue so that there is a basis of eigenvectors. But the number of linearly independent solutions is determined by the rank of the system. Since $A - dI$ is real, the standard method of solving the linear system $(A - dI)X = 0$ will yield real solutions, and the number of linearly independent solutions is equal to the nullity of the system. Thus, the eigenvectors of A can all be chosen with real coordinates. If several linearly independent eigenvectors correspond to one eigenvalue, the Gram-Schmidt procedure can be applied and it will yield real orthonormal eigenvectors. As before, eigenvectors corresponding to distinct eigenvalues will be orthogonal. Thus, the matrix of transition to this orthonormal basis will be real; it will be an orthogonal matrix.

Theorem 7.1 *A real matrix A is orthogonal similar to a diagonal matrix if and only if it is symmetric.* □

The procedure for finding an orthonormal basis of eigenvectors and the eigenvalues is the same as for a normal matrix. The reasoning in the proof of Theorem 7.1 assures us that this procedure will yield real eigenvalues and a basis of eigenvectors. Examples 1 and 3 in Section 7–6 are examples of real symmetric matrices.

318 • Orthogonal and Unitary Spaces

If an orthogonal matrix is not symmetric, it is not orthogonal similar to a diagonal matrix even though it is normal. However, the characteristic values, both real and complex, of an orthogonal matrix give important information about the transformation represented by the matrix.

Let σ be a linear transformation of the real plane into itself which is a rotation about the origin through an angle of θ. This transformation has been discussed in Section 4-2, and we found that it was represented by the matrix

$$A = \begin{bmatrix} \cos \theta & -\sin \theta \\ \sin \theta & \cos \theta \end{bmatrix}. \tag{7.1}$$

It is easily checked that A is an orthogonal matrix. The characteristic polynomial for A is $f(x) = (x - \cos \theta)^2 + \sin^2 \theta$, and the characteristic values are $d_1 = \cos \theta + i \sin \theta$ and $d_2 = \cos \theta - i \sin \theta$. A slightly more compact notation would be more convenient. We write cis θ for $\cos \theta + i \sin \theta$. (Read "cis" as "cosine-i-sine.") In this notation, $d_1 = $ cis θ and $d_2 = $ cis$(-\theta)$. Notice that both d_1 and d_2 are of absolute value 1, which is also implied by Theorem 4.1.

Conversely, assume A is a real normal matrix with complex characteristic values cis θ and cis$(-\theta)$. For notational convenience, let cis $\theta = a + bi$. Since A is normal, there exists a unitary matrix U such that $U^*AU = D$ is a diagonal matrix with $a + bi$ and $a - bi$ in the main diagonal. Then $A = UDU^*$ and

$$\begin{aligned} A + A^T = A + A^* &= UDU^* + (UDU^*)^* \\ &= UDU^* + UD^*U^* \\ &= U(D + D^*)U^* \\ &= U \begin{bmatrix} 2a & 0 \\ 0 & 2a \end{bmatrix} U^* \\ &= 2aUU^* = 2aI. \end{aligned} \tag{7.2}$$

This means A is of the form

$$A = \begin{bmatrix} a & -b \\ b & a \end{bmatrix} \text{ or } A = \begin{bmatrix} a & b \\ -b & a \end{bmatrix}. \tag{7.3}$$

Thus, A represents a rotation through an angle of θ or $-\theta$.

Let us describe all orthogonal transformations of a plane into itself. Let σ be an orthogonal transformation of the plane represented by the orthogonal matrix A. By Theorem 4.1, the characteristic values of A are of absolute value 1. The product of the characteristic values is the value of the determinant of A which is a real number. Thus, the product of the characteristic values is $+1$ or -1.

First, consider the situation when the product of the characteristic values is $+1$. If one is complex (non-real), the other is its complex conjugate and we are in the situation described above; the orthogonal transformation is a rotation. If one characteristic value is real, both are real. If both are $+1$, the transformation is the identity. If both are -1, the transformation is a rotation through 180°. If we consider the identity as a rotation of 0, then all orthogonal transformations of the plane into itself where the product of the characteristic values is $+1$ are rotations.

Second, consider the situation when the product of the characteristic values is -1. In this case one characteristic value must be -1 and the other must be $+1$. Since they are real both are eigenvalues. Let ξ_1 be an eigenvector corresponding to $+1$ and let ξ_2 be an eigenvector corresponding to -1. The entire line from the origin through the end point of ξ_1 is left fixed. The vector ξ_2 is orthogonal to this line and direction is reversed by the transformation. Thus, the transformation is a reflection with respect to the line through ξ_1. Thus, all orthogonal transformations of the plane into itself, where the product of the characteristic values is -1, are re-

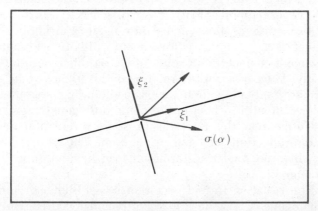

Figure 7-7.1

flections. Notice that the rotations are orientation preserving and the reflections are orientation reversing.

Now let us turn our attention to orthogonal transformations of a real 3-dimensional space into itself. Let the orthogonal transformation σ be represented by the orthogonal matrix A. As before, the product of the characteristic values of A is real and either $+1$ or -1. First, consider the situation when the product of the characteristic values is $+1$. In any case one of the characteristic values must be real. If two characteristic values are complex, their product is $+1$ and the real characteristic value must be $+1$. If all three are real at least one of them is $+1$. Thus, if the product of the characteristic values is $+1$, one of the characteristic values is $+1$. Let ξ_1 be an eigenvector corresponding to $+1$. The line through ξ_1 is left fixed by σ. Since σ preserves the inner product, the plane orthogonal to ξ_1 is mapped onto itself by σ. Since σ maps an orthonormal set onto an orthonormal set, even within this plane, it is an orthogonal transformation of this plane into itself. The product of the characteristic values for this orthogonal transformation is $+1$, since the product of all characteristic values is $+1$ and one characteristic value of $+1$ has been accounted for. The previous discussion applies to this plane. Hence σ acts like a rotation of this plane. Then in the entire space, ξ_1 is in the axis of the rotation and the plane of the rotation is orthogonal to this axis.

Now, consider the situation when the product of the characteristic values is -1. We can conclude, using reasoning similar to that in the previous paragraph, that one of the characteristic values is -1. Let ξ_1 be an eigenvector corresponding to -1. Since σ preserves the inner product, the plane orthogonal to ξ_1 is mapped onto itself. Again, σ acts like an orthogonal transformation of this plane into itself. Furthermore, the product of the characteristic values for this transformation is $+1$ since a characteristic value of -1 has been accounted for already. Thus, σ acts like a rotation of this plane. In the entire space, σ is a rotation about an axis containing ξ_1 together with a reflection with respect to the plane orthogonal to ξ_1.

Notice that the rotations preserve orientation and the rotation-reflections reverse orientation. The general nature of each orthogonal transformation can be identified by the characteristic values.

The situation in real vector spaces of higher dimensions is analogous. Let A be a $n \times n$ orthogonal matrix. The characteristic polynomial for A is real, and the characteristic values are of absolute value 1. Those that are complex can be grouped into conjugate pairs. To each pair there is a 2-dimensional subspace in which the orthogonal transformation represented by A is a rotation. The real characteristic values are $+1$'s and -1's. An even number of the -1's can be grouped into pairs. To each pair there is a 2-dimensional subspace in which the orthogonal transformation represented by A is a 180° rotation. To the $+1$'s there are fixed axes. If there is a -1 left, there is a reflection, in which case the orthogonal transformation is orientation reversing. If there is no -1 left after all pairings, the orthogonal transformation is orientation preserving.

EXERCISES 7-7

1. The matrix

$$\frac{1}{3}\begin{bmatrix} 1 & 2 & 2 \\ 2 & 1 & -2 \\ 2 & -2 & 1 \end{bmatrix}$$

is both orthogonal and symmetric. Show that the characteristic values are real and of absolute value 1. Without finding a diagonal form, describe the geometric properties of the transformation represented by this matrix.

2. Identify the rotation represented by $\begin{bmatrix} 1/2 & -\sqrt{3}/2 \\ \sqrt{3}/2 & 1/2 \end{bmatrix}$.

3. The matrix

$$\begin{bmatrix} 0 & 0 & 1 \\ 1 & 0 & 0 \\ 0 & 1 & 0 \end{bmatrix}$$

is easily seen to represent a rotation σ that rotates α_1 to α_2, α_2 to α_3, and α_3 to α_1. With this interpretation the axis is easy to identify. Find an orthonormal basis with respect to which σ is represented in the form of formula (5.12) of Chapter 6.

4. Consider the matrix $A = \dfrac{1}{3}\begin{bmatrix} 2 & 2 & 1 \\ 1 & -2 & 2 \\ 2 & -1 & -2 \end{bmatrix}$.

Verify, by considering the orthonormality of the columns of A, that A is an orthogonal matrix and represents an orthogonal transformation σ. This means there is an orthonormal basis such that σ can be represented in the form of (5.12) or (5.16) of Chapter 4. How do we find out which it is? Any matrix similar to A has the same determinant as A. Determine whether A is orthogonal similar to a matrix of the form of (5.12) or the form of (5.16).

5. Now, how do we find the angle θ (or the $\cos \theta$) that appears in (5.12) or (5.16)? In Section 6-3 the trace of a matrix is defined. Show that $\text{Tr}(AB) = \text{Tr}(BA)$ for any two square matrices of the same order. Use this to show that if P is non-singular, then $\text{Tr}(A) = \text{Tr}(P^{-1}AP)$. Use this information to compute $\cos \theta$ from A.

(Note: In the discussion suggested in Exercise 4, it is not even necessary to evaluate the determinant exactly. We know from A that the value of the determinant is real. We know from Theorem 4.1 that the product of the characteristic values of A is of absolute value 1. Thus, the value of the determinant can be only $+1$ or -1. A crude estimate for the value of the determinant is sufficient.)

Summary

This chapter is concerned with vector spaces over the real and complex numbers in which an inner product is defined. An *inner product* is a scalar-valued function of pairs of vectors $\alpha, \beta \in \mathscr{V}$ such that

1. $(\alpha, \alpha) \geq 0$ and $(\alpha, \alpha) = 0$ if and only if $\alpha = 0$.
2. $(\alpha, \beta) = \overline{(\beta, \alpha)}$.
3. The inner product is linear in the second vector variable. (This means that it is linear in the first variable in the real case, and conjugate linear in the first variable in the complex case.)

The fundamental inequality for the inner product is Schwarz' inequality.

$$|(\alpha, \beta)| \leq \|\alpha\| \|\beta\|. \tag{S.1}$$

where $\|\alpha\|^2 = (\alpha, \alpha)$.

Two vectors $\alpha, \beta \in \mathscr{V}$ are orthogonal if $(\alpha, \beta) = 0$. A set vectors is *orthonormal* if the vectors in the set are mutually orthogonal and of length 1.

Let $\mathscr{A} = \{\alpha_1, \ldots, \alpha_r\}$ be a linearly independent set of vectors. The Gram-Schmidt orthonormalization procedure is used to obtain an orthonormal set $\mathscr{X} = \{\xi_1, \ldots, \xi_r\}$ such that for each k, $1 \leq k \leq r$,

$$\xi_k = \sum_{i=1}^{k} a_{ik} \alpha_i. \tag{S.2}$$

This chapter is concerned primarily with changes of basis from one orthonormal basis to another orthonormal basis. If A represents a linear transformation σ with respect to an orthonormal basis $\mathscr{X}$ and A' represents σ with respect to an orthonormal basis $\mathscr{X}'$, then A and A' are *unitary similar* in the complex case, *orthogonal similar* in the real case. This is a special kind of similarity, studied in Chapter 6, where there was no restriction on the kind of basis.

Over the complex numbers, every matrix is unitary similar to a matrix in upper triangular form.

The principal theorem of this chapter is that, in the complex case, the necessary and sufficient condition for a linear transformation to have an orthonormal basis of eigenvectors is that the linear transformation be normal. This requires the definition of the adjoint. The linear transformation σ^* is the *adjoint* of σ if $(\sigma^*(\alpha), \beta) = (\alpha, \sigma(\beta))$ for all $\alpha, \beta \in \mathscr{V}$. The linear transformation σ is *normal* if and only if $\sigma \sigma^* = \sigma^* \sigma$.

If A is a matrix, A^* is the conjugate transpose of A. If A represents a linear transformation σ with respect to an orthonormal basis, then A^* represents σ^* with respect to the same basis. A matrix A is *normal* if and only if $AA^* = A^*A$. The principal theorem says that A is unitary similar to a diagonal matrix if and only if it is normal.

Note that the principal theorem, whether stated for linear transformations or for matrices, refers to unitary similarity or the complex case. In the real case a linear transformation σ has an orthonormal basis of eigenvectors if and only if $\sigma^* = \sigma$, in which case we say σ is *self adjoint*. In terms of matrices, a real matrix A is orthogonal similar to a diagonal matrix if and only if it is symmetric.

A *unitary transformation* preserves the inner product. That is, $(\sigma(\alpha), \sigma(\beta)) = (\alpha, \beta)$ for all $\alpha, \beta \in \mathscr{V}$. A unitary transformation is normal. In the real case a linear transformation that preserves the inner product is said to be orthogonal. An orthogonal linear transformation is normal, but not necessarily self adjoint. Thus, an orthogonal linear transformation may not have an orthonormal basis of eigenvectors.

APPENDIX

In this appendix we shall discuss some topics—set, indexed sets, elementary logic, summation notation, algebra—that we consider to be outside the subject orientation of this book. While we cannot presume that these topics are already familiar to every reader, it is a great convenience to use these ideas. The discussion of these topics here is neither logically nor pedagogically complete. It is intended to be merely sufficient to make these ideas useful to the reader for the purpose of reading the material in this text.

A-1 SETS AND LOGIC

A careful treatment of set theory from first principles is a rather lengthy affair. Furthermore, there are some subtle logical problems that are of great mathematical interest. For example, there are several paradoxes that require understanding either to see that they are not really problems or to find ways to avoid them. However, our use of set theory in this book does not involve these subtleties. We use set theory in a naïve way and all we need is an understanding of the terminology and notation of set theory.

We assume the reader to have an intuitive idea of what a set is, and what it means to say that something is an element of a set. We assume there to be one set $\mathcal{U}$, a "universe" or "universe of discourse," that contains all elements that are considered. The sets under consideration contain elements taken from this universe. For each such set it must be possible, in principle, to ascertain for each element of the universe whether it is or is not an element of this set. Even a statement as innocent as this is not entirely simple. For example, suppose $\mathcal{S}$ is the set of all numbers whose digits appear in order in the decimal expansion of π. If a is the number 1234567890, and this sequence of digits does appear in any of the known computed approximations of π we can decide that a is an element of $\mathcal{S}$. However, if this sequence of digits does not appear in any known decimal approximation of π, we cannot conclude that a is not an element of $\mathcal{S}$. It may be that the known approximations of π have not been calculated far enough to include the given sequence of digits. In fact, it is not possible to conclude, in a *finite* number of steps, that a is not an element of $\mathcal{S}$. Despite this kind of problem, "a is an element of $\mathcal{S}$" and "a is not an element of $\mathcal{S}$" are both meaningful statements.

The notation $a \in \mathcal{S}$ is used to stand for the statement, "a is an element of the set $\mathcal{S}$." The notation $a \notin \mathcal{S}$ means that "a is not an element of the set $\mathcal{S}$." We assume for each set $\mathcal{S}$ and each element a, either $a \in \mathcal{S}$ or $a \notin \mathcal{S}$ is true and both are not true at the same time. This kind of assumption evades the problem mentioned in the previous paragraph.

Elements and sets are considered to be on different "levels." That is, an element and the set consisting of that single element are not considered to be the same thing. This distinction is consistent with common usage. For example, we would certainly distinguish between a student and a class in a college, even if the class contained only one student. This idea also raises the possibility of considering collections of sets. For example, we might wish to refer to all the courses offered by the mathematics department. Thus, we can have elements, sets of elements, and collections of sets; the hierarchy of levels could be continued indefinitely.

There are several ways a set can be specified or described. If a set contains just a few elements it is possible to list the elements of the set. Thus, $\{a, b, c, d, e\}$ denotes the set consisting of the first five letters of the Latin alphabet. We shall consistently use braces to denote sets where the items enclosed within the braces are the elements (or the names of the elements) of the set.

Most of the sets we will consider have too many elements to allow them to be listed. Often a set can be specified by a defining criterion. For example, we might refer to "the set of all even integers," or "the set of all solutions of the equation $x^5 = 1$." In some cases, we can describe the set verbally, as we have done here. If P is the defining property, we can denote the set by $\{x \mid P\}$. This notation should be read as "the set of all x satisfying property P." For example, $\{x \mid x^5 = 1\}$ is the set of all x satisfying the equation $x^5 = 1$. Similarly, $\{(x, y) \mid x = 2y\}$ is the set of all pairs of numbers (or whatever the elements of the universe of discourse are) for which the first is twice the second.

A fundamental relation between sets is set inclusion. We say a set $\mathcal{A}$ is a *subset* of a set $\mathcal{B}$, or $\mathcal{A}$ is contained in $\mathcal{B}$, if every element in $\mathcal{A}$ is also in $\mathcal{B}$. In this case we write $\mathcal{A} \subset \mathcal{B}$ or $\mathcal{B} \supset \mathcal{A}$, whichever is more convenient. Often we wish to show that two sets which are described differently are, in fact, the same. Clearly, two sets are the same if and only if they contain exactly the same elements. Usually, we show $\mathcal{A} = \mathcal{B}$ by showing $\mathcal{A} \subset \mathcal{B}$ and $\mathcal{B} \subset \mathcal{A}$.

It is sometimes desirable to consider a situation in which $\mathcal{A} \subset \mathcal{B}$ but $\mathcal{A} \neq \mathcal{B}$. In this case we say $\mathcal{A}$ is a *proper* subset of $\mathcal{B}$.

Two fundamental operations with sets are set union and set intersection. It is traditional to illustrate these concepts by means of a Venn diagram, which is shown in Figure A–1.1. In Figure A–1.1, the large rectangle represents the universe of discourse, the universe $\mathcal{U}$. $\mathcal{A}$ and $\mathcal{B}$ are sets, represented by circles. The *intersection* of $\mathcal{A}$ and $\mathcal{B}$ is the set

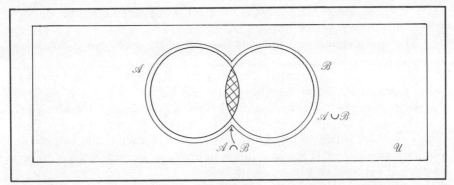

Figure A-1.1 A Venn diagram. $\mathcal{U}$ is the universal set. The cross-hatched region is the intersection of the sets $\mathcal{A}$ and $\mathcal{B}$. The color-outlined region is the union of $\mathcal{A}$ and $\mathcal{B}$.

of all points (elements) in both $\mathcal{A}$ and $\mathcal{B}$. It is shown as the lens-shaped region that is cross-hatched. The *union* of $\mathcal{A}$ and $\mathcal{B}$ is the set of all points in either $\mathcal{A}$ or $\mathcal{B}$ or both. In Figure A-1.1 it is represented by the binocular-shaped region. The intersection of $\mathcal{A}$ and $\mathcal{B}$ is denoted by $\mathcal{A} \cap \mathcal{B}$, and the union of $\mathcal{A}$ and $\mathcal{B}$ is denoted by $\mathcal{A} \cup \mathcal{B}$.

An element is in the intersection of $\mathcal{A}$ and $\mathcal{B}$ if and only if it is in $\mathcal{A}$ and $\mathcal{B}$. An element is in the union of $\mathcal{A}$ and $\mathcal{B}$ if and only if it is in $\mathcal{A}$ or in $\mathcal{B}$. In fact, in mathematical logic, the intersection of sets and the logical connection "and" are considered to be equivalent. The interpretation is as follows. Each element in $\mathcal{U}$ is considered to be a possible outcome, a possible state of whatever is under consideration. Then $\mathcal{U}$ is the set of all possible states. The set $\mathcal{A}$ consists of those states for which a certain statement or assertion is true. For simplicity, we will use the same letter $\mathcal{A}$ to denote the statement, whatever it is. Similarly, let $\mathcal{B}$ be the set for which the statement $\mathcal{B}$ is true. In this context, set inclusion $\mathcal{A} \subset \mathcal{B}$ has a logical interpretation. Every state for which $\mathcal{A}$ is true is a state for which $\mathcal{B}$ is true. In other words, every time $\mathcal{A}$ is true, $\mathcal{B}$ is true. In this case we say, "$\mathcal{A}$ implies $\mathcal{B}$." Thus, $\mathcal{A} \subset \mathcal{B}$ is equivalent to "$\mathcal{A}$ implies $\mathcal{B}$."

Now $\mathcal{A} \cap \mathcal{B}$ is the set for which both $\mathcal{A}$ and $\mathcal{B}$ are true. The union $\mathcal{A} \cup \mathcal{B}$ is the set for which $\mathcal{A}$ or $\mathcal{B}$ is true. Notice that the word "or" is used here in a sense which is different from the way it is usually used in common discourse. Suppose you are asked if you intend to go out to the movies this evening or do something else. In the sense in which "or" is used in mathematical logic, the correct answer is, "Yes." If $\mathcal{A}$ represents the states in which you intend to go to the movies, the rest of the universe $\mathcal{U}$ contains the states in which you do not intend to go to the movies. Thus, the union of these two sets is the universe. In other words, $\mathcal{A}$ or "not $\mathcal{A}$" is true for all states. The answer is "Yes" regardless of the state.

From the Venn diagram it is easily seen that $\mathcal{A} \subset \mathcal{A} \cup \mathcal{B}$ and

$\mathcal{A} \cap \mathcal{B} \subset \mathcal{A}$. These are equivalent to the logical statements, "$\mathcal{A}$ implies $\mathcal{A}$ or $\mathcal{B}$," and "$\mathcal{A}$ and $\mathcal{B}$ implies $\mathcal{A}$."

The *complement* of a set $\mathcal{A}$ is the part of the universe $\mathcal{U}$ that is not in $\mathcal{A}$. Let $\sim \mathcal{A}$ denote the complement of $\mathcal{A}$. Then $a \in \sim\mathcal{A}$ if and only if $a \notin \mathcal{A}$. For this reason, the symbol "$\sim$" is often read as "not." In other words, $\sim\mathcal{A}$ is the set where "not $\mathcal{A}$" is true.

In Figure A-1.2 the complement of $\mathcal{A}$ is indicated by horizontal lines, and the complement of $\mathcal{B}$ is indicated by vertical lines. Thus, $(\sim\mathcal{A}) \cap (\sim\mathcal{B})$ is the part that is cross-hatched. Notice that it is the complement of $\mathcal{A} \cup \mathcal{B}$. This is a fundamental law known as *DeMorgan's law*. As illustrated in Figure A-1.2 it is

$$\sim(\mathcal{A} \cup \mathcal{B}) = (\sim\mathcal{A}) \cap (\sim\mathcal{B}). \quad (1.1)$$

Although the truth of this law is evident from Figure A-1.2, it deserves to be established on a logical basis.

If $a \in \sim(\mathcal{A} \cup \mathcal{B})$, then $a \notin \mathcal{A} \cup \mathcal{B}$. Since $a \in \mathcal{A} \cup \mathcal{B}$ if and only if $a \in \mathcal{A}$ or $a \in \mathcal{B}$, $a \notin \mathcal{A}$ and $a \notin \mathcal{B}$. That is, $a \in (\sim\mathcal{A}) \cap (\sim\mathcal{B})$. This shows $\sim(\mathcal{A} \cup \mathcal{B}) \subset (\sim\mathcal{A}) \cap (\sim\mathcal{B})$.

If $a \in (\sim\mathcal{A}) \cap (\sim\mathcal{B})$, then $a \in \sim\mathcal{A}$ and $a \in \sim\mathcal{B}$. That is, $a \notin \mathcal{A}$ and $a \notin \mathcal{B}$. Thus $a \notin \mathcal{A} \cup \mathcal{B}$, or $a \in \sim(\mathcal{A} \cup \mathcal{B})$. This shows $(\sim\mathcal{A}) \cap (\sim\mathcal{B}) \subset \sim(\mathcal{A} \cup \mathcal{B})$. The two inclusions show that $\sim(\mathcal{A} \cup \mathcal{B}) = (\sim\mathcal{A}) \cap (\sim\mathcal{B})$.

It should be evident that $\sim(\sim\mathcal{A}) = \mathcal{A}$, that the complement of the complement is the original set. This idea can be used to give another equivalent form of DeMorgan's law. Replace $\mathcal{A}$ by $\sim\mathcal{A}$ and $\mathcal{B}$ by $\sim\mathcal{B}$ in Formula (1.1). This gives

$$\sim[(\sim\mathcal{A}) \cup (\sim\mathcal{B})] = (\sim(\sim\mathcal{A})) \cap (\sim(\sim\mathcal{B})) = \mathcal{A} \cap \mathcal{B}, \quad (1.2)$$

or

$$\sim(\mathcal{A} \cap \mathcal{B}) = (\sim\mathcal{A}) \cup (\sim\mathcal{B}). \quad (1.3)$$

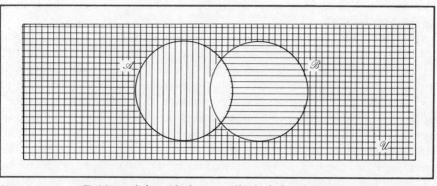

Figure A-1.2 DeMorgan's law. The horizontally shaded region is the complement of $\mathcal{A}$. The vertically shaded region is the complement of $\mathcal{B}$. The intersection of the complements is the complement of $\mathcal{A} \cup \mathcal{B}$.

Although formulas (1.1) and (1.3) are equivalent, it is customary to list both forms and refer to them as DeMorgan's laws.

In Figure A–1.1, the possibility that $\mathscr{A}$ and $\mathscr{B}$ have no points in common must be considered. It would be awkward if the intersection $\mathscr{A} \cap \mathscr{B}$ was a set if $\mathscr{A}$ and $\mathscr{B}$ had points in common, and we had to make some other kind of statement if $\mathscr{A}$ and $\mathscr{B}$ had no points in common. For reasons of this type we define the *empty set* as the set with no points, with no elements. We will denote the empty set by the symbol $\varnothing$. Since for all a in the universe $\mathscr{U}$, $a \notin \varnothing$, the empty set is the complement of $\mathscr{U}$.

The connection between the sets concepts discussed here and elementary mathematical logic is direct. As already pointed out, $\mathscr{A} \subset \mathscr{B}$ is equivalent to the logical statement "$\mathscr{A}$ implies $\mathscr{B}$." It is clear that $\mathscr{A} \subset \mathscr{B}$ if and only if $\sim\mathscr{A} \supset \sim\mathscr{B}$, which in turn is equivalent to the logical statement "Not $\mathscr{B}$ implies not $\mathscr{A}$." In logic, "$\mathscr{A}$ implies $\mathscr{B}$" and "Not $\mathscr{B}$ implies not $\mathscr{A}$" are equivalent and one is called the *contrapositive* of the other. Quite often a theorem is proved by proving its contrapositive.

Suppose we wish to show that "$\mathscr{A}$ implies $\mathscr{B}$" is false. This is equivalent to showing that $\mathscr{A}$ is not a subset of $\mathscr{B}$. To do this we must produce at least one element in $\mathscr{A}$ that is not in $\mathscr{B}$. In theorems and proofs the situation is generally of this type, but not so simply phrased. A typical situation is to show (or show false) that every $a \in \mathscr{A}$ is an element of $\mathscr{B}$. For example, "All men are green" is a typical statement involving an implication. Here, $\mathscr{A}$ is the set of men and $\mathscr{B}$ is the set of green things. To negate this statement we would have to show, "There is a man who is not green." That is, there is an element of $\mathscr{A}$ that is not in $\mathscr{B}$.

DeMorgan's laws also have direct logical meaning and are often used in mathematical arguments. Formula (1.1), $\sim(\mathscr{A} \cap \mathscr{B}) = (\sim\mathscr{A}) \cup (\sim\mathscr{B})$ is equivalent to: "$\mathscr{A}$ and $\mathscr{B}$" is false if and only if $\mathscr{A}$ is false or $\mathscr{B}$ is false. Formula (1.3), $\sim(\mathscr{A} \cup \mathscr{B}) = (\sim\mathscr{A}) \cap (\sim\mathscr{B})$ is equivalent to: "$\mathscr{A}$ or $\mathscr{B}$" is false if and only if $\mathscr{A}$ is false and $\mathscr{B}$ is false.

A–2 INDEXED SETS

It is very useful to be able to construct "lists" in mathematics. To the extent that $\{6, -1, 3, 5\}$ is a list, it is called an ordered set. This terminology refers to the fact that the items listed are given in a certain order and means that we regard $\{-1, 5, 6, 3\}$ as a different ordered set even though it contains the same elements. It is just as effective and more convenient to think of $\{6, -1, 3, 5\}$ as the set $\{a_1, a_2, a_3, a_4\}$ where $a_1 = 6$, $a_2 = -1$, $a_3 = 3$, $a_4 = 5$. The subscripts are called *indices*. The *index set* is the set from which the indices are chosen. In the example given, the index set is $\{1, 2, 3, 4\}$. The fact that the index set $\{1, 2, 3, 4\}$ is an ordered set (in the usual ordering) induces the order of the elements listed in the ordered set $\{a_1, a_2, a_3, a_4\}$.

In some applications of indexing the order of the index set is important, but in many other situations it is not. For example, suppose Nancy and Douglas are two children in a family and R_N is Nancy's room while R_D is Douglas' room. Then $\{R_N, R_D\}$ is a set of two rooms, one associated with each child. There is no need to think of either the rooms or the children in any particular order. All that is really important is the association of each one of the children with a room. In the same sense, what is usually important is the association of each element i in the index set with an element a_i.

Definition. *Let $\mathcal{A}$ and $\mathcal{I}$ be any sets. If for every element $i \in \mathcal{I}$ there is associated an element $a_i \in \mathcal{A}$, we call this association an* **indexing***. The set $\mathcal{I}$ is called the* **index set** *of the indexing. We write $\{a_i \mid a_i \in \mathcal{A}, i \in \mathcal{I}\} = \{a_i\}_{i \in \mathcal{I}}$ to denote the indexing. The set $\{a_i \mid a_i \in \mathcal{A}, i \in \mathcal{I}\}$ is the* **indexed set***. Notice that we do not require that every element of $\mathcal{A}$ be indexed, and we do not require that $a_i \neq a_j$ for $i \neq j$.*

Common index sets are the positive integers $\mathcal{N}$ and finite sets of positive integers $\{1, 2, \ldots, n\}$. In this notation $\{a_i \mid a_i \in \mathcal{R}, i \in \mathcal{N}\}$ is a sequence of real numbers. A more common notation for a sequence is $\{a_1, a_2, a_3, \ldots\}$, where the continuation of the indexing is implied.

The most frequently occurring index set in this book is a finite set of integers $\{1, 2, \ldots, n\} = (n)$. In Section 1 of Chapter 1 we consider a linear equation of the form

$$a_1 x_1 + a_2 x_2 + \cdots + a_n x_n = b. \tag{2.1}$$

Here, there are two indexed sets; the set $\{a_1, a_2, \ldots, a_n\}$ of coefficients and the set $\{x_1, x_2, \ldots, x_n\}$ of unknowns. The index set is (n). In these expressions the three dots indicate that the indices are intended to run through the index set from the least to the largest values shown.

An equation like (2.1) could also be indicated by the notation

$$ax + by + \cdots + cz = d, \tag{2.2}$$

but using index notation offers several advantages. Not only does it commit fewer available typographical symbols to each equation, it allows us to use the same kind of symbol for all quantities or objects of the same type. For example, a_i is a coefficient and x_i is an unknown.

Index notation also allows us to write rather complicated expressions in more compact form. Notations for sums and products are examples of this use of index notation.

$$\sum_{i=1}^{4} a_i x_i \qquad (2.3)$$

stands for "the sum, as i runs through the index set $\{1, 2, 3, 4\}$, of all terms of the form $a_i x_i$." The large sigma is the sign that a summation is to be performed. The part of the symbol $_{i=1}^{4}$ tells us the name of the index (i in this case) and the range of the index set. The implication here is that the index runs through integral values from the least to the largest value shown (from 1 to 4 in this case). The expression that the summation symbol acts upon, $a_i x_i$, is to be written down once for each value of the index and the resulting terms are to be summed. Written out in full expression (2.3) becomes

$$a_1 x_1 + a_2 x_2 + a_3 x_3 + a_4 x_4. \qquad (2.4)$$

In a similar way, equation (2.1) could be written in the form

$$\sum_{i=1}^{n} a_i x_i = b. \qquad (2.5)$$

Examine the following examples which illustrate the use of summation notation

$$\sum_{i=1}^{3} i = 1 + 2 + 3 = 6 \qquad (2.6)$$

$$\sum_{i=1}^{4} i^2 = 1^2 + 2^2 + 3^2 + 4^2 = 30 \qquad (2.7)$$

$$\sum_{k=0}^{3} P(1+i)^k = P + P(1+i) + P(1+i)^2 + P(1+i)^3 \qquad (2.8)$$

$$\sum_{k=1}^{4} \frac{1}{1+k} = \frac{1}{1+1} + \frac{1}{1+2} + \frac{1}{1+3} + \frac{1}{1+4} \qquad (2.9)$$

$$\sum_{k=0}^{n} ar^k = a + ar + ar^2 + \cdots + ar^n \qquad (2.10)$$

$$\sum_{k=1}^{n} kx^k = x + 2x^2 + \cdots + nx^n \qquad (2.11)$$

$$\sum_{j=1}^{n} a_{ij} x_j = a_{i1} x_1 + a_{i2} x_2 + \cdots + a_{in} x_n \qquad (2.12)$$

In formula (2.12) the symbol a_{ij} has two indices. In this situation there are two index sets. For example, consider $\{1, 2, \ldots, m\} = (m)$ and $\{1, 2, \ldots, n\} = (n)$. In Section 1–1 there is a system of linear equations indicated with the following notation:

$$\sum_{j=1}^{n} a_{ij}x_j = b_i: i = 1, 2, \ldots, m. \qquad 1\text{–}(1.5)$$

Here, there is one equation for each index i in the index set (m). For each equation, the summation is over the index j which runs through the index set (n).

For a matrix $A = [a_{ij}]$ it is also useful to form a single index set $(m) \times (n)$ consisting of all pairs (i, j) where $i \in (m)$ and $j \in (n)$. This index set has mn pairs. The arranging of the elements of a matrix into rows and columns is merely a way of implying the index sets without cumbersome notation.

When two or more indices appear, several indices may occur actively in the summation formulas. For example, consider

$$\sum_{k=1}^{3} \sum_{i=1}^{4} (k + 2i). \qquad (2.13)$$

The convention of this notation is that the inner summation (using the index i) is to be performed before the outer summation (using the index k). Thus, the expression (2.13) would be expanded in the following way.

$$\sum_{k=1}^{3} \sum_{i=1}^{4} (k + 2i) = \sum_{k=1}^{3} \{(k+2) + (k+4) + (k+6) + (k+8)\}$$
$$= \{(1+2) + (1+4) + (1+6) + (1+8)\} \qquad (2.14)$$
$$+ \{(2+2) + (2+4) + (2+6) + (2+8)\}$$
$$+ \{(3+2) + (3+4) + (3+6) + (3+8)\}.$$

The expression in (2.13) can also be unravelled from the outside in the following way.

$$\sum_{k=1}^{3} \sum_{i=1}^{4} (k + 2i) = \sum_{i=1}^{4} (1 + 2i) + \sum_{i=1}^{4} (2 + 2i) + \sum_{i=1}^{4} (3 + 2i). \qquad (2.15)$$

When the three terms on the right side of formula (2.15) are expanded we obtain the right side of (2.14) with the terms obtained in the same order.

Compare the following expansion with formula (2.14).

$$\sum_{i=1}^{4} \sum_{k=1}^{3} (k+2i) = \sum_{i=1}^{4} \{(1+2i) + (2+2i) + (3+2i)\}$$

$$\begin{aligned}
&= \{(1+2) + (2+2) + (3+2)\} \\
&\quad + \{(1+4) + (2+4) + (3+4)\} \\
&\quad + \{(1+6) + (2+6) + (3+6)\} \\
&\quad + \{(1+8) + (2+8) + (3+8)\}.
\end{aligned} \quad (2.16)$$

Because of the associative and commutative laws for addition, the right sides of formulas (2.14) and (2.16) are equal. Thus,

$$\sum_{k=1}^{3} \sum_{i=1}^{4} (k+2i) = \sum_{i=1}^{4} \sum_{k=1}^{3} (k+2i). \quad (2.17)$$

Reversing the order of summation, which is illustrated in formula (2.17), occurs very frequently in this book. Most of the time such a step presents no problem. Justifying the step is no more involved than an application of the associative and commutative laws. However, we want to be able to write down the reversal of the order of summation in one step, as it appears in formula (2.17), without obtaining the full expansions that are given in (2.14) and (2.16). This requires a clear understanding of what is involved in reversing the order of summation.

The important thing in formula (2.17) is that the index sets for i and k are independent. The actual expression which is summed is irrelevant. To make this statement clearer, let us illustrate a situation in which the index sets are not independent. Consider

$$\sum_{k=1}^{3} \sum_{i=1}^{k} (k+2i). \quad (2.18)$$

For each k, the index set for i is $\{1, \ldots, k\}$. The index set for i depends on the value of k. When the expression in (2.18) is expanded fully we obtain

$$\begin{aligned}
\sum_{k=1}^{3} \sum_{i=1}^{k} (k+2i) &= \sum_{i=1}^{1} (1+2i) + \sum_{i=1}^{2} (2+2i) + \sum_{i=1}^{3} (3+2i) \\
&= (1+2) + \{(2+2) + (2+4)\} \\
&\quad + \{(3+2) + (3+4) + (3+6)\} \\
&= \{(1+2) + (2+2) + (3+2)\} \\
&\quad + \{(2+4) + (3+4)\} + (3+6) \\
&= \sum_{k=1}^{3} (k+2) + \sum_{k=2}^{3} (k+4) + \sum_{k=3}^{3} (k+6) \\
&= \sum_{i=1}^{3} \sum_{k=i}^{3} (k+2i).
\end{aligned} \quad (2.19)$$

To direct our attention to those terms in formula (2.19) that we wish to emphasize, let us write down just the first and final expressions. Also, let us suppress the terms to be summed and indicate their presence with asterisks.

$$\sum_{k=1}^{3} \sum_{i=1}^{k} *** = \sum_{i=1}^{3} \sum_{k=i}^{3} ***. \tag{2.20}$$

Notice the presence of the index k in the limits of the inner summation on the left and the presence of the index i in the limits of the inner summation on the right.

If the principle at work in formula (2.20) looks obscure, the following interpretation may be helpful. Consider the pairs (i, k) as points plotted on a coordinate plane, as in Figure A-2.1. The points in Figure A-2.1 that are included in the sum (2.20) are indicated in color. For the inner sum on the left side of formula (2.20), the index set is $\{1, \ldots, k\}$. For this inner sum the summation is over the index i. That is, the index k is a constant for this sum. The pairs that are summed in this way are shown in Figure A-2.1 by horizontal line segments. Each segment extends from the vertical line $i = 1$ to the 45° diagonal line $i = k$. Notice that $i = 1$ and $i = k$ are the lower and upper limits for the inner sum on the left.

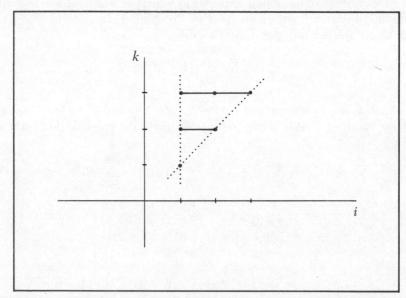

Figure A-2.1 The grouping of terms in a sum given by the left side of formula (2.20). The points on a horizontal line have a common value of k. For each such line the left and right end points correspond to the values of the index i indicated by the summation limits of the inside sum.

A-2 Indexed Sets • 335

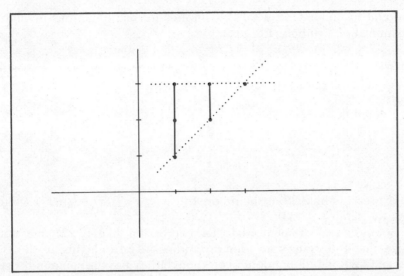

Figure A–2.2   The grouping of terms in a sum given by the right side of formula (2.20). The points on a vertical line have a common value of i. For each such line the bottom and top end points correspond to the values of the index k indicated by the summation limits of the inside sum.

In Figure A–2.2 the same points are shown in color. They are grouped into vertical line segments. Each line segment extends from the 45° diagonal line $k = i$ to the horizontal line $k = 3$. Notice that $k = i$ and $k = 3$ are the lower and upper limits for the inner sum on the right side of formula (2.20).

Finally, notice that the outer sum on the left side of formula (2.20) has the index k running from $k = 1$ to $k = 3$. In Figure A–2.1, this is interpreted as summing the horizontal line segments from the lowest segment to the top segment. Similarly, the outer sum on the right side of formula (2.20) has the index i running from $i = 1$ to $i = 3$. In Figure A–2.2, this is interpreted as summing the vertical line segments from the left to the right.

Although reversing the order of summation occurs often in the main body of this book, none of the occurrences is as complicated as the illustration given here. Though simpler, the same principles apply and they should be understood.

The indices appearing in expression (2.3) are called "dummy indices." This means the value of the expression does not depend on what the indices are called. That is,

$$\sum_{k=1}^{4} a_k x_k \quad \text{and} \quad \sum_{i=1}^{4} a_i x_i \qquad (2.21)$$

also denote the same sum, the sum shown in expression (2.4). It is important to understand what this implies for manipulations involving the summation symbols. For example,

$$\left(\sum_{i=1}^{3} a_i\right)\left(\sum_{i=1}^{3} x_i\right) \tag{2.22}$$

is not equal to

$$\sum_{i=1}^{3} a_i x_i = a_1 x_1 + a_2 x_2 + a_3 x_3. \tag{2.23}$$

The expression in (2.22) is the product $(a_1 + a_2 + a_3)(x_1 + x_2 + x_3)$, which is not the same as (2.23).

Before one can manipulate an expression like (2.22) one must realize that the indices are dummy indices. The indexing in the two sums of (2.22) is independent even though they use the same index set. The expression in (2.22) should be expanded by first renaming some of the indices so that every index has a different name. For example, in (2.22) we have

$$\left(\sum_{i=1}^{3} a_i\right)\left(\sum_{i=1}^{3} x_i\right) = \left(\sum_{i=1}^{3} a_i\right)\left(\sum_{j=1}^{3} x_j\right) \tag{2.24}$$

$$= \sum_{i=1}^{3} a_i \left(\sum_{j=1}^{3} x_j\right) = \sum_{i=1}^{3} \sum_{j=1}^{3} a_i x_j.$$

Write out the expression on the right side of (2.24) in full and show that it is the correct equivalent of expression (2.22).

Index notation can be used for purposes other than sums. For example, Π is used to denote a product. We define

$$\prod_{i=1}^{n} a_i = a_1 a_2 \cdots a_n. \tag{2.25}$$

The rules for handling indexed products are quite similar to those for indexed sums.

In Section A-1 we defined the intersection and union of two sets. If we need to consider the intersection and union of larger collections of sets, indexing allows convenient handling of rather complex expressions. For example,

$$\bigcap_{i=1}^{n} \mathscr{S}_i = \mathscr{S}_1 \cap \mathscr{S}_2 \cap \cdots \cap \mathscr{S}_n \tag{2.26}$$

denotes the intersection of an indexed collection of sets. An indexed union can be handled in a similar way.

In Section 4 of Chapter 2 the notation

$$\cap_i \mathscr{S}_i \tag{2.27}$$

appears. There is no indication of what the index set is. The expression (2.27) denotes the intersection of the entire indexed collection, and it does not matter whether the index set is finite, ordered, or infinite. One cannot be so casual about indexed sums. The difference is caused by the fact that addition is a binary operation, while the intersection (or union) involves the whole indexed collection at once. In fact, we shall restate the definitions of the intersection and union to make this point more evident.

The symbol ∃ stands for the words "there is" or "there exists," and ∀ stands for the words "for each" or "for all." The symbol ∃ is the *existential quantifier* and ∀ is called the *universal quantifier*. Existential and universal statements are routine in mathematics, and the use of these quantifiers allows us to express such statements more compactly and systematizes our handling of them. For example, in this notation axiom 4) for a vector space (See Section 2-2.) becomes, "∀$\alpha \in \mathscr{V}$, ∃$(-\alpha) \in \mathscr{V}$, such that $(-\alpha) + \alpha = 0$."

The quantifiers defined in the previous paragraph are closely related to the ideas of set unions and intersections. Let $\{\mathscr{S}_i \mid i \in \mathscr{I}\}$ be any indexed collection of subsets of the universal set $\mathscr{U}$. Then

$$\cap_{i \in \mathscr{I}} \mathscr{S}_i = \{x \mid x \in \mathscr{U}, \forall\, i \in \mathscr{I}, x \in \mathscr{S}_i\} \tag{2.28}$$

$$\cup_{i \in \mathscr{I}} \mathscr{S}_i = \{x \mid x \in \mathscr{U}, \exists\, i \in \mathscr{I}, x \in \mathscr{S}_i\} \tag{2.29}$$

In other words, an element belongs to the intersection of an indexed collection of sets if and only if it belongs to every set in the collection. An element belongs to the union of an indexed collection of sets if and only if there is a set in the collection to which it belongs. This is the idea behind the expression (2.27): we have merely suppressed the identity of the index set.

DeMorgan's laws can be easily expressed for indexed collections of sets.

$$\sim(\cap_{i \in \mathscr{I}} \mathscr{S}_i) = \cup_{i \in \mathscr{I}} (\sim \mathscr{S}_i) \tag{2.30}$$

$$\sim(\cup_{i \in \mathscr{I}} \mathscr{S}_i) = \cap_{i \in \mathscr{I}} (\sim \mathscr{S}_i) \tag{2.31}$$

It is instructive to prove these laws by use of the existential and universal quantifiers.

A-3 FIELDS, RINGS, AND GROUPS

In several places in this book we use algebraic terminology which may be unfamiliar to many readers. In most cases the use of these terms does not require a deep understanding of the concepts involved. In fact, every reader of this book has extensive experience with examples of these concepts. It is mostly a question of relating this experience to the abstract terminology. An intuitive feeling for these concepts based on such experience should be quite sufficient to handle the terminology in this book.

Definition. A *field* is a non-empty set $\mathscr{F}$ satisfying the following conditions.
(1) For all $a, b \in \mathscr{F}$, there is associated another element in $\mathscr{F}$ which is designated by $a + b$.
(2) For all $a, b, c \in \mathscr{F}$, $(a + b) + c = a + (b + c)$.
(3) There is a $0 \in \mathscr{F}$ such that for all $a \in \mathscr{F}$, $0 + a = a$.
(4) For all $a \in \mathscr{F}$, there is a $(-a) \in \mathscr{F}$ such that $(-a) + a = 0$.
(5) For all $a, b \in \mathscr{F}$, $a + b = b + a$.
(6) For all $a, b \in \mathscr{F}$, there is associated another element in $\mathscr{F}$ which is designated by ab.
(7) For all $a, b, c \in \mathscr{F}$, $(ab)c = a(bc)$.
(8) There is a $1 \in \mathscr{F}$ such that for all $a \in \mathscr{F}$, $1a = a$.
(9) For all $a \in \mathscr{F}$, $a \neq 0$, there is a $a^{-1} \in \mathscr{F}$ such that $a^{-1}a = 1$.
(10) For all $a, b \in \mathscr{F}$, $ab = ba$.
(11) For all $a, b, c \in \mathscr{F}$, $a(b + c) = ab + ac$.

Familiar examples of fields are the rational numbers, the real numbers, and the complex numbers. The field axioms include the familiar associative and commutative laws for addition, the associative and commutative laws for multiplication, and the distributive law. They also assert the existence of a zero and negatives, a unit and inverses for non-zero numbers.

A sophisticated and interesting theory of fields can be erected on the basis of these axioms. There are many fields, even many that are much less familiar than those cited in the previous paragraph. However, neither a deep understanding of field theory nor an acquaintance with esoteric examples is needed to read this book.

The arithmetic operations of addition, subtraction, multiplication, and division are *rational operations*. These four operations are mentioned in the field axioms (axioms (1), (4), (6), and (9)). In other words these operations can be performed in any field. These are the operations that must be available to perform the elementary row operations and pivot operations described in Chapter 1. In fact, throughout the book there are just two places where non-rational operations are involved. One occurrence involves the solutions of the characteristic equation in Chapter 6. The other involves taking a square root to normalize a vector in Chapter 7. The fact that only rational operations are required throughout the rest of the book means that the theory of linear algebra under discussion works in any field. Thus, the theory of linear algebra applies to a vector space over any field.

Sometimes we may be considering a vector space over one field while a particular problem may be defined over a smaller field. For example, a particular system of linear equations may have the property that all coefficients that appear in the system are from a field smaller than the field over which the vector space is defined. The arithmetic operations that we perform to solve the system of equations involve only rational operations with the coefficients that are given and that can be computed from those coefficients. Thus, the numbers that appear in the generators for the solution will be contained in the smallest field that contains all the given coefficients. For example, if all the coefficients for a system of linear equations are integers, the generators for the solution will be given in terms of rational numbers. If the problem was posed for a vector space over the real numbers, the part of the solution set that involves real non-rational numbers will be obtained by assigning real non-rational values to the parameters.

For the most part this is all that one needs to know about fields to read this book: to know that one can carry out the four rational operations in a field. The only place where a deeper understanding is required is in the discussions of eigenvalues in Chapters 6 and 7. What is required there is an understanding of the solvability of polynomial equations with real and with complex coefficients to the extent that this topic is usually covered in a college algebra course. Wherever the word "field" appears (and a particular field is not specified) it is quite safe to substitute in your mind your favorite field, perhaps the rational number field or the real number field. The generality that is lost is insignificant and no problem is caused by such a substitution. The proofs, however, are quite general and are no easier (or even different) regardless of which field is chosen.

A *ring* is a non-empty set of elements that satisfies all the field axioms, except possibly axioms (9) and (10). Thus, a field is also a ring. A familiar ring that is not a field is the set of all integers with addition and multiplication specified in the usual way. Another ring that the reader will probably meet in this book for the first time is the set of all $n \times n$ matrices with coefficients in a field $\mathscr{F}$. For this consideration the binary operations that are involved are addition of matrices and multiplication of matrices. As a ring we choose to ignore multiplication of matrices by scalars.

The ring of integers also satisfies axiom (10), and it is a *commutative* ring. The ring of n-th order matrices is a non-commutative ring.

The only place in this book where a ring is referred to is in Exercise 10 of Section 2-2. The point of that exercise is that the coefficient set $\mathscr{Z}$ is a ring but not a field. Therefore, $\mathscr{L}_n$ is not a vector space.

A *group* is a non-empty set of elements satisfying the first four axioms for a field. However, it is traditional to write the group operation with multiplicative notation. A group that also satisfies axiom (5) is called a *commutative* group or an *abelian* group.

Because of the relation between the axioms for fields, rings, and groups, every field is a commutative group under addition. Every ring is a commutative group under addition.

Although no use of group theory occurs in this book, several groups have been discussed. A vector space is a commutative group under the operation of vector addition. The set of all non-singular n-th order matrices over a given field is a group under matrix multiplication. Equivalently, the set of all non-singular linear transformations of a vector space into itself is a group under the composition of mappings.

In Chapter 5 we discussed permutations of a finite set $\{1, 2, \ldots, n\}$ in connection with the determinant function. A permutation is a one-to-one mapping of a set $\mathscr{S}$ onto itself. The set of all permutations of the set $\mathscr{S}$ is a group under the composition of mappings.

We have not made use of the theory of groups in any way in this book. There are several places where the discussion could have been shorter if we could have assumed the reader to have some previous knowledge of elementary group theory. In some places we have given proofs of these elementary properties. In others we have assumed these properties to be either evident, easily proved by the reader, or intuitively acceptable. The examples of groups that most students know about are so familiar, and encountered so early, that most students accept these properties without critical examination. To be entirely honest, we should prove these properties from the axioms. We devote the remainder of this section to doing this.

Remember, a group satisfies the first four of the field axioms, but we shall write the group operation in multiplicative notation. Let $\mathscr{G}$ be a group, and let e be an element in $\mathscr{G}$ such that $ea = a$ for all $a \in \mathscr{G}$ (axiom (3)). For each $a \in \mathscr{G}$, there is an element $a^{-1} \in \mathscr{G}$ such that $a^{-1}a = e$ (axiom (4)). There is an $(a^{-1})^{-1} \in \mathscr{G}$ such that $(a^{-1})^{-1}(a^{-1}) = e$. Then

$$\begin{aligned}a(a^{-1}) &= ea(a^{-1}) = (a^{-1})^{-1}(a^{-1})a(a^{-1})\\ &= (a^{-1})^{-1}e(a^{-1}) = (a^{-1})^{-1}(a^{-1}) = e.\end{aligned} \quad (3.1)$$

This shows that an inverse on the left is also an inverse on the right. Then

$$ae = a(a^{-1})a = ea = a. \quad (3.2)$$

This shows that the unit on the left is also a unit on the right. If the group were commutative, these two arguments would be unnecessary.

It is now possible to show that the unit and inverses are unique. Let e' be an element in $\mathscr{G}$ such that $e'a = a$ for at least one element $a \in \mathscr{G}$ (not necessarily for all elements in $\mathscr{G}$ as is assumed for e). Then

$$e' = e'e = e'(aa^{-1}) = (e'a)a^{-1} = aa^{-1} = e. \tag{3.3}$$

Now, let $(a^{-1})'$ be an element in $\mathscr{G}$ such that $(a^{-1})'a = e$. Then

$$(a^{-1})' = (a^{-1})'e = (a^{-1})'aa^{-1} = ea^{-1} = a^{-1}. \tag{3.4}$$

In Section 2–4, we discussed the conditions that a subset of a vector space must satisfy to be a subspace. There we asserted (without proof) that $0\alpha = 0$. This is what one expects, but it should have been proved. By the distributive law,

$$0\alpha = (0 + 0)\alpha = 0\alpha + 0\alpha. \tag{3.5}$$

Thus, $\beta = \beta + 0\alpha$ for at least one β. As shown above, it then follows that $0\alpha = 0$. We also asserted (again, without proof) that $(-1)\alpha = -\alpha$. We have

$$0 = 0\alpha = (1 + (-1))\alpha = 1\alpha + (-1)\alpha = \alpha + (-1)\alpha. \tag{3.6}$$

Since the group inverse is unique, it follows that $(-1)\alpha = -\alpha$.

In the main body of the book, the question of the inverse of a matrix was handled in a different way. In Section 2–2, we assumed that a square matrix had an inverse on each side (possibly different on each side). We then showed these two inverses to be the same, and the inverse of a matrix (if it has one) to be unique. In Section 4–4, we gave an argument based on the concept of rank. Since this, in turn, was dependent on the concept of the inverse of a linear transformation, that argument is really rather sophisticated. On the basis of the group-theoretic discussion given in this section, the question of the inverse of a matrix can be handled along the following lines. Start with the set of all $n \times n$ matrices over a field $\mathscr{F}$. Consider the subset $\mathscr{G}$ consisting of the n-th order matrices of rank n. If A is an n-th order matrix of rank n, the discussion in Section 3–3 shows there is a matrix B such that $BA = I$. Furthermore, the matrix B that is obtained by the method shown is also of rank n. This shows that $\mathscr{G}$ satisfies axiom (4) for groups. Clearly, $I \in \mathscr{G}$. The associative law (axiom (2)) is satisfied for $\mathscr{G}$ since it is satisfied for the set of all n-th order matrices. Assume A and C are in $\mathscr{G}$. Then there is a B such that $BA = I$ and there is a D such that $DC = I$. Then $(DB)(AC) = D(BA)C = DIC =$

$DC = I$. This argument shows that AC is of rank n, and axiom (1) for groups is satisfied. We have shown, then, that $\mathscr{G}$ is a group. Now, the argument given above shows that these matrices have inverses and they are unique. Incidentally, Theorem 4–1 of Chapter 4 for matrices shows that $\mathscr{G}$ is the largest group under multiplication in the set of all n-th order matrices.

A-4 GREEK ALPHABET

The symbolism of mathematics makes it desirable to have available a large number of symbols. The Latin alphabet provides 26 lower case letters and 26 upper case letters. In type, the number of available symbols can be increased by using different type fonts and styles. However, this method is not convenient for handwritten work. To increase the number of symbols available, mathematicians commonly use Greek letters, German script, and a few Hebrew letters. We make extensive use of the Greek alphabet in this book. A list of the letters of the Greek alphabet is given here for your convenience.

Letter	Upper case	Lower case
alpha	A	α
beta	B	β
gamma	Γ	γ
delta	Δ	δ
epsilon	E	ϵ
zeta	Z	ζ
eta	H	η
theta	Θ	θ
iota	I	ι
kappa	K	κ
lambda	Λ	λ
mu	M	μ
nu	N	ν
xi	Ξ	ξ
omicron	O	o
pi	π	π
rho	P	ρ
sigma	Σ	σ
tau	T	τ
upsilon	Y	υ
phi	Φ	ϕ
chi	X	χ
psi	Ψ	ψ
omega	Ω	ω

SELECTED ANSWERS TO EXERCISES

This is a reasonably complete set of answers to the exercises. However, most "show" type exercises are not provided with answers. Many exercises have several correct answers or several correct forms. In such cases no attempt has been made to describe all alternatives.

EXERCISES 1–1

1. a) $2A + B = \begin{bmatrix} 3 & 3 & 6 \\ 10 & 10 & 10 \end{bmatrix}$

   b) $(A+B)C = \begin{bmatrix} -3 & 11 \\ 0 & 18 \end{bmatrix}$

   c) $AC + BC = \begin{bmatrix} -3 & 11 \\ 0 & 18 \end{bmatrix} = (A+B)C$

   d) $(AC)B = \begin{bmatrix} 20 & 0 & -20 \\ 44 & 0 & -44 \end{bmatrix}$

3. $B = \begin{bmatrix} 5 & 2 \\ 2 & 1 \end{bmatrix}$

4. $AB = \begin{bmatrix} 2 & 2 \\ 7 & 3 \end{bmatrix}$, $BA = \begin{bmatrix} -3 & -3 & -4 \\ 2 & 4 & 6 \\ 3 & 3 & 4 \end{bmatrix}$

5. a) $AD = \begin{bmatrix} 2 & -6 & 3 \\ 8 & -15 & 6 \\ 14 & -24 & 9 \end{bmatrix}$, $DA = \begin{bmatrix} 2 & 4 & 6 \\ -12 & -15 & -18 \\ 7 & 8 & 9 \end{bmatrix}$

   The columns of AD are multiples of the columns of A. The rows of DA are multiples of the rows of A.

   b) $D'A = \begin{bmatrix} -2 & 8 & 9 \\ -8 & 20 & 18 \\ -14 & 32 & 27 \end{bmatrix}$, $AD' = \begin{bmatrix} -2 & -4 & -6 \\ 16 & 20 & 24 \\ 21 & 24 & 27 \end{bmatrix}$

6. A and B are not square matrices.

7. $AD = \begin{bmatrix} 1 & 2 & 0 \\ 4 & 5 & 0 \\ 7 & 8 & 0 \end{bmatrix}$, $DA = \begin{bmatrix} 1 & 2 & 3 \\ 4 & 5 & 6 \\ 0 & 0 & 0 \end{bmatrix}$

$D^2 = D^{10} = D$

9. $AB = BA$

10. AB is an $m \times r$ matrix.

11. $\begin{bmatrix} 2 & 3 & 3 \\ 1 & -1 & 0 \\ -1 & 1 & 3 \end{bmatrix}$ is the coefficient matrix. $\begin{bmatrix} 2 & 3 & 3 & 16 \\ 1 & -1 & 0 & 6 \\ -1 & 1 & 3 & 18 \end{bmatrix}$

is the augmented matrix.

12. $0.1x_1 + 0.3x_2 + 0.4x_3 = 8$

13. $5x_1 = x_2$
$6x_1 = x_3$

14. $0.1x_1 + 0.3x_2 + 0.4x_3 = 8$
$5x_1 - x_2 = 0$
$6x_1 - x_3 = 0$

EXERCISES 1-2

3. $(1 + 2x_3) + 2(3 - 2x_3) + 2x_3 = 7$
$-(1 + 2x_3) + (3 - 2x_3) + 4x_3 = 2$

8. Add the two equations to obtain $3x_2 + 6x_3 = 9$, or $x_2 = 3 - 2x_3$. Add the first equation to -2 times the second to obtain $3x_1 - 6x_3 = 3$, or $x_1 = 1 + 2x_3$.

EXERCISES 1-3

1, 3, 4, 5, 6 are in basic form.

7. 1) $x_2 = 3 - 3x_1 + x_4$
$x_3 = 5 + x_1$
$x_5 = 4 - 2x_1 - 3x_4$

2) is not in basic form.

3) $x_1 = 6 - x_4$
$x_2 = 4 - x_4$
$x_3 = 2 - x_4$

Selected Answers to Exercises • 345

6) $x_3 = 2 - x_1$
$x_5 = 3 - x_1 - x_2 + x_4$
$x_6 = 1 - x_1 - x_2 - x_4$

8. $x_1 = 4 - x_2 - x_3 - x_4$
$x_2 = 4 - x_1 - x_3 - x_4$
$x_3 = 4 - x_1 - x_2 - x_4$
$x_4 = 4 - x_1 - x_2 - x_3$

9. $x_1 = 1 - x_3$
$x_4 = 2 \quad - x_2$
$x_5 = 3 - x_3$

$x_1 = 1 - x_3$
$x_2 = 2 \quad - x_4$

$x_5 = 3 - x_3$

EXERCISES 1-4

1. $(-1, 2)$ is the unique solution, a general solution with no parameters.

2. $x_1 = 2 + x_3$
$x_2 = -1 - 2x_3$

3. $(9, -10, -7)$ is the unique solution.

4. $x_1 = 3 - x_2 - 3x_4$
$x_3 = -1$

5. $(2, 1)$ is the unique solution.

6. $x_1 = 1 + 3x_2$
$x_3 = 2$
$x_4 = 3$

7. $x_2 = -3 + x_1 + 2x_4$
$x_3 = 4 - 2x_1 - 5x_4$

8. $x_1 = 3 + x_2 - 2x_4$
$x_3 = -2 - 2x_2 - x_4$

13. We cannot convert one system into the other by means of a sequence of elementary operations.

EXERCISES 1-5

1. A is obtained by multiplying the second row by 3. B is obtained by interchanging rows one and three. C is obtained by adding -2 times the third row to the second row.

2. $$AD = \begin{bmatrix} 1 & 2 & 3 \\ 6 & 9 & -3 \\ -1 & 2 & 1 \end{bmatrix}$$

   $$BD = \begin{bmatrix} -1 & 2 & 1 \\ 2 & 3 & -1 \\ 1 & 2 & 3 \end{bmatrix}$$

   $$CD = \begin{bmatrix} 1 & 2 & 3 \\ 4 & -1 & -3 \\ -1 & 2 & 1 \end{bmatrix}$$

3. The answers are not unique. Possible answers are given.

   a) $\begin{bmatrix} 1 & -2 & 0 \\ 0 & 0 & 0 \\ 0 & 2 & 1 \end{bmatrix}$

   b) $\begin{bmatrix} 1 & 0 & 3 \\ 0 & 1 & -1 \\ 0 & 0 & 0 \end{bmatrix}$

   c) $\begin{bmatrix} 1 & 0 & 2 & 1 & 1 \\ 0 & 1 & -1 & 1 & 2 \\ 0 & 0 & 0 & 0 & 0 \end{bmatrix}$

   d) $\begin{bmatrix} 0 & 0 & 9/2 & 1 \\ 1 & 1 & -11 & 0 \end{bmatrix}$

   e) $\begin{bmatrix} 1 & 0 & 0 \\ 0 & 1 & 0 \\ 0 & 0 & 1 \end{bmatrix}$

5. The basic elements are underscored.

   a) $\begin{bmatrix} \underline{-1} & 0 & \underline{1} & 2 & 0 \\ 1 & \underline{-1} & 0 & 1 & \underline{1} \end{bmatrix}$

   b) $\begin{bmatrix} \underline{1} & 0 & \underline{1} & 0 & 0 \\ 0 & \underline{1} & 0 & 1 & 0 \\ 1 & 0 & 0 & 1 & \underline{1} \end{bmatrix}$

c) $\begin{bmatrix} 0 & 0 & 0 & 0 & 0 & 0 & 1 \\ 0 & 0 & 0 & 0 & 1 & 1 & 0 \\ 0 & 0 & 1 & 2 & 0 & 2 & 0 \\ 1 & -1 & 0 & 0 & 0 & 3 & 0 \end{bmatrix}$

d) $\begin{bmatrix} 1 & 0 & 0 & 0 & 1 & 1 & 0 & 1 \\ 1 & 0 & 0 & 0 & 0 & 1 & 1 & 0 \\ 0 & 0 & -1 & 1 & 1 & 0 & 0 & 0 \\ 1 & 1 & 0 & 0 & 0 & 0 & 0 & 0 \end{bmatrix}$

EXERCISES 1–6

1. (−4, 18) is the exact solution. If .333 is used to represent 1/3, the solution is (−4.04, 18.07) to an accuracy justified by the approximation of 1/3.

2. $x_1 = -100 + 43x_4$
    $x_2 = 15 - 6x_4$
    $x_3 = 10 - 4x_4$
    $x_5 = 3$

    Actually, only six multiplicative steps are required. Three of the steps where a multiplication might be required involved a multiplication by 1.

3. (3, 1, 2, −2) is the exact solution. If Gaussian elimination is used, with the pivots all selected in the main diagonal, and decimal approximations rounded to 4 significant digits, the answer is (3.002, 1.00, 2.000, −2.001). The procedure does not yield an accurate value for the 4-th digit of x_2.

4. Pivoting on .000100, we obtain $(0.00 \times 10^4, 1.00)$. The notation 0.00×10^4 means that x_1 is 0 to an accuracy that does not determine the 1's digit. Pivoting on the coefficient of x_1 in the second equation, we obtain (1.00, 1.00) as the solution. If the computation is carried out exactly, the answer is (10000/9999, 9998/9999).

5. (.00716, .0259, −.0584, 4.42)

EXERCISES 2–1

2. A and B intersect at (2,1).

3. A and C intersect at (5,3).

4. A and D do not intersect.

5. A, B, and C do not have a common point.

6. 2–1.4c

7. 2–1.4b

8. 2–1.4d

9. 2–1.5a

10. 2–1.5c
11. 2–1.5b

12. Two of the three planes are parallel.

EXERCISES 2–2

6. If we multiply a function that takes on at least one positive value by a negative number we obtain a function that is not in $\mathscr{E}$. Thus, condition 6 is not satisfied for $\mathscr{E}$.

9. An integral n-tuple, multiplied by an irrational element of $\mathscr{R}$ is not an integral n-tuple.

10. The coefficient domain $\mathscr{L}$ is not a field. (See the Appendix for relevant definitions.)

EXERCISES 2–3

2. $2(3+3t)+3(-2t)=6$

3. $2(x_1)+3(5/3-2t/3)=5$
   The parametrizations describe parallel lines.

5. $(3+s+2t)+2(3+s-t)+3(-1-s)=6$

6. $(2+s+2t)+2(3+s-t)+3(1-x)=11$

9. $x_1 = 2+2x_4$
   $x_2 = -1+3x_4$
   $x_3 = 3-4x_4$

10. $x_1 = 1 + 6x_3 + 9x_5$
    $x_2 = 2 + 7x_3 + 7x_5$
    $x_4 = 4 + 9x_3 + 3x_5$

EXERCISES 2-4

7. $(x_1, x_2, x_3) = x_1(1, 0, -7/4) + x_2(0, 1, -2)$

8. $(x_1, x_2, x_3) = x_3(-4, 3, 1)$

9. $(x_1, x_2, x_3, x_4) = x_1(1, -4, 0, 5) + x_3(0, 7, 1, -12)$

10. $(x_1, x_2, x_3, x_4) = x_4(-3, 1, -2, 1)$

11. $(x_1, x_2, x_3, x_4) = x_2(-2, 1, -1, 0)$

12. $(x_1, x_2, x_3, x_4) = x_3(1, -2, 1, 0) + x_4(-2, -3, 0, 1)$

13. a) If $(x_1, \ldots, x_n)$ is in the subset $2(x_1, \ldots, x_n)$ is not.
    b) $(1, 0, \ldots, 0)$ is in the subset but $(-1)(1, 0, \ldots, 0)$ is not.
    c) $(1, 0, \ldots, 0)$ is in the subset but $101(1, 0, \ldots, 0)$ is not.
    d) $(1, 0, \ldots, 0)$ is in the subset but $\sqrt{2}\,(1, 0, \ldots, 0)$ is not.
    e) $(1, 0, \ldots, 0)$ and $(0, 1, \ldots, 1)$ are in the subset, but their sum is not.

EXERCISES 2-5

1. $\{(1, -3, 1, 0, -2), (0, 1, 0, 1, -3)\}$

2. ϕ is the set of generators. $\{(0, 0)\}$ is the solution subspace.

3. $\{(1, -2, 1)\}$

4. $\{(0, 2, 1)\}$

5. $\{(-1, 1, 0, 0), (22, 0, 2, -9)\}$

6. $\{(3, 1, 0, 0)\}$

7. $\{(-2, -3, 1)\}$

8. $\{(-3, -2, 1, 0, 0), (-2, 0, 0, -2, 1, 1)\}$

9. $(-11, 3, 6) = -2(6, -1, -3) + (2, 0, -1) + (-1, 1, 1) \in \mathscr{S}$

10. $\{(6, -1, -3), (2, 0, -1), (-1, 1, 1)\}$ spans $\mathscr{S}$. $(14, -2, -7) = 2(6, -1, -3) + (2, 0, -1)$. $(7, -2, -4) = (6, -1, -3) - (-1, 1, 1)$

11. $(1, 1, 1, 2) \notin \mathscr{S}$. $(1, 1, 2, 2) = -(0, 1, 0, 0) + (1, 0, -2, 0) + 2(0, 1, 2, 1) \in \mathscr{S}$

12. $t^3 + t^2 + t + 2$ is represented by $(1, 1, 1, 2)$ with respect to the basis $\{t^3, t^2, t, 1\}$. $t^3 + t^2 + t + 2 \notin \mathscr{S}$. $t^3 + t^2 + 2t + 2 = -(t^2) + (t^3 - 2t) + 2(t^2 + 2t + 1)$

13. $2x_1 + x_3 - 2x_4 = 0$

EXERCISES 2-6

1. $(-3, 2, -1) = -2(3, 1, 2) + (3, 4, 3)$; $(9, 9, 8) = (3, 1, 2) + 2(3, 4, 3)$; $(4, -4, 1) = 3(3, 1, 2) - 2(3, 4, 3) + (1, 1, 1)$
   All other such linear relations are linear combinations of these.

2. $(8, 5, 2) = 3(3, 3, 1) - (1, 4, 1)$; $(12, -7, 0) = 5(3, 3, 1) - 7(1, 4, 1) + 2(2, 3, 1)$; $(4, 7, 2) = (3, 3, 1) + (1, 4, 1)$

3. $(-2, 6, 2, 8) = -2(1, -2, 1, -3) + 2(0, 1, 2, 1)$; $(1, 0, 4, 0) = (1, -2, 1, -3) + 2(0, 1, 2, 1) - (0, 0, 1, -1)$; $(4, -10, -1, -10) = 4(1, =2, 1, -3) - 2(0, 1, 2, 1) - (0, 0, 1, -1) + 3(0, 0, 0, 1)$

4. $\{(2, 1, 1, 2), (1, 2, -2, -1), (-1, 3, 0, 1), (3, 4, 2, 2)\}$ is a maximal linearly independent subset. There are others.

5. $(0, -1, 1, -1) = 2(1, 1, 1, 1) - (2, 3, 1, 3)$; $(1, 1, 3, 5) = -(1, 1, 1, 1) + 2(2, 3, 1, 3) + 2(-1, -2, 1, 0)$

6. All three-element subsets except $\{(1, 1, 1, 1), (2, 3, 1, 3), (0, -1, 1, -1)\}$ are linearly independent.

7. All four-element subsets except $\{(1, 2, 1, 3), (1, 1, 0, 2), (2, 4, 3, 6), (1, 3, 4, 4)\}$ and the sets containing $\{(1, 2, 1, 3), (1, 2, 1, 4), (4, 8, 4, 11)\}$ are linearly independent.

8. $1 \cdot 0 = 0$

Selected Answers to Exercises • 351

9. If $\{\alpha\}$ is linearly independent, there exists an $a \neq 0$ such that $a\alpha = 0$. Then $\alpha = a^{-1}(a\alpha) = a^{-1}0 = 0$.

10. Let $A_1, \ldots, A_6$ denote the columns of the matrix. Then $\{A_1, A_2, A_3\}$, $\{A_1, A_4, A_5\}$, and $\{A_2, A_3, A_4, A_5\}$ are minimal linearly dependent subsets.

EXERCISES 2–7

1. $\{(1, -3, 1, 0, -2), (0, 1, 0, 1, -3)\}$; dim $\mathscr{S} = 2$

2. ϕ; dim $\mathscr{S} = 0$

3. $\{(1, -2, 1)\}$; dim $\mathscr{S} = 1$

4. $\{(0, 2, 1)\}$; dim $\mathscr{S} = 1$

5. $\{(-1, 1, 0, 0), (22, 0, 2, -9)\}$; dim $\mathscr{S} = 2$

6. $\{(3, 1, 0, 0)\}$; dim $\mathscr{S} = 1$

7. $\{(-2, -3, 1)\}$; dim $\mathscr{S} = 1$

8. $\{(-3, -2, 1, 0, 0, 0), (-2, 0, 0, -2, 1, 1)\}$; dim $\mathscr{S} = 2$

9. $\{(-3, -2, 1, 0, 0, 0), (-2, 0, 0, -2, 1, 1), (1, 0, 0, 0, 0, 0), (0, 1, 0, 0, 0, 0), (0, 0, 0, 1, 0, 0), (0, 0, 0, 0, 1, 0)\}$

10. The set is linearly independent in a 3-dimensional space. Hence, it is a basis.

11. The set is linearly dependent.

12. Let $A_1, \ldots, A_5$ denote the vectors. Then $\{A_1, A_2, A_3\}$ and $\{A_1, A_4, A_5\}$ are linearly dependent. All other three-element subsets are linearly independent.

EXERCISES 2–8

1. $\{[1 \ 0 \ 7 \ -4], [0 \ 1 \ -8 \ 3]\}$ represents a basis for the set of equations. That is, every equation is a linear combination of $x_1 + 7x_3 - 4x_4 = 0$ and $x_2 - 8x_3 + 3x_4 = 0$. A basis for the solution set is $\{(-7, 8, 1, 0), (4, -3, 0, 1)\}$. The two given vectors also form a basis.

2. $\{[1\ 0\ 0\ 0\ -1],\ [0\ 1\ 0\ -1\ 0]\}$ is a basis of the set of equations. $\{(1, 0, 0, 0, 1),\ (0, 1, 0, 1, 0),\ (0, 0, 1, 0, 0)\}$ is a basis of the solution set.

3. $\{[0\ 1\ 1\ -1\ 0]\}$ is a basis of the set of equations. $\{(1,0,0,0,0),\ (0,-1,1,0,0),\ (0,1,0,1,0),\ (0,0,0,0,1)\}$ is a basis of the solution set.

4. $\begin{bmatrix} 2 & -3 \\ -1 & -2 \\ 1 & 0 \\ 0 & -1 \\ 0 & 1 \end{bmatrix}$

5. $\begin{bmatrix} 1 & 0 & 0 & 0 \\ 0 & 1 & 0 & 0 \\ -1 & -2 & 2 & -1 \\ 0 & 0 & 1 & 0 \\ 0 & 0 & 0 & 1 \\ -2 & -1 & 1 & -2 \end{bmatrix}$

6. $\begin{bmatrix} 1 & 0 & 0 \\ 2 & 2 & -1 \\ -1 & -3 & -2 \\ 0 & 1 & 0 \\ 0 & 0 & 1 \\ 0 & -1 & -1 \end{bmatrix}$

7. $\begin{bmatrix} 1 & 0 & 0 & 0 \\ 0 & 1 & 0 & 0 \\ 0 & 0 & 1 & 0 \\ 0 & 0 & 0 & 1 \\ -1 & -1 & -1 & -1 \end{bmatrix}$

8. $x_1\ \ \ \ \ -\ x_3 = 0$
$\ \ \ \ x_2 + 2x_3 = 0$

9. $x_1\ \ \ \ \ \ \ \ \ \ \ = 0$
$\ \ \ \ x_2 - 2x_3 = 0$

10. $x_1\ \ \ \ \ \ + 3x_3\ \ \ \ \ \ \ \ \ \ + x_6 = 0$
$\ \ \ \ x_2 + 2x_3\ \ \ \ \ \ \ \ \ \ \ \ \ \ \ = 0$
$\ \ \ \ \ \ \ \ \ \ \ \ \ \ \ \ \ x_4\ \ \ \ + 2x_6 = 0$
$\ x_5 - x_6 = 0$

EXERCISES 3-1

1. (2.2)

2. Yes

3. $(-1/3, 1/3),\ (1/3, 1/6)$

4. $(3, 5, -7)$; 3; $3\alpha_1$; $(9, 15, 21)$; $\xi + \eta$ is represented by $(8, 5, -4)$; $(21, 10, -5)$

5. α_1 is represented by $(1, 0, -1)$. $\beta_1 + 2\beta_2 + 3\beta_3 = (11, 20, 6)$

7. $-3x_1 + x_3 = 0$ is the equation for $\mathscr{S}_1$. $x_1 + 2x_2 - 5x_3 = 0$ is the equation for $\mathscr{S}_2$. The two equations together is the system for $\mathscr{S}_1 \cap \mathscr{S}_2$.

EXERCISES 3-2

1. $\begin{bmatrix} 6 & 2 & -1 \\ -1 & 0 & 1 \\ -3 & -1 & 1 \end{bmatrix}, \begin{bmatrix} 1 & -1 & 2 \\ -2 & 3 & -5 \\ 1 & 0 & 2 \end{bmatrix}$

2. $(9, -22, 8)$

3. $\beta_1 = 2\alpha_1 + 3\alpha_2 + 4\alpha_3$, $\beta_2 = 2\alpha_1 + 5\alpha_2 + 3\alpha_3$, $\beta_3 = \alpha_1 + 2\alpha_2 + 2\alpha_3$
   $\alpha_1 = 4\beta_1 + 2\beta_2 - 11\beta_3$, $\alpha_2 = -\beta_1 + 2\beta_3$, $\alpha_3 = -\beta_1 - \beta_2 + 4\beta_3$

4. a) $\begin{bmatrix} 5 & -8 \\ -3 & 5 \end{bmatrix}$

   b) $\begin{bmatrix} 43 & -5 & -25 \\ 10 & -1 & -6 \\ -7 & 1 & 4 \end{bmatrix}$

   c) $\begin{bmatrix} -73 & 43 & -25 \\ -17 & 10 & -6 \\ 29 & -17 & 10 \end{bmatrix}$

5. $(1, 0, 0)$; (β is the first column of the matrix of transition.)

6. $\begin{bmatrix} 7 & -3 \\ -2 & 1 \end{bmatrix}$

7. $\begin{bmatrix} -1 & 1 & -1 \\ 2 & 1 & 1 \\ 1 & 0 & 0 \end{bmatrix}, \begin{bmatrix} 0 & 0 & 1 \\ \frac{1}{2} & \frac{1}{2} & -\frac{1}{2} \\ -\frac{1}{2} & \frac{1}{2} & -\frac{3}{2} \end{bmatrix}$

9. $\xi = 4\beta_1 + \beta_2 - 2\beta_3$
   $\beta_1 = 3\alpha_1 + 3\alpha_2 + \alpha_3$
   $\beta_2 = \alpha_1 + 4\alpha_2 + \alpha_3$
   $\beta_3 = 2\alpha_1 + 3\alpha_2 + \alpha_3$
   $X = PY = (9, 10, 3)$
   $= 9\alpha_1 + 10\alpha_2 + 3\alpha_3$
   $Q = \begin{bmatrix} 1 & 1 & -5 \\ 0 & 1 & -3 \\ -1 & -2 & 9 \end{bmatrix}$
   $QX = (4, 1, -2) = Y$

10. $Y = (3, -2, 1)$; $X = PY = (4, -4, 1)$; $Q = \begin{bmatrix} 1 & 0 & -1 \\ 1 & 1 & -2 \\ -5 & -3 & 9 \end{bmatrix}$

EXERCISES 3-3

1. $\begin{bmatrix} 1 & 4 & 7 & 10 \\ 2 & 5 & 8 & 11 \\ 3 & 6 & 9 & 12 \end{bmatrix}$

2. β_2 or β_3

3. $\{\alpha_1, \beta_4, \alpha_3\}$; $(-1, 2, -7)$; $(-1, 1, -2)$; $\{\beta_2, \beta_4, \alpha_3\}$; $\{\beta_2, \beta_1, \alpha_3\}$; $(1, -1, 0)$; $\beta_4 = \beta_2 - \beta_1$

5. $\begin{bmatrix} 5 & 8 & 1 \\ -15 & -15 & -1 \\ -6 & -10 & -1 \end{bmatrix}$, $\dfrac{1}{3}\begin{bmatrix} 15 & -2 & 17 \\ -9 & 1 & -10 \\ 0 & 2 & -5 \end{bmatrix}$

6. $\begin{bmatrix} 1 & 0 & 2 & 0 \\ 0 & 1 & 1 & 0 \\ 0 & 0 & 3 & 0 \\ 0 & 0 & 2 & 1 \end{bmatrix}$

7. $\begin{bmatrix} 1 & 0 & 0 & -2 \\ 0 & 1 & 0 & 1 \\ 0 & 0 & 1 & 3 \\ 0 & 0 & 0 & 1 \end{bmatrix}$

8. α_2 is replaced by $2\alpha_1 - \alpha_2 + \alpha_3 + 3\alpha_4$.

9. α_3 is replaced by $4\alpha_1 - \alpha_2 + 2\alpha_3 + 7\alpha_4$.

10. The basis is $\{\alpha_7, \alpha_2, \alpha_4, \alpha_8\}$. The new basis is $\{\alpha_7, \alpha_2, \alpha_4, \alpha_1\}$. The matrix of transition is

    $\begin{bmatrix} 1 & 0 & 0 & 2 \\ 0 & 1 & 0 & 1 \\ 0 & 0 & 1 & 3 \\ 0 & 0 & 0 & 1 \end{bmatrix}.$

    After the second pivot the new basis is $\{\alpha_7, \alpha_3, \alpha_4, \alpha_1\}$. The second matrix of transition is

    $\begin{bmatrix} 1 & 7 & 0 & 0 \\ 0 & 2 & 0 & 0 \\ 0 & 7 & 1 & 0 \\ 0 & -2 & 0 & 1 \end{bmatrix}.$

EXERCISES 3-4

1. $\begin{bmatrix} 1 & 0 & 0 \\ 0 & 1 & -1 \\ 0 & 0 & 1 \end{bmatrix} \begin{bmatrix} 1 & 0 & -1 \\ 0 & 1 & 0 \\ 0 & 0 & 1 \end{bmatrix} \begin{bmatrix} 1 & 0 & 0 \\ 0 & 1 & 0 \\ 0 & -1 & 1 \end{bmatrix} \begin{bmatrix} 1 & 0 & 0 \\ 2 & 1 & 0 \\ 0 & 0 & 1 \end{bmatrix}$ is one possibility.

2. a) Interchange first and third rows.
   b) Multiply row two by 3.
   c) Add -2 time row two to row one.
   d) Add row three to row one.

4. $P = \begin{bmatrix} 1 & 0 & 0 & 0 \\ 0 & 3 & 0 & 0 \\ 0 & 0 & 1 & 0 \\ 0 & 0 & 0 & 1 \end{bmatrix} \begin{bmatrix} 1 & 0 & 0 & 0 \\ 0 & 1 & 0 & 0 \\ 0 & 0 & 1 & 0 \\ 0 & 7 & 0 & 1 \end{bmatrix} \begin{bmatrix} 1 & 0 & 0 & 0 \\ 0 & 1 & 0 & 0 \\ 0 & -1 & 1 & 0 \\ 0 & 0 & 0 & 1 \end{bmatrix} \begin{bmatrix} 1 & 2 & 0 & 0 \\ 0 & 1 & 0 & 0 \\ 0 & 0 & 1 & 0 \\ 0 & 0 & 0 & 1 \end{bmatrix}$

6. b) $\begin{bmatrix} -3 & 0 & 0 \\ 0 & 1 & 0 \\ 0 & 0 & 1 \end{bmatrix} \begin{bmatrix} 1 & 0 & 0 \\ 2 & 1 & 0 \\ 0 & 0 & 1 \end{bmatrix} \begin{bmatrix} 1 & 0 & 0 \\ 0 & 1 & 0 \\ 5 & 0 & 1 \end{bmatrix} \begin{bmatrix} 1 & 0 & 0 \\ 0 & 1 & 0 \\ 0 & -3 & 1 \end{bmatrix} \begin{bmatrix} 1 & 0 & 0 \\ 0 & 1 & 0 \\ 0 & 0 & 9 \end{bmatrix}$

   d) $\begin{bmatrix} 1 & 0 & 0 \\ 0 & 1 & 0 \\ 0 & 0 & -1 \end{bmatrix} \begin{bmatrix} 1 & 0 & 0 \\ 0 & 1 & 6 \\ 0 & 0 & 1 \end{bmatrix} \begin{bmatrix} 1 & 0 & 0 \\ 0 & 2 & 0 \\ 0 & 0 & 1 \end{bmatrix} \begin{bmatrix} 1 & 0 & 7 \\ 0 & 1 & 0 \\ 0 & 0 & 1 \end{bmatrix} \begin{bmatrix} 1 & -4 & 0 \\ 0 & 1 & 0 \\ 0 & 0 & 1 \end{bmatrix} \begin{bmatrix} 3 & 0 & 0 \\ 0 & 1 & 0 \\ 0 & 0 & 1 \end{bmatrix}$

EXERCISES 3-5

1. $\dfrac{n(n-1)(n-2)}{6}$

2. $\dfrac{n(n+1)(n+2)}{6}$

4. $\dfrac{n(n^2-1)}{3}$

5. $\dfrac{n(n^2-1)}{3} + \dfrac{n(n-1)(n-2)}{6} + \dfrac{n(n+1)(n+2)}{6} + \dfrac{n(n^2-1)}{3} = n^3$

6. $\begin{bmatrix} 2 & 0 & 0 \\ -3 & 4 & 0 \\ 4 & 1 & -2 \end{bmatrix} \begin{bmatrix} 1 & 2 & -3 \\ 0 & 1 & -2 \\ 0 & 0 & 1 \end{bmatrix}$

7. $\begin{bmatrix} 1 & 0 & -1 \\ 0 & 1 & 0 \\ 0 & -2 & 1 \end{bmatrix} \begin{bmatrix} 0 & 0 & 4 \\ 2 & -1 & -8 \\ 0 & -2 & 1 \end{bmatrix}$

8. $\begin{bmatrix} 2 & 0 & 0 & 0 \\ 3 & 4 & 0 & 0 \\ -2 & -2 & 1 & 0 \\ -3 & -3 & 2 & 11 \end{bmatrix} \begin{bmatrix} 1 & -2 & 3 & -4 \\ 0 & 1 & -2 & 2 \\ 0 & 0 & 1 & -6 \\ 0 & 0 & 0 & 1 \end{bmatrix}$

9. $(1, -3, 2, 1)$

10. $(x_1, x_2, x_3, x_4) = (2, -2, 0, 0) + x_3(1, -2, 1, 0) + x_4(2, -3, 0, 1)$

11. $(2, 1, -1, -2)$

12. a) $\begin{bmatrix} 1 & 0 & 0 \\ -a & 1 & 0 \\ -b & 0 & 1 \end{bmatrix}$ b) $\begin{bmatrix} 1 & 0 & 0 & 0 \\ -a & 1 & 0 & 0 \\ -b & 0 & 1 & 0 \\ -c & 0 & 0 & 1 \end{bmatrix}$

 c) $\begin{bmatrix} 1 & 0 & 0 \\ -a & 1 & 0 \\ -b+ac & -c & 1 \end{bmatrix}$ d) $\begin{bmatrix} 1 & - & 0 & 0 \\ -a & 1 & 0 & 0 \\ -b+ad & -d & 1 & 0 \\ -c+ae-adf+bf & -e+df & -f & 1 \end{bmatrix}$

 e) $\begin{bmatrix} a^{-1} & 0 & 0 \\ -a^{-1}bd^{-1} & d^{-1} & 0 \\ a^{-1}(bd^{-1}e-c)f^{-1} & -d^{-1}ef^{-1} & f^{-1} \end{bmatrix}$

13. $\begin{bmatrix} 1 & 0 & 0 \\ -2 & 1 & 0 \\ 5 & -4 & 1 \end{bmatrix}$

14. $\begin{bmatrix} 1 & -6 & 37 \\ 0 & 1 & -5 \\ 0 & 0 & 1 \end{bmatrix}$

15. $\begin{bmatrix} 198 & -154 & 37 \\ -27 & 21 & -5 \\ 5 & -4 & 1 \end{bmatrix}$

16. $\begin{bmatrix} 1 & 0 & 0 & 0 \\ 2 & 1 & 0 & 0 \\ -5 & -2 & 1 & 0 \\ -4 & -3 & 1 & 1 \end{bmatrix}$

17. $\begin{bmatrix} 1 & 1 & -3 & -4 \\ 0 & 1 & -2 & -5 \\ 0 & 0 & 1 & 2 \\ 0 & 0 & 0 & 1 \end{bmatrix}$

18. $\begin{bmatrix} 34 & 19 & -7 & -4 \\ 32 & 20 & -7 & -5 \\ -13 & -8 & 3 & 2 \\ -4 & -3 & 1 & 1 \end{bmatrix}$

EXERCISES 4–1

1. e), f), and h) are not linear. The others are linear.

2. a) $f(x+y) = (x+y)^2 = x^2 + 2xy + y^2 = f(x) + f(y) + 2xy \neq f(x) + f(y)$
   b) $f(x+y) = \sin(x+y) = \sin x \cos y + \cos x \sin y \neq f(x) + f(y)$
   c) $f(3x) = \log(3x) = \log 3 + \log x \neq 3f(x)$
   d) $f(x+y) = \tan(x+y) = \dfrac{\tan x + \tan y}{1 - \tan x \tan y} \neq f(x) + f(y)$

5. $\sigma_a(\alpha + \beta) = a(\alpha + \beta) = a\alpha + a\beta = \sigma_a(\alpha) + \sigma_a(\beta)$ $\sigma_a(b\beta) = a(b\beta) = b(a\beta) = b\sigma_a(\beta)$

14. (2, 9, 9)

15. (0, 0, 0)

EXERCISES 4–2

1. $\sigma(\xi)$ is represented by $(-17, -41)$.

2. $\sigma(\xi_1)$ is represented by $(1, 1)$. $\sigma(\xi_2)$ is represented by $(-1, 1)$.

3. $\sigma(\xi)$ is represented by $(4, 7)$. $\sigma(\xi_1)$ is represented by $(2, 2)$. $\sigma(\xi_2)$ is represented by $(2, 1)$.

4. $\sigma((2, 5)) = (2, 5)$, $\sigma((1, 3)) = (0, 0)$; Thus, σ is represented by $\begin{bmatrix} 1 & 0 \\ 0 & 0 \end{bmatrix}$ with respect to the basis $\{(2, 5), (1, 3)\}$. σ is a projection.

5. $\sigma((2, 5)) = (1, 3)$, $\sigma((1, 3)) = (2, 5)$: Thus, σ is represented by $\begin{bmatrix} 0 & 1 \\ 1 & 0 \end{bmatrix}$ with respect to the basis $\{(2, 5), (1, 3)\}$. σ is a reflection.

6. $\sigma((2, 5)) = (2, 5)$, $\sigma((1, 3)) = (3, 8) = (1, 3) + (2, 5)$. Thus, σ is represented by $\begin{bmatrix} 1 & 1 \\ 0 & 1 \end{bmatrix}$. σ is a shear.

7. This transformation is a projection of the space onto a 2-dimensional subspace.

8. $\sigma(\alpha_1)$ is represented by $(-2, 0, 1)$. That is, $\sigma(\alpha_1) = \alpha_1$. $\sigma(\alpha_2)$ is represented by $(0, 3, -2)$. That is, $\sigma(\alpha_2) = \alpha_2$. $\sigma(\alpha_3)$ is represented by $(0, 0, 0)$. That is, $\sigma(\alpha_3) = 0$. σ is a projection, like that described in Exercise 7.

9. Let $\{\alpha_1, \alpha_2, \alpha_3\}$ be the basis used in the representation of σ. Then σ leaves the plane spanned by $\{\alpha_1, \alpha_2\}$ fixed and is a reflection in that plane parallel to the line generated by α_3.

EXERCISES 4–3

1. $\dim K(\sigma) = 4 - \dim \text{Im}(\sigma) \geq 4 - 3 = 1$. $\{(-6, -5, 5, 4)\}$ is a basis of $K(\sigma)$. Thus, $\text{rank } \sigma = \dim \text{Im}(\sigma) = 4 - \dim K(\sigma) = 3$.

2. $\{(1, 2, 3), (2, 1, 4)\}$ is a basis for $\text{Im}(\sigma)$.

3. $\{(-2, 1, 1, 0)\}$ is a basis for $K(\sigma)$. $\{(1, 2, -1, 3), (2, 1, 2, 1), (3, 2, 1, 2)\}$ is a basis for $\text{Im}(\sigma)$.

4. $\dim K(\sigma) = 6 - \dim \text{Im}(\sigma) \geq 6 - \dim \mathscr{V} = 3$.

5. $\dim \text{Im}(\sigma) = 3 - \dim K(\sigma) \leq 3 < 6 = \dim \mathscr{V}$. Thus, $\text{Im}(\sigma)$ is a proper subspace of $\mathscr{V}$.

6. $\{1, x\}$ is a basis for $K(D)$.

7. $\{\alpha_2, \alpha_3\}$ is a basis for $K(\sigma)$. $\{\alpha_1\}$ is a basis of $\text{Im}(\sigma)$.

8. $\{\alpha_1 - \alpha_2, \alpha_1 - \alpha_3\}$ is a basis of $K(\tau)$. $\{\alpha_1\}$ is a basis of $\text{Im}(\tau)$.

9. $\{1\}$ is a basis for $K(D)$.

10. $\text{Im}(G)$ is the set of all polynomials with zero constant term.

EXERCISES 4-4

1. σ^2 is represented by $A^2 = \begin{bmatrix} 6 & -2 \\ 15 & -5 \end{bmatrix} = A$. $K(\sigma) = K(\sigma^2)$ has $\{(1, 3)\}$ for a basis. $\mathrm{Im}(\sigma) = \mathrm{Im}(\sigma^2)$ has $\{(2, 5)\}$ for a basis.

2. $\tau\sigma$ is represented by $BA = \begin{bmatrix} -1 & 0 \\ 0 & 1 \end{bmatrix}$. $\tau\sigma$ is a reflection.

3. $(\tau\sigma)^2$ is represented by I. Thus, $(\tau\sigma)^2 = 1$ and $\tau\sigma\tau$ is the inverse of σ. $\tau\sigma\tau$ is represented by $\begin{bmatrix} -7 & 3 \\ -16 & 7 \end{bmatrix}$.

4. σ^2 is represented by $A^2 = A$. Thus, $\sigma^2 = \sigma$. $\{(1, -2, 1)\}$ is a basis for $K(\sigma) = K(\sigma^2)$. $\{(-2, 6, -3), (-4, 9, -4)\}$ is a basis for $\mathrm{Im}(\sigma)$.

5. τ^2 is represented by $A^2 = 0$. rank $\tau^2 = 0$. (Here, $\mathrm{Im}(\tau) = K(\tau)$.)

6. If σ has an inverse, σ is surjective and injective. Hence, dim codomain = dim domain = dim $\mathrm{Im}(\sigma)$ = dim $\mathrm{Im}(\sigma^{-1})$.

7. rank $\tau \geq$ rank $\sigma\tau \geq$ rank $\sigma^{-1}(\sigma\tau) =$ rank τ.

8. rank $\tau \geq$ rank $\tau\sigma \geq$ rank $(\tau\sigma)\sigma^{-1} =$ rank τ.

9. $\mathrm{Im}(\tau\sigma) = \mathrm{Im}\, \tau$ since σ is an epimorphism.

10. Since τ is injective, $\tau\sigma(\alpha) = 0$ if and only if $\sigma(\alpha) = 0$.

11. If $\tau_1 \neq \tau_2$, there is a β such that $\tau_1(\beta) \neq \tau_2(\beta)$. Since σ is an epimorphism, there is an α such that $\sigma(\alpha) = \beta$. Then $\tau_1\sigma(\alpha) = \tau_1(\beta) \neq \tau_2(\beta) = \tau_2\sigma(\alpha)$.

12. If $\sigma_1 \neq \sigma_2$, there is an α such that $\sigma_1(\alpha) \neq \sigma_2(\alpha)$. Then, since τ is a monomorphism, $\tau\sigma_1(\alpha) \neq \tau\sigma_2(\alpha)$.

13. $(1 - \pi)^2 = 1 - 2\pi + \pi^2 = 1 - 2\pi + \pi = 1 - \pi$.

14. $\pi(1 - \pi) = \pi - \pi^2 = \pi - \pi = 0$. Thus, $\mathrm{Im}(1 - \pi) \subset K(\pi)$. Also, $(1 - \pi)\pi = 0$ so that $\mathrm{Im}(\pi) \subset K(1 - \pi)$. Thus, dim $\mathrm{Im}(1 - \pi) =$ rank$(1 - \pi) = n -$ nullity$(1 - \pi) = n -$ dim $K(1 - \pi) \geq n -$ dim $\mathrm{Im}(\pi) =$ dim $K(\pi) \geq$ dim $\mathrm{Im}(1 - \pi)$. It then follows that all inequalities are equalities. Hence, $\mathrm{Im}(\pi) = K(1 - \pi)$ and $\mathrm{Im}(1 - \pi) = K(\pi)$.

15. If $\alpha \in K(\pi) \cap \mathrm{Im}(\pi)$, there is a β such that $\alpha = \pi(\beta)$. But then, $\alpha = \pi^2(\beta) = \pi(\alpha) = 0$.

16. Since $A^2 = A$, $\sigma^2 = \sigma$. Thus, σ is a projection. $K(\sigma) = \langle (1, 3) \rangle$, $\text{Im}(\sigma) = \langle (2, 5) \rangle$.

17. For $\alpha \in \mathcal{V}$, let $\alpha_1 = \pi(\alpha)$ and let $\alpha_2 = \alpha - \alpha_1$. Then $\pi(\alpha_2) = \pi(\alpha) - \pi(\alpha_1) = \alpha_1 - \pi(\pi(\alpha)) = \alpha_1 - \pi(\alpha) = 0$. Thus, $\alpha_2 \in K(\pi)$. Clearly, $\alpha_1 \in \text{Im}(\pi)$. Since $\mathcal{V} = \text{Im}(\pi) \oplus K(\pi)$, the representation is unique.

EXERCISES 4–5

1. $\begin{bmatrix} \cos\theta & 0 & -\sin\theta \\ 0 & 1 & 0 \\ \sin\theta & 0 & \cos\theta \end{bmatrix}$

2. The product $\begin{bmatrix} 0 & 0 & -1 \\ 1 & 0 & 0 \\ 0 & -1 & 0 \end{bmatrix}$ leaves $(-1, -1, 1)$ fixed. This vector, therefore, lines in the axis of the rotation.

3. The composition of distance preserving, orientation preserving transformations is distance preserving and orientation preserving.

4. $\begin{bmatrix} \cos\theta & -\sin\theta & 0 \\ \sin\theta & \cos\theta & 0 \\ 0 & 0 & 1 \end{bmatrix} \begin{bmatrix} -1 & 0 & 0 \\ 0 & -1 & 0 \\ 0 & 0 & 1 \end{bmatrix} = \begin{bmatrix} -\cos\theta & \sin\theta & 0 \\ -\sin\theta & -\cos\theta & 0 \\ 0 & 0 & 1 \end{bmatrix}$

Let $\zeta = \theta + \pi$.

EXERCISES 4–6

1. $\begin{bmatrix} 2 & 0 \\ 0 & -1 \end{bmatrix}$ 2. $\begin{bmatrix} 1 & 0 \\ 0 & 0 \end{bmatrix}$

3. $\begin{bmatrix} 0 & 1 \\ 1 & 0 \end{bmatrix}$, with $\begin{bmatrix} 2 & 1 \\ 5 & 3 \end{bmatrix}$ as the matrix of transition.

4. $\begin{bmatrix} 1 & 1 \\ 0 & 1 \end{bmatrix}$ with $\begin{bmatrix} 2 & 1 \\ 5 & 3 \end{bmatrix}$ as the matrix of transition.

5. $\begin{bmatrix} 1 & 0 & 0 \\ 0 & 1 & 0 \\ 0 & 0 & 0 \end{bmatrix}$

7. $A^{10} = P \begin{bmatrix} 2 & 0 \\ 0 & -1 \end{bmatrix}^{10} P^{-1} = \begin{bmatrix} 1 & -2 \\ 1 & -1 \end{bmatrix} \begin{bmatrix} 2^{10} & 0 \\ 0 & 1 \end{bmatrix} \begin{bmatrix} -1 & 2 \\ -1 & 1 \end{bmatrix}$

$= \begin{bmatrix} -2^{10} + 2 & 2^{11} - 2 \\ -2^1 + 1 & 2^{11} - 1 \end{bmatrix}$

EXERCISES 4-7

1. $f(x) = -\cos x + C$

3. For $A = \begin{bmatrix} 2 & -3 & -8 \\ -1 & 2 & 5 \\ 3 & 4 & 5 \end{bmatrix}$, rank $A = 2$. Thus, dim Im$(\sigma) = 2$ so that Im(σ) is a proper subspace of $\mathscr{V}$.

4. $10y_1 + 17y_2 - y_3 = 0$

5. $y_1 - y_2 - y_3 = 0$

6. $66y_1 + 13y_2 - 35y_3 - 46y_4 + 3y_5 = 0$. $(5, 8, 8, 4, 10)$ satisfies this equation and is, therefore, in the image. That is, the system of equations has a solution.

EXERCISES 5-1

1. 1
2. 3
3. d
4. $1 + 0 = 1$
5. 1
6. 1
7. d
8. ad
9. $-bc$
10. $ad - bc$
11. abc
12. $-abc$
13. $a_1 bc$
14. $a_1 b_2 c - a_2 b_1 c$
15. $(a_1 b_2 - a_2 b_1) c$
16. $(a_1 b_2 - a_2 b_1) c$
17. $(a_1 b_2 - a_2 b_1) c_3 - (a_1 b_3 - a_3 b_1) c_2 + (a_2 b_3 - a_3 b_2) c_1$

EXERCISES 5-2

1. odd
2. even

3. $\pi\sigma$ maps $\{1, 2, 3, 4, 5\}$ to $\{1, 2, 4, 5, 3\}$. $\sigma\pi$ maps $\{1, 2, 3, 4, 5\}$ to $\{1, 4, 3, 5, 2\}$.

4. Suppose π leaves k fixed. For $i < k$, we have an inversion involving k if and only if $\pi(i) > \pi(k) = k$. Similarly, for $j > k$, we have an inversion involving k if and only if $\pi(j) < \pi(k) = k$. But for every $i < k$ with $\pi(i) > k$ there is exactly one $j > k$ for which $\pi(j) < k$. Thus, the total number of inversions involving k is even.

5. Such a permutation must either permute all five elements cyclically, or it must interchange two elements and cyclically permute the remaining three. There are 24 of the first kind and 20 of the second kind.

6. One permutation, the identity, leaves all five elements fixed. None leaves four fixed. Ten leave three elements fixed (interchanging the other two elements). Fifteen interchange two pairs. Thirty cyclically permute four elements. The remaining two types are described in the answer to Exercise 5.

7. When permuting n objects, there are n possible choices for the destiny of the first element. After that choice is made there are $n-1$ choices for the destiny of the second element. After these choices are made there are $n-2$ choices for the destiny of the third element. Since these choices can be made independently, the total number of choices is $n(n-1)(n-2) \cdots 2 \cdot 1 = n!$.

8. The permutation described sends 1 to 5, 5 to 4, 4 to 2, 2 to 3, and 3 to 1.

9. A convenient notation for the permutation described in Exercise 8 is (15423). In this notation, each element is mapped into the following element, except the last, which is mapped onto the first. In this notation the permutations identified as cyclic can be written in the form (1234), (1243), (1324), (1342), (1423), (1432).

EXERCISES 5–3

1. a) 1; b) 23; c) 0; d) 13; e) −95; f) 135

2. $17; n(n!) - 1$

3. $(n-1)(n!) - 1$

4. $(n-2)(n!) - 1$

EXERCISES 5-4

1. a) 1; b) -7; c) 77; d) 0; e) -191; f) 196

3. The product $a_{\pi(1),1} a_{\pi(2),2} \cdots a_{\pi(n),n}$ vanishes for a lower triangular matrix unless $\pi(i) \geq i$. But this can be true for all i if and only if $\pi(i) = i$ for all i.

EXERCISES 5-5

1. $1; -5; -7$

2. $-\dfrac{1}{2}\begin{bmatrix} 5 & -3 \\ -4 & 2 \end{bmatrix}$

5. $-22; 162$

EXERCISES 6-1

1. $\mathscr{S}_1 \cup \mathscr{S}_2$ generates $\langle \mathscr{S}_1 \rangle + \langle \mathscr{S}_2 \rangle$.

2. $\mathscr{S}_1 \cap \mathscr{S}_2 = \langle (1, -2, -2, -12) \rangle$, $\mathscr{S}_1 + \mathscr{S}_2 = \langle (1, -2, -2, -12), (1, 1, 10, 3), (1, 1, 7, 9) \rangle$.

3. Since $\mathscr{W}_1 + \mathscr{W}_2 = \mathscr{R}^4$, $\mathscr{W}_1 \cap \mathscr{W}_2 = \{0\}$.

4. If $\mathscr{W}_1 \subset \mathscr{W}_2$, then $\mathscr{W}_1 \cup \mathscr{W}_2 = \mathscr{W}_2$ is a subspace. If neither $\mathscr{W}_1 \subset \mathscr{W}_2$ nor $\mathscr{W}_1 \supset \mathscr{W}_2$, there is an $\alpha_1 \in \mathscr{W}_1$, $\alpha_1 \notin \mathscr{W}_2$ and there is an $\alpha_2 \in \mathscr{W}_2$, $\alpha_2 \notin \mathscr{W}_1$. If $\mathscr{W}_1 \cup \mathscr{W}_2$ is a subspace, then $\alpha_1 + \alpha_2 \in \mathscr{W}_1 \cup \mathscr{W}_2$. Thus, $\alpha_1 + \alpha_2 \in \mathscr{W}_1$ or $\alpha_1 + \alpha_2 \in \mathscr{W}_2$. If $\alpha_1 + \alpha_2 \in \mathscr{W}_1$, then $\alpha_2 = (\alpha_1 + \alpha_2) - \alpha_1 \in \mathscr{W}_1$, which is a contradiction. There is a similar contradiction if $\alpha_1 + \alpha_2 \in \mathscr{W}_2$. Thus, $\mathscr{W}_1 \cup \mathscr{W}_2$ is not a subspace.

5. $\langle (1, 0, 0, 0, 0), (0, 1, 0, 0) \rangle$ is a complementary subspace.

6. Since $\mathscr{W}_1 \subset \mathscr{W}_2$, a basis for $\mathscr{W}_1$ is a linearly independent subset of $\mathscr{W}_2$. It can be extended to a basis of $\mathscr{W}_2$. Since $\dim \mathscr{W}_1 = \dim \mathscr{W}_2$, this linearly independent subset is already a basis of $\mathscr{W}_2$. Thus, $\mathscr{W}_1 = \mathscr{W}_2$.

7. $\dim \mathscr{W}_1 = \dim(\mathscr{S} + \mathscr{W}_1) + \dim(\mathscr{S} \cap \mathscr{W}_1) - \dim \mathscr{S} = \dim(\mathscr{S} + \mathscr{W}_2) + \dim(\mathscr{S} \cup \mathscr{W}_2) - \dim \mathscr{S} = \dim \mathscr{W}_2$. Thus, $\mathscr{W}_1 = \mathscr{W}_2$.

8. (a) For $\mathscr{W}_1 = \langle (1, 0) \rangle$, $\mathscr{W}_2 = \langle (0, 1) \rangle$, and $\mathscr{S} = \langle (1, 1) \rangle$, we have $\mathscr{W}_1 + \mathscr{S} = \mathscr{R}^2 = \mathscr{W}_2 + \mathscr{S}$ and $\mathscr{W}_1 \cap \mathscr{S} = \{0\} = \mathscr{W}_2 \cap \mathscr{S}$. (b) For $\mathscr{W}_1 = \langle (1, 0) \rangle$,

$W_2 = \mathscr{R}^2$, and $\mathscr{S} = W_1$, we have $W_1 \subset W_2$ and $W_1 \cap \mathscr{S} = \mathscr{S} = W_2 \cap \mathscr{S}$.
(c) For $W_1 = \{(1, 0)\}$, $W_2 = \mathscr{R}^2$, and $\mathscr{S} = W_2$, we have $W_1 \subset W_2$ and $W_1 + \mathscr{S} = W_2 = W_2 + \mathscr{S}$.

EXERCISES 6-2

1. $\{(1, 3), (2, 5)\}$ is a basis of eigenvectors and $P = \begin{bmatrix} 1 & 2 \\ 3 & 5 \end{bmatrix}$ diagonalizes A.

2. $\{(3, -5), (1, -2)\}$ is a basis of eigenvectors and $P = \begin{bmatrix} 3 & 1 \\ -5 & -2 \end{bmatrix}$ diagonalizes B.

3. $\{(-2, 0, 1), (1, -2, 1), (0, -3, 2)\}$ is a basis of eigenvectors and
$P = \begin{bmatrix} -2 & 1 & 0 \\ 0 & -2 & -3 \\ 1 & 1 & 2 \end{bmatrix}$ diagonalizes C.

4. $\{(-2, 1, 0), (0, -1, 1), (-1, 0, 1)\}$ is a basis of eigenvectors and
$P = \begin{bmatrix} -2 & 0 & -1 \\ 1 & -1 & 0 \\ 0 & 1 & 1 \end{bmatrix}$ diagonalizes D.

5. $\{(-1, 1, 0), (-1, 0, 2), (4, 2, 11)\}$ is a basis of eigenvectors and
$P = \begin{bmatrix} -1 & -1 & 4 \\ 1 & 0 & 2 \\ 0 & 2 & 11 \end{bmatrix}$ diagonalizes E.

6. Using the given matrix of transition, we find $P^{-1}FP = \begin{bmatrix} 1 & 1 & 0 \\ 0 & 1 & 0 \\ 0 & 0 & 2 \end{bmatrix}$.
This tells us that $(-1, -1, 4)$ is an eigenvector with eigenvalue 1, and $(4, 2, -11)$ is an eigenvector with eigenvalue 2. Since $P^{-1}FP$ is similar to F it has the same eigenvalues. Thus, 1 and 2 are the eigenvalues of F. Since there is only one linearly independent eigenvector with eigenvalue 1, there are at most two linearly independent eigenvectors for F.

EXERCISES 6-3

1. $(x-1)(x-2)^2 = 0$ is the characteristic equation for both A and B. Both have 1 and 2 as eigenvalues.

2. Both C and D have the same characteristic equation and eigenvalues as do A and B in Exercise 1.

3. For any matrix P such that $P^{-1}DP = B$ we also have $P^{-1}CP = A$. However, it does not follow from $P^{-1}CP = A$ that $P^{-1}DP = B$. The reason for this situation is that the eigenspaces of C are also invariant subspaces of D. The eigenspace corresponding to 1 is the same for both C and D. The 2-dimensional eigenspace of C corresponding to the eigenvalue 2 is also an invariant subspace of D. Thus, any matrix of transition that diagonalizes C will reduce D to a matrix of the form $\begin{bmatrix} 1 & 0 & 0 \\ 0 & * & * \\ 0 & * & * \end{bmatrix}$, where the 2×2 matrix in the lower-right corner is not uniquely determined.

4. D has only two eigenvalues. Each produces just one linearly independent eigenvector. Since there are no other eigenvectors for D, D is not similar to a diagonal matrix.

5. Let $P^{-1}AP = B$. Then $\det(B - xI) = \det(P^{-1}AP - xI) = \det P^{-1}(A - xI)P = \det P^{-1} \det(A - xI) \det P = \det P^{-1} \det P \det(A - xI) = \det I \det(A - xI) = \det(A - xI)$. Since A and B have the same characteristic polynomial, they have the same characteristic equation.

6. C has a set of three linearly independent eigenvectors. D has at most two linearly independent eigenvectors.

EXERCISES 6-4

1. a) Reflexivity: $AI = A$. b) Symmetry: If $AP = B$, then $BP^{-1} = A$.
   c) Transitivity: If $AP = B$ and $BQ = C$, then $A(PQ) = C$.

2. Write $a \mid b$ to stand for "a divides b." a) Reflexivity: $m \mid (a - a) = 0$.
   b) Symmetry: If $m \mid (a - b)$, then $m \mid (a - b)$. c) Transitivity: If $m \mid (a - b)$ and $m \mid (b - c)$, then $m \mid (a - c) = (a - b) + (b - c)$. There are m equivalence classes. $\{0, 1, 2, \ldots, m - 1\}$ is a complete set of representatives of the equivalence classes.

3. a) $f(1) = f(1)$. b) If $f(1) = g(1)$, then $g(1) = f(1)$. c) If $f(1) = g(1)$ and $g(1) = h(1)$, then $f(1) = h(1)$. Let $f(1) = c$. Then g is equivalent to f if and only if $g(1) = c$. That is, there is one equivalence class for each value of c.

4. Any reasonable estimate of the number of hairs on a person's head would be considerably smaller than the total population of the world. According to the "pigeon-hole principle," when there are more items than there are classifications, at least one classification must

contain two or more items. The application here is that there are more people than there are equivalence classes. Thus, at least one class contains at least two people.

5. a) $\sigma(\alpha) = \sigma(\alpha)$. b) If $\sigma(\alpha) = \sigma(\beta)$, then $\sigma(\beta) = \sigma(\alpha)$. c) If $\sigma(\alpha) = \sigma(\beta)$ and $\sigma(\beta) = \sigma(\gamma)$, then $\sigma(\alpha) = \sigma(\gamma)$.

6. $\sigma(\alpha) = \sigma(\beta)$ if and only if $\sigma(\alpha - \beta) = 0$, that is, if and only if $\alpha - \beta \in K(\sigma)$.

7. a) $\alpha - \alpha \in \mathscr{W}$ since a subspace contains the zero vector. b) If $\alpha - \beta \in \mathscr{W}$, then $\beta - \alpha \in \mathscr{W}$ since $(-1)(\alpha - \beta) \in \mathscr{W}$. c) If $\alpha - \beta \in \mathscr{W}$ and $\beta - \gamma \in \mathscr{W}$, then $\alpha - \gamma \in \mathscr{W}$ since $(\alpha - \beta) + (\beta - \gamma) \in \mathscr{W}$. Notice how the crucial subspace properties, closure under addition and scalar multiplication, are used here.

8. We must show that addition and scalar multiplication defined for equivalence classes are, in fact, really defined. Specifically, α' and β' might be other elements in $\mathscr{V}$ such that $\mathscr{C}(\alpha') = \mathscr{C}(\alpha)$ and $\mathscr{C}(\beta') = \mathscr{C}(\beta)$. Then $\mathscr{C}(\alpha) + \mathscr{C}(\beta)$ could also have been defined to be $\mathscr{C}(\alpha' + \beta')$. But since $(\alpha + \beta) - (\alpha' + \beta') = (\alpha - \alpha') + (\beta - \beta') \in \mathscr{W}$, the equivalence class $\mathscr{C}(\alpha' + \beta')$ is the same class as the equivalence class $\mathscr{C}(\alpha + \beta)$. That is, $\mathscr{C}(\alpha) + \mathscr{C}(\beta) = \mathscr{C}(\alpha + \beta)$ defined the addition of two equivalence classes regardless of which $\alpha' \in \mathscr{C}(\alpha)$ and $\beta' \in \mathscr{C}(\beta)$ are used to define the sum. Similarly, $a\alpha - a\alpha' = a(\alpha - \alpha') \in \mathscr{W}$. That is, $\mathscr{C}(a\alpha') = \mathscr{C}(a\alpha)$ so that $a\mathscr{C}(\alpha)$ is well defined.

The definition of addition and scalar multiplication of equivalence classes is of the same form as the statement that the mapping of $\alpha \in \mathscr{V}$ onto $\mathscr{C}(\alpha)$ is linear. On this basis alone the ten axioms for a vector space can be verified.

9. The linearity was established in the answer to Exercise 8. The kernel of mapping is $\mathscr{W}$.

EXERCISES 6-6

1. The set $\{I, A, A^2, \ldots, A^{n^2}\}$ contains $n^2 + 1$ matrices. Thus, this set is linearly dependent and there is non-trivial linear relation of the form $a_{n^2} A^{n^2} + \ldots + a_1 A + a_0 I = 0$.

2. The Hamilton-Cayley theorem shows that there is a polynomial equation of degree n which A satisfies. Thus, the polynomial equation of lowest degree is n or less.

3. Let $g(x) = 0$ be a polynomial equation of lowest degree which A satisfies, and let $f(x)$ be any other polynomial which A satisfies. By the division algorithm, $f(x) = g(x)q(x) + r(x)$, where $r = 0$ or the degree of $r(x)$ is less than the degree of $g(x)$. Substituting A for x we have $f(A) = 0 = g(A)q(A) + r(A) = r(A)$. If $r \neq 0$, this would contradict the assumption that $g(x)$ is a polynomial of lowest degree for A. Thus, $r = 0$ and $g(x)$ divides $f(x)$. In particular, $g(x)$ divides the characteristic polynomial for A.

4. Let $f(x) = x^n + a_{n-1}x^{n-1} + \cdots + a_1x + a_0$ be the characteristic polynomial for P. Let $P^m = [p_{ij}{}^{(m)}]$. Since $f(P) = 0$, $P^{n+1} = -(a_{n-1}P^n + \cdots + a_1P^2 + a_0P)$. If $p_{ij}{}^{(m)} = 0$ for $m = 1, 2, \ldots, n$, it then follows that $p_{ij}{}^{(n+1)} = 0$. By iteration of this argument it then follows that $p_{ij}{}^{(m)} = 0$ for all $m \geq 1$.

EXERCISES 7-1

1. (1) If a continuous function is not identically zero, it must be non-zero throughout a small interval. It then follows that $(f, f) = \int_0^1 (f(t))^2 \, dt > 0$. This shows that $(f, f) > 0$ unless $f = 0$. (2) $(f, g) = \int_0^1 f(t)g(t) \, dt = (g, f)$. (3) $(f, ag + bh) = \int_0^1 f(t)[ag(t) + bh(t)] \, dt = a(f, g) + b(f, h)$.

2. $|(\alpha, \beta)| \leq \|\alpha\| \cdot \|\beta\| = 0$ implies $(\alpha, \beta) = 0$.

3. $(\alpha, \beta) = 2$

4. $(\alpha, \beta) = 0$

5. $\|\alpha - \beta\|^2 = (\alpha - \beta, \alpha - \beta) = (\alpha, \alpha) - (\alpha, \beta) - (\beta, \alpha) + (\beta, \beta) = \|\alpha\|^2 + \|\beta\|^2 - 2(\alpha, \beta) = \|\alpha\|^2 + \|\beta\|^2 - 2\|\alpha\|\|\beta\| \cos \theta$.

EXERCISES 7-2

3. $(\alpha + \beta, \alpha - \beta) = (\alpha, \alpha) + (\beta, \alpha) - (\alpha, \beta) - (\beta, \beta) = \|\alpha\|^2 - \|\beta\|^2 = 0$.

4. $\left\{ \frac{1}{2}(1, 1, 1, 1), \frac{1}{\sqrt{2}}(1, -1, 0, 0), \frac{1}{2\sqrt{19}}(-1, -1, 7, -5) \right\}$

5. Since $g_{ij} = (\alpha_i, \alpha_j) = (\alpha_j, \alpha_i) = g_{ji}$, G is symmetric. If G were singular, the columns of G would be linearly dependent. That is,

there would be a non-trivial linear relation of the form $\sum_{j=1}^{n} a_j g_{ij} = 0$ for all i. Let $\beta = \sum_{j=1}^{n} a_j \alpha_j$. Then for each i $(\alpha_i, \beta) = \left(\alpha_i, \sum_{j=1}^{n} a_j \alpha_j\right) = \sum_{j=1}^{n} a_j(\alpha_i, \alpha_j) = \sum_{j=1}^{n} a_j g_{ij} = 0$. Thus, $(\beta, \beta) = \left(\sum_{i=1}^{n} a_i \alpha_i, \beta\right) = \sum_{i=1}^{n} \bar{a}_i(\alpha_i, \beta) = 0$. This means $\beta = 0$ and contradicts the linear independence of $\{\alpha_1, \ldots, \alpha_n\}$.

6. If $\alpha_1, \alpha_2 \in \mathscr{S}^\perp$, then $(a_1\alpha_1 + a_2\alpha_2, \beta) = \bar{a}_1(\alpha_1, \beta) + \bar{a}_2(\alpha_2, \beta) = 0$ for all a_1, a_2. Thus, $a_1\alpha_1 + a_2\alpha_2 \in \mathscr{S}^\perp$ and $\mathscr{S}^\perp$ is a subspace.

7. For each $\beta \in \mathscr{S}$, for all $\alpha \in \mathscr{S}^\perp$, $(\beta, \alpha) = \overline{(\alpha, \beta)} = 0$. Thus, $\beta \in (\mathscr{S}^\perp)^\perp$ and $\mathscr{S} \subset \mathscr{S}^{\perp\perp}$.

8. Let $\beta \in \mathscr{T}^\perp$. For all $\alpha \in \mathscr{S}$, $(\alpha, \beta) = 0$ since $\alpha \in \mathscr{T}$. Thus, $\beta \in \mathscr{S}^\perp$ and $\mathscr{T}^\perp \subset \mathscr{S}^\perp$.

9. From Exercise 7, $\mathscr{S} \subset \mathscr{S}^{\perp\perp}$. From Exercise 8, $\mathscr{S}^\perp \supset \mathscr{S}^{\perp\perp\perp}$. Also, from Exercise 7, $\mathscr{S}^\perp \subset (\mathscr{S}^\perp)^{\perp\perp} = \mathscr{S}^{\perp\perp\perp}$.

10. If $\alpha \in \mathscr{S} \cap \mathscr{S}^\perp$, then $(\alpha, \alpha) = 0$ because the first $\alpha \in \mathscr{S}$ and the second $\alpha \in \mathscr{S}^\perp$. Thus, $\alpha = 0$. If $\mathscr{S}$ is a subspace, $0 \in \mathscr{S} \cap \mathscr{S}^\perp$.

11. Let $\{\alpha_1, \ldots, \alpha_m\}$ be a basis of $\mathscr{S}$. This linearly independent set can be extended to a basis of $\mathscr{V}$. Apply the Gram-Schmidt procedure to obtain the orthonormal set $\{\beta_1, \ldots, \beta_n\}$. $\{\beta_1, \ldots, \beta_m\}$ spans $\mathscr{S}$, and $\{\beta_{m+1}, \ldots, \beta_n\}$ spans a subspace $\mathscr{T} \subset \mathscr{S}^\perp$.

12. Since $\mathscr{S} \cap \mathscr{S}^\perp = \{0\}$, $\dim \mathscr{S} + \dim \mathscr{S}^\perp \leq n$. Since $\dim \mathscr{S} + \dim \mathscr{T} = n$ and $\mathscr{T} \subset \mathscr{S}^\perp$, $\mathscr{T} = \mathscr{S}^\perp$.

13. Exercise 12 shows that if $\mathscr{S}$ is a subspace, then $\mathscr{S}^{\perp\perp} = \mathscr{S}$. This assertion can be arrived at as follows. If $\mathscr{S}$ is a subspace, then $\mathscr{S} \oplus \mathscr{S}^\perp = \mathscr{V}$. Since $\mathscr{S} \subset \mathscr{S}^{\perp\perp}$ and $\dim \mathscr{S} = \dim \mathscr{S}^{\perp\perp}$ we have $\mathscr{S} = \mathscr{S}^{\perp\perp}$. Now since $\mathscr{S} + \mathscr{T} \supset \mathscr{S}$, $(\mathscr{S} + \mathscr{T})^\perp \subset \mathscr{S}^\perp$. Similarly, $(\mathscr{S} + \mathscr{T})^\perp \subset \mathscr{T}^\perp$. Thus, $(\mathscr{S} + \mathscr{T})^\perp \subset \mathscr{S}^\perp \cap \mathscr{T}^\perp$. Also, since $\mathscr{S} \cap \mathscr{T} \subset \mathscr{S}$, $(\mathscr{S} \cap \mathscr{T})^\perp \supset \mathscr{S}^\perp$. Similarly, $(\mathscr{S} \cap \mathscr{T})^\perp \supset \mathscr{T}^\perp$. Since $(\mathscr{S} \cap \mathscr{T})^\perp$ is a subspace, $(\mathscr{S} \cap \mathscr{T})^\perp \supset \mathscr{S}^\perp + \mathscr{T}^\perp$. This implies $(\mathscr{S}^\perp \cap \mathscr{T}^\perp)^\perp \supset \mathscr{S}^{\perp\perp} + \mathscr{T}^{\perp\perp} = \mathscr{S} + \mathscr{T}$. Hence, $(\mathscr{S}^\perp \cap \mathscr{T}^\perp) = (\mathscr{S}^\perp \cap \mathscr{T}^\perp)^{\perp\perp} \subset (\mathscr{S} + \mathscr{T})^\perp$. This shows $(\mathscr{S}^\perp \cap \mathscr{T}^\perp) = (\mathscr{S} + \mathscr{T})^\perp$.

EXERCISES 7-3

1. If A and B are orthogonal, then $A^T = A^{-1}$ and $B^T = B^{-1}$. Then $(AB)^T = B^T A^T = B^{-1} A^{-1} = (BA)^{-1}$.

2. If A and B are unitary matrices, then $A^* = A^{-1}$ and $B^* = B^{-1}$. Then $(AB)^* = B^* A^* = B^{-1} A^{-1} = (AB)^{-1}$.

3. 1) $A = I^T A I$. 2) If $B = P^T A P$, then $A = (P^{-1})^T B P^{-1}$. 3) If $B = P^T A^P$ and $C = Q^T B^Q$, then $C = Q^T P^T A P Q = (PQ)^T A (PQ)$.

4. Read the answer to Exercise 3 with P^* in place of P^T and Q^* in place of Q^T.

5. $\begin{bmatrix} 1 & 0 \\ 0 & 2 \end{bmatrix} \begin{bmatrix} 1 & 2 \\ 2 & 3 \end{bmatrix} = \begin{bmatrix} 1 & 2 \\ 4 & 6 \end{bmatrix}$

6. The inverse of an orthogonal matrix is its transpose. All the matrices given in this exercise are orthogonal.

7. The inverse of a unitary matrix is its conjugate transpose. All are unitary.

EXERCISES 7-4

1. If σ and τ are distance preserving, then their composition $\tau\sigma$ is also distance preserving.

2. The argument is the same as that for Exercise 1.

3. The inverse of a distance preserving transformation is also distance preserving.

4. The argument is the same as that for Exercise 3.

5. $((\sigma\tau)^*(\alpha), \beta) = (\alpha, (\sigma\tau)(\beta)) = (\alpha, \sigma(\tau(\beta))) = (\sigma^*(\alpha), \tau(\beta)) = (\tau^*(\sigma^*(\alpha)), \beta) = (\tau^*\sigma^*(\alpha), \beta)$

6. If $\alpha \in K(\sigma^*)$, then $(\alpha, \sigma(\beta)) = (\sigma^*(\alpha), \beta) = (0, \beta) = 0$ for all β. That is, $\alpha \in (\text{Im}(\sigma))^\perp$. If $\alpha \in (\text{Im}(\sigma))^\perp$, then for all β, $(\alpha, \sigma(\beta)) = 0 = (\sigma^*(\alpha), \beta)$. Thus, $\sigma^*(\alpha) = 0$ and $\alpha \in K(\sigma^*)$. Hence, $K(\sigma^*) = \text{Im}(\sigma)^\perp$.

7. If σ is a scalar transformation, $(\sigma^*(\alpha), \beta) = (\alpha, \sigma(\beta)) = (\alpha, a\beta) = a(\alpha, \beta) = (\bar{a}\alpha, \beta)$. Thus, $\sigma^*(\alpha) = \bar{a}\alpha$.

8. Let $\alpha \in \mathscr{W}^\perp$, then for all $\beta \in \mathscr{W}$, $(\sigma^*(\alpha), \beta) = (\alpha, \sigma(\beta))$. Since $\sigma(\beta) \in \mathscr{W}$, $(\alpha, \sigma(\beta)) = 0$. Thus, $\sigma^*(\alpha) \in \mathscr{W}^\perp$.

9. Use Exercise 8 and the fact that $\sigma^{**} = \sigma$.

10. Use Exercise 8 and the fact that $\mathscr{W}^{\perp\perp} = \mathscr{W}$.

11. $0 = (\alpha + \beta, \sigma(\alpha + \beta)) = (\alpha, \sigma(\alpha)) + (\alpha, \sigma(\beta)) + (\beta, \sigma(\alpha)) + (\beta, \sigma(\beta)) = (\alpha, \sigma(\beta)) + (\beta, \sigma(\alpha))$.

12. $(\alpha + i\beta, \sigma(\alpha + i\beta)) = (\alpha, \sigma(\alpha)) + (i\beta, \sigma(\alpha)) + (\alpha, i\sigma(\beta)) + (i\beta, \sigma(i\beta)) = -i(\beta, \sigma(\alpha)) + i(\alpha, \sigma(\beta)) = 0$.

13. From Exercises 11 and 12, $(\beta, \sigma(\alpha)) + (\alpha, \sigma(\beta)) = 0$ and $(\beta, \sigma(\alpha)) - (\alpha, \sigma(\beta)) = 0$. Thus $(\alpha, \sigma(\beta)) = 0$.

14. Take $\alpha = \sigma(\beta)$. Then $(\sigma(\beta), \sigma(\beta)) = 0$.

15. The matrix representing σ^* is $\begin{bmatrix} 0 & 1 \\ -1 & 0 \end{bmatrix}$, which is the negative of the matrix representing σ. Thus, $\sigma^* = -\sigma$. Hence $(\alpha, \sigma(\alpha)) = (\sigma^*(\alpha), \alpha) = (-\sigma(\alpha), \alpha) = -(\alpha, \sigma(\alpha))$ for a real vector space.

EXERCISES 7-5

1. Take the basis used in the proof of Theorem 5.2 in the reverse order.

2. The matrix of transition from the given orthonormal basis to the basis used in Exercise 1 is unitary.

4. σ^* is represented by A^*. If $\sigma^* = \sigma$, then $A^* = A$.

5. $A^* = A$.

EXERCISES 7-6

1. The columns of the matrices of transition given are the eigenvectors.
   (a) real symmetric, $P = \dfrac{1}{\sqrt{2}} \begin{bmatrix} 1 & -1 \\ 1 & 1 \end{bmatrix}$
   (b) real symmetric, $P = \dfrac{1}{5} \begin{bmatrix} 4 & -3 \\ 3 & 4 \end{bmatrix}$

(c) real symmetric, $P = \begin{bmatrix} 1/3\sqrt{2} & 1/\sqrt{2} & 2/3 \\ -4/3\sqrt{2} & 0 & 1/3 \\ -1/3\sqrt{2} & 1/\sqrt{2} & -2/3 \end{bmatrix}$

(d) real symmetric, $P = \dfrac{1}{3}\begin{bmatrix} 1 & -2 & -2 \\ -2 & 1 & -2 \\ -2 & -2 & 1 \end{bmatrix}$

(e) real symmetric, $P = \begin{bmatrix} 1/\sqrt{3} & 1/\sqrt{2} & -1/\sqrt{6} \\ -1/\sqrt{3} & 0 & -2/\sqrt{6} \\ -1/\sqrt{3} & 1/\sqrt{2} & 1/\sqrt{6} \end{bmatrix}$

(f) Hermitian, $P = \dfrac{1}{\sqrt{2}}\begin{bmatrix} 1 & i \\ i & 1 \end{bmatrix}$

(g) Hermitian, $P = \dfrac{1}{\sqrt{3}}\begin{bmatrix} 1 & 1-i \\ 1+i & -1 \end{bmatrix}$

(h) Hermitian, $P = \begin{bmatrix} 1/\sqrt{2} & -1/2 & -1/2 \\ 0 & i/\sqrt{2} & -i/\sqrt{2} \\ 1/\sqrt{2} & 1/2 & 1/2 \end{bmatrix}$

2. Suppose $\alpha \in K(\sigma)$. Then $\|\sigma^*(\alpha)\|^2 = (\sigma^*(\alpha), \sigma^*(\alpha)) = (\sigma\sigma^*(\alpha), \alpha) = (\sigma^*\sigma(\alpha), \alpha) = (\sigma(\alpha), \sigma(\alpha)) = 0$. Thus, $K(\sigma) \subset K(\sigma^*)$. A symmetric argument shows that $K(\sigma^*) \subset K(\sigma)$.

3. From Exercise 7-4.6, $K(\sigma^*) = \operatorname{Im}(\sigma)^\perp$. From Exercise 2, $K(\sigma) = \operatorname{Im}(\sigma)^\perp$.

4. $\operatorname{Im}(\sigma)^\perp = K(\sigma) = \operatorname{Im}(\sigma^*)^\perp$. Thus, $\operatorname{Im}(\sigma) = \operatorname{Im}(\sigma^*)$.

5. $\alpha \in \mathscr{S}^\perp$ if and only if $(\xi, \alpha) = 0$ for all $\xi \in \mathscr{S}$. Then $(\xi, \sigma(\alpha)) = (\sigma^*(\xi), \alpha) = (d\xi, \alpha) = d(\xi, \alpha) = 0$. Hence, $\sigma(\alpha) \in \mathscr{S}^\perp$ and $\mathscr{S}^\perp$ is invariant under σ.

6. Since $\mathscr{W}$ is invariant under σ, σ acts like a linear transformation of $\mathscr{W}$ into itself. Furthermore, σ acts like a normal linear transformation on $\mathscr{W}$. Thus, $\mathscr{W}$ has a basis of eigenvectors of σ. Let $\{\xi_1, \ldots, \xi_m\}$ be a basis of eigenvectors of $\mathscr{W}$. Then $\mathscr{W}^\perp = \{\xi_1, \ldots, \xi_m\}^\perp$. By Exercise 5, $\mathscr{W}^\perp$ is invariant under σ.

EXERCISES 7-7

1. The eigenvalues are 1, 1, −1. The linear transformation is a reflection with respect to the plane containing the eigenvectors corresponding to 1.

2. This is a rotation through θ, where $\theta = \arccos 1/2$.

3. A vector in the axis is $(1, 1, 1)$. Any two vectors orthogonal to this axis can be used as basis vectors.

4. We can calculate $\det A = 1$. Thus, A is similar to a matrix of the form of 4–(5.12).

5. $\text{Tr}(A) = -2/3$ and the trace of a matrix of the form of 4–(5.12) is $2 \cos \theta + 1$. Thus, $\cos \theta = -5/6$, $\sin \theta = \pm\sqrt{11}/6$. In this way the matrix of the form 4–(5.12) to which A is orthogonal similar can be found without computing the characteristic values or eigenvectors for A.

Index

A

abelian, 340
adjoint, 305
adjunct, 237
algebra, 176
algebraic multiplicity, 262
annihilate, 95
associated homogeneous problem, 200
augmented matrix, 7
automorphism, 199

B

back substitution, 36
backward course, 36
backward pivot, 38
basic coefficient, 29
basic element, 29
basic form, 15, 27
basic variable, 15
basis, 83
bijective, 177
bilinear, 282
binary relation, 177

C

canonical form, 196, 266
characteristic equation, 257
characteristic matrix, 257
characteristic polynomial, 257
characteristic value, 258
closed under scalar multiplication, 62
closed under vector addition, 62
codomain, 165
coefficient matrix, 7
cofactor, 234
commutative group, 340
commutative ring, 339
commutes (a diagram), 111
complement, 328
complementary subspace, 244
complementary systems, 95
complete matrix algebra, 181
components, 104
composition, 174
congruent modulo m, 267
conjugate biliner, 283
conjugate symmetric, 283
coordinates, 104
coordinate space, 43, 53
coordinate-wise, 104
coset, 203, 204
cyclic permutation, 220

D

DeMorgan's law, 328
determinant, 215, 238
diagonal matrix, 5
diagonalizable, 248
dimension, 84
direct sum, 241
distance function, 286
distance preserving, 188
domain, 165
dot product, 281

E

eigenspace, 248
eigenvalue, 247
eigenvector, 247
element (of a matrix), 3
elementary matrix, 126
elementary operation, 19
elementary row operation, 25
empty set, 329
epimorphism, 177
equivalence class, 265
equivalence relation, 265
equivalent (system of equations), 19

373

even permutation, 219
existential quantifier, 337

F

factor set, 268
factor space, 268
field, 338
finite dimensional, 84
form, 282
forward course, 36
forward elimination, 36
forward pivot, 38
function space, 65

G

Gauss-Jordan elimination, 33
Gaussian elimination, 36
general solution, 16, 43
generates, 69
geometric multiplicity, 248
Gram-Schmidt process, 291
group, 340

H

Hamilton-Cayley Theorem, 275
Hermitian bilinear, 283
Hermitian matrix, 310
Hermitian symmetric, 283
homogenous linear equations, 63
homomorphism, 177

I

identity matrix, 6
identity transformation, 176
image, 165, 167
implies, 327
inconsistent system of equations, 11
index, 329
index set, 329, 330
indexed set, 330
indexing, 330
injective mapping, 176
inner product, 281
intersection, 326
introduce coordinate, 104
invariant subspace, 246
inverse, 176
inverse matrix, 6, 113
inverse matrix, right, 114
inversion (of a permutation), 218
isomorphism, 104, 177

J

Jordan (Wilhelm) elimination, 33
Jordan (Camille) normal form, 267

K

kernel, 167
Kronecker delta, 6

L

length, 282
linear algebra, 176
linear combination, 62, 67
linear device, 149
linear mapping, 147
linear problem, 199
linear relation, 77
linear transformation, 147
linearly dependent, 77
linearly independent, 77
lower triangular form, 131, 145, 307

M

main diagonal, 5
matrix, 3
matrix of transition, 108
maximal set, 81
mean value, 148
minimal set, 81
minimum polynomial, 276
minimum polynomial equation, 273
minor, 235
monomorphism, 177

N

n-dimensional coordinate space, 43
n-tuple, 42
natural basis, 86
negative definite, 282
non-negative semi-definite, 282
non-singular, 128
non-trivial linear relation, 77
norm, 282
normal form, 196, 266
normal linear transformation, 311
normal matrix, 310
normalization, 291
null space, 168
nullity, 168, 171

O

odd permutation, 219
one-to-one, 176
onto, 177
order of a matrix, 3
orientation, 188
orthogonal, 288
orthogonal matrix, 297
orthogonal similar, 298
orthogonal transformation, 304
orthonormal, 288

P

parameter, 15
parametric solution, 69
permutation, 218
pivot element, 26, 29
pivot operation, 26
position vector, 51
positive definite, 282
pre-image, 165
probability function, 148
product of matrices, 4
product of vector by scalar, 52
projection, 162, 163, 183
proper subset, 326

R

random variable, 148
rank, 167
rational operations, 338
real coordinate space, 53
reduced form, 28
reduction to basic form, 28
reflection, 159, 160
relation, 170, 264
relation, equivalence, 265
represents a linear function, 154
represents (a vector space), 100
represents with respect to a basis, 101
resolution of the identity, 270
right-hand rule, 222
right inverse, 114
ring, 339
rotation, 162
row echelon form, 28
row equivalence, 266

S

sample space, 148
scalars, 4
scalar multiple, 4

scalar multiplication, 53
scalar product, 281
scalar transformation, 152, 176
Schwarz's inequality, 283
semi-definite, 282
shear, 160, 161
similar, 197, 267
similarity, 267
singular, 128
Skew-Hermitian, 310
Skew-symmetry, 310
solution, 12
solutions set, 12, 43
span, 69
spectral resolution, 271
spectrum, 271
Steinitz replacement principle, 83
stochastic matrix, 273
submatrix, 234
subset, 326
subset, proper, 326
subspace, 62
sum of matrices, 4
sum of vectors, 52
super diagonal form, 307
surjective, 177
symmetric (form), 282
symmetric (matrix), 294

T

trace, 257
transpose, 93, 227
triangle inequality, 283
trivial linear relation, 77
tuple, 42

U

union, 327
unitary matrix, 297
unitary similar, 298
unitary transformation, 304
universal quantifier, 337
upper triangular form, 135, 307

V

vector addition, 53
vector space, 54
Venn diagram, 326

W

with respect to a basis, 101